METAL IONS IN LIFE SCIENCES

VOLUME 9

Structural and Catalytic Roles of Metal Ions in RNA

Metal Ions in Life Sciences

edited by

Astrid Sigel,[1] Helmut Sigel,[1] and Roland K. O. Sigel[2]

[1] *Department of Chemistry*
Inorganic Chemistry
University of Basel
Spitalstrasse 51
CH-4056 Basel, Switzerland

[2] *Institute of Inorganic Chemistry*
University of Zürich
Winterthurerstrasse 190
CH-8057 Zürich, Switzerland

Volume 9

Structural and Catalytic Roles of Metal Ions in RNA

DE GRUYTER

First published by the Royal Society of Chemistry in 2011. Publication Details:

ISBN: 978-1-84973-094-5
ISSN: 1559-0836
DOI: 10.1039/9781849732512

A cataloque record for this book is available from the British Library

ISBN 978-3-11-044282-3
e-ISBN (PDF) 978-3-11-043664-8
Set-ISBN (Print + Ebook) 978-3-11-043665-5

Library of Congress Cataloging-in-Publication Data
A CIP catalog record for this book has been applied for at the Library of Congress.

Bibliographic information published by the Deutsche Nationalbibliothek
The Deutsche Nationalbibliothek lists this publication in the Deutsche Nationalbibliografie; detailed bibliographic data are available on the Internet at http://dnb.dnb.de.

Cover image: The figure on the dust cover shows Figure 5(b) of Chapter 7 by Daniela Donghi and Joachim Schnabl.

www.degruyter.com

Historical Development and Perspectives of the Series *Metal Ions in Life Sciences*[*]

It is an old wisdom that metals are indispensable for life. Indeed, several of them, like sodium, potassium, and calcium, are easily discovered in living matter. However, the role of metals and their impact on life remained largely hidden until inorganic chemistry and coordination chemistry experienced a pronounced revival in the 1950s. The experimental and theoretical tools created in this period and their application to biochemical problems led to the development of the field or discipline now known as *Bioinorganic Chemistry*, *Inorganic Biochemistry*, or more recently also often addressed as *Biological Inorganic Chemistry*.

By 1970 *Bioinorganic Chemistry* was established and further promoted by the book series *Metal Ions in Biological Systems* founded in 1973 (edited by H.S., who was soon joined by A.S.) and published by Marcel Dekker, Inc., New York, for more than 30 years. After this company ceased to be a family endeavor and its acquisition by another company, we decided, after having edited 44 volumes of the *MIBS* series (the last two together with R.K.O.S.) to launch a new and broader minded series to cover today's needs in the *Life Sciences*. Therefore, the Sigels new series is entitled

Metal Ions in Life Sciences.

After publication of the first four volumes (2006–2008) with John Wiley & Sons, Ltd., Chichester, UK, we are happy to join forces now in this still new endeavor with the Royal Society of Chemistry, Cambridge, UK; a most experienced Publisher in the *Sciences*.

---

[*]Reproduced with some alterations by permission of John Wiley & Sons, Ltd., Chichester, UK (copyright 2006) from pages v and vi of Volume 1 of the series *Metal Ions in Life Sciences* (MILS-1).

The development of *Biological Inorganic Chemistry* during the past 40 years was and still is driven by several factors; among these are (i) the attempts to reveal the interplay between metal ions and peptides, nucleotides, hormones or vitamins, etc., (ii) the efforts regarding the understanding of accumulation, transport, metabolism and toxicity of metal ions, (iii) the development and application of metal-based drugs, (iv) biomimetic syntheses with the aim to understand biological processes as well as to create efficient catalysts, (v) the determination of high-resolution structures of proteins, nucleic acids, and other biomolecules, (vi) the utilization of powerful spectroscopic tools allowing studies of structures and dynamics, and (vii), more recently, the widespread use of macromolecular engineering to create new biologically relevant structures at will. All this and more is and will be reflected in the volumes of the series *Metal Ions in Life Sciences*.

The importance of metal ions to the vital functions of living organisms, hence, to their health and well-being, is nowadays well accepted. However, in spite of all the progress made, we are still only at the brink of understanding these processes. Therefore, the series *Metal Ions in Life Sciences* will endeavor to link coordination chemistry and biochemistry in their widest sense. Despite the evident expectation that a great deal of future outstanding discoveries will be made in the interdisciplinary areas of science, there are still "language" barriers between the historically separate spheres of chemistry, biology, medicine, and physics. Thus, it is one of the aims of this series to catalyze mutual "understanding".

It is our hope that *Metal Ions in Life Sciences* proves a stimulus for new activities in the fascinating "field" of *Biological Inorganic Chemistry*. If so, it will well serve its purpose and be a rewarding result for the efforts spent by the authors.

Astrid Sigel, Helmut Sigel
Department of Chemistry
Inorganic Chemistry
University of Basel
CH-4056 Basel
Switzerland

Roland K. O. Sigel
Institute of Inorganic Chemistry
University of Zürich
CH-8057 Zürich
Switzerland

October 2005
and October 2008

Preface to Volume 9

Structural and Catalytic Roles of Metal Ions in RNA

This volume is solely devoted to the vibrant research area indicated in the book title. It opens with three general chapters: The first one presents an overview on the binding of metal ions to RNA focusing on monovalent and divalent ions, but providing also information on higher charged cations. The roles of metal ions in the formation of RNA structures, in RNA function, and in RNA catalysis are closely intertwined; therefore, precise knowledge on metal ion-binding motifs is required to understand these processes. In Chapter 2 methods to detect and to characterize metal ion-binding sites in RNA are discussed. Spectroscopic methods encompassing X-ray crystallography, NMR, EPR, lanthanide(III) luminescense, etc. as well as chemical and biochemical methods including sulfur-rescue experiments, nucleotide analogue interference mapping (NAIM), hydrolytic and Fenton cleavage experiments to probe metal ion-binding sites are considered. The importance of computational methods involving databases as well as the calculation of binding constants are emphasized. The chapter highlights the advantages, but lists also the caveats and possible difficulties associated with each method. The majority of metal ions surrounding nucleic acids are used for charge screening of the phoshate-sugar backbone, yet an estimated 10% of the divalent metal ions are bound more specifically, taking over structural as well as catalytic roles. In Chapter 3 the importance of 'diffuse' metal ion binding to RNA is pointed out. This binding type provides a significant stabilizing force and is critical in the folding kinetics of RNA.

The following chapters are devoted to specific RNAs, like RNA quadruplexes, which are dealt with in Chapter 4. The quadruplexes involve four

Metal Ions in Life Sciences, Volume 9 Edited by Astrid Sigel, Helmut Sigel, and Roland K. O. Sigel

Published by the Royal Society of Chemistry, www.rsc.org DOI: 10.1039/9781849732512FP007 guanine residues leading to four-stranded nucleic acid conformations. K^+, due to its size, fits most suitably into the center of the G-quartet. These higher order nucleic acid conformations have been proposed to be involved in various biological functions, such as gene regulation, nucleosome positioning, recombination, and genomic maintenance. The roles of metal ions in gene regulation by riboswitches are discussed in Chapter 5. Classes of signal-responsive RNA elements have been discovered recently to function as metalloregulating agents. This suggests that RNA-based regulatory strategies exist which precisely tune the intracellular metal ion level.

Nearly 30 years ago several small RNA motifs capable of chemical catalysis have been discovered. They play vital roles in the replication of subviral and viral pathogens, gene regulation in prokaryotes, and have recently been found in non-coding eukaryotic RNAs as is reviewed in Chapter 6. Although the interactions between small ribozymes and metal ions are relatively non-specific, the activity of these ribozymes is quite sensitive to the types and concentrations of metal ions present in solution, suggesting a close evolutionary relationship between cellular metal ion homeostasis and cation requirements of catalytic RNAs. Chapter 7 describes the multiple roles of metal ions in large ribozymes, that is, in group I and group II introns as well as in ribonuclease P (RNase P). Metal ions affect RNA folding and catalysis; regarding the last point in particular, the 'two-metal-ion mechanism', which has been confirmed repeatedly, is emphasized

The spliceosome, discussed in Chapter 8, is a massive complex of 5 RNAs and many proteins that associate to catalyze precursor mRNA splicing, which is an essential step in eukaryotic gene expression. A wealth of data indicates that the core of the spliceosome is comprised of RNA, suggesting that the spliceosome is a ribozyme. There is also growing evidence for a magnesium-dependence with at least two metal ions in the active site. "Metal ions are the salt in the soup of essentially every biological system", as the authors of Chapter 9 point out. Also in the ribosome, the largest natural ribozyme that produces all proteins in every living cell, metal ions have been found to contribute significantly to the highly dynamic and accurate process of translation. Indeed, the catalytic center and surrounding domains are densely packed with divalent metal ions and direct roles in mRNA decoding and reading frame maintenance are likely.

The means of *in vitro* selection has yielded a number of artificial ribozymes with functions that have not been discovered yet in modern biological systems. The metal ion requirements of two such artificial ribozymes, which catalyze aminoacylations and redox reactions, are discussed in Chapter 10. In Chapter 11 the fundamental structure and metal ion-binding properties of four naturally occurring RNA enzymes, i.e., the hammerhead, hairpin, hepatitis delta virus, and *glms*-metabolite sensing ribozyme, are described. In addition, the fold and metal ion coordination of three artificial ribozymes are discussed, namely of the leadzyme, the flexizyme, and the Diels-Alder ribozyme. An emergent theme is that natural and artificial ribozymes that catalyze single-step reactions often possess a pre-formed active site and that multivalent ions facilitate RNA-active site formation.

In the terminating Chapter 12 several aspects of the kinetically inert platinum(II) are highlighted, focusing on Pt(II)-RNA adducts and the possibility that they influence RNA-based processes in cells. Nucleic acids are targets of metal-based therapeutic agents, the most extensively studied being the Pt(II) anticancer compounds, especially *cis*-diamminedichloro-Pt(II) (*Cisplatin*). Indeed, the formation of Pt(II)-RNA adducts provides a rationale for the observation that the treatment with Pt(II) compounds disrupts RNA-based processes including enzymatic processing, splicing, and translation.

Astrid Sigel
Helmut Sigel
Roland K. O. Sigel

Contents

Metal Ions in Life Sciences, Volume 9 Edited by Astrid Sigel, Helmut Sigel, and Roland K. O. Sigel

Published by the Royal Society of Chemistry, www.rsc.org DOI: 10.1039/9781849732512FP011

Contributors to Volume 9

Numbers in parentheses indicate the pages on which the authors' contributions begin.

Pascal Auffinger Architecture et Reactivité de l'ARN, IBMC-CNRS, Université de Strasbourg, 15, Rue René Descartes, F-67084 Strasbourg Cédex, France, Fax: +33-38860-2218 <p.auffinger@ibmc.u-strasbg.fr> (1)

Samuel E. Butcher Department of Biochemistry, University of Wisconsin-Madison, 433 Babcock Drive, Madison, WI 53706-1544, USA <butcher@biochem.wisc.edu> (235)

Erich G. Chapman Department of Chemistry, University of Oregon, Eugene, OR 97403-1253, USA (347)

Shi-Jie Chen Department of Physics & Astronomy and Department of Biochemistry, 223 Physics Building, University of Missouri-Columbia, Columbia, MO 65211, USA, Fax: +1-573-882-4195 <chenshi@missouri.edu> (101)

Victoria J. DeRose Department of Chemistry, University of Oregon, Eugene, OR 97403-1253, USA <derose@uoregon.edu> (347)

Daniela Donghi Institute of Inorganic Chemistry, University of Zürich, Winterthurerstrasse 190, CH-8057 Zürich, Switzerland <daniela.donghi@aci.uzh.ch> (197)

Michèle C. Erat Department of Biochemistry, University of Oxford, South Parks Road, Oxford, OX1 3Qu, UK <michele.erat@bioch.ox.ac.uk> (37)

Adrian Ferré-D'Amaré Howard Hughes Medical Institute and Division of Basic Sciences, Fred Hutchinson Cancer Research Center, Seattle, WA 98109-1024, USA <aferre@fhcrc.org> (141)

Kazuki Futai Research Center for Advanced Science and Technology, The University of Tokyo, 4-6-1 Komaba, Meguro-ku, Tokyo 153-8904, Japan (277)

Neena Grover Department of Chemistry and Biochemistry, The Colorado College, Colorado Springs, CO 80903, USA <ngrover@coloradocollege.edu> (1)

Kangkan Halder Department of Chemistry, University of Konstanz, Universitätsstrasse 10, D-78457 Konstanz, Germany (125)

Jörg S. Hartig Department of Chemistry, University of Konstanz, Universitätsstrasse 10, D-78457 Konstanz, Germany, Fax: +49-7531-88-5140 <joerg.hartig@uni-konstanz.de> (125)

Alethia A. Hostetter Department of Chemistry, University of Oregon, Eugene, OR 97403-1253, USA (347)

Koichiro Jin Department of Chemistry and Biotechnology, School of Engineering, The University of Tokyo, 7-3-1 Hongo, Bunkyo-ku, 113-0033 Tokyo, Japan <jin@cbl.rcast.u-tokyo.ac.jp> (277)

Alexander E. Johnson-Buck Department of Chemistry, University of Michigan, 930 N. University, Ann Arbor, MI 48109-1055, USA <alebuck@umich.edu> (175)

Sarah E. McDowell Department of Chemistry, University of Michigan, 930 N. University, Ann Arbor, MI 48109-1055, USA <semcdowe@umich.edu> (175)

Amanda L. Miller Department of Chemistry, University of Oregon, Eugene, OR 97403-1253, USA (347)

Maire F. Osborn Department of Chemistry, University of Oregon, Eugene, OR 97403-1253, USA (347)

Norbert Polacek Innsbruck Biocenter, Dividion of Genomics and RNomics, Medical University Innsbruck, Fritz-Pregl-Strasse 3, A-6020 Innsbruck, Austria, Fax: +43-512-9003-73100 <norbert.polacek@i-med.ac.at> (253)

Joachim Schnabl Institute of Inorganic Chemistry, University of Zürich, Winterthurerstrasse 190, CH-8057 Zürich, Switzerland, <joachim.schnabl@aci.uzh.ch> (197)

Roland K. O. Sigel Institute of Inorganic Chemistry, University of Zürich, Winterthurerstrasse 190, CH-8057 Zürich, Switzerland, Fax: +41-44-635-6802 <roland.sigel@aci.uzh.ch> (37)

Hiroaki Suga Department of Chemistry, School of Science, The University of Tokyo, 7-3-1 Hongo, Bunkyo-ku, 113-0033 Tokyo, Japan, Department of Chemistry and Biotechnology, School of Engineering, The University of Tokyo, 7-3-1 Hongo, Bunkyo-ku, 113-0033 Tokyo, Japan, and Research Center for Advanced Science and Technology, The University of Tokyo, 4-6-1 Komaba, Meguro-ku, Tokyo 153-8904, Japan <hsuga@rcast.u-tokyo.ac.jp> (277)

Zhi-Jie Tan Department of Physics and Key Laboratory of Artificial Micro- and Nano-Structures of the Ministry of Education, School of Physics and Technology, Wuhan University, Wuhan, China 430072 (101)

Krista Trappl Innsbruck Biocenter, Dividion of Genomics and RNomics, Medical University Innsbruck, Fritz-Pregl-Strasse 3, A-6020 Innsbruck, Austria <krista.trappl@i-med.ac.at> (253)

Nils G. Walter Department of Chemistry, University of Michigan, 930 N. University, Ann Arbor, MI 48109-1055, USA, Fax: +1-734-647-4865 <nwalter@umich.edu> (175)

Joseph E. Wedekind Department of Biochemistry and Biophysics and Center for RNA Biology, University of Rochester School of Medicine and Dentistry, 601 Elmwood Avenue, Rochester, NY 14642, USA, Fax: +1-585-275-6007 <joseph.wedekind@rochester.edu>(299)

Eric Westhof Architecture et Reactivité de l'ARN, IBMC-CNRS, Université de Strasbourg, 15, Rue René Descartes, F-67084 Strasbourg Cédex, France, Fax: +33-38860-2218 <eric.westhof@ibmc.u-strasbg.fr> (1)

Wade C. Winkler Department of Biochemistry, The University of Texas Southwestern Medical Center, 5323 Harry Hines Blvd, Dallas, TX 75390-9038, USA, Fax: +1-214-648-8856 <wade.winkler@utsouthwestern.edu> (141)

Titles of Volumes 1–44 in the *Metal Ions in Biological Systems* Series

edited by the SIGELs
and published by Dekker/Taylor & Francis (1973–2005)

Volume 1: **Simple Complexes**
Volume 2: **Mixed-Ligand Complexes**
Volume 3: **High Molecular Complexes**
Volume 4: **Metal Ions as Probes**
Volume 5: **Reactivity of Coordination Compounds**
Volume 6: **Biological Action of Metal Ions**
Volume 7: **Iron in Model and Natural Compounds**
Volume 8: **Nucleotides and Derivatives: Their Ligating Ambivalency**
Volume 9: **Amino Acids and Derivatives as Ambivalent Ligands**
Volume 10: **Carcinogenicity and Metal Ions**
Volume 11: **Metal Complexes as Anticancer Agents**
Volume 12: **Properties of Copper**
Volume 13: **Copper Proteins**
Volume 14: **Inorganic Drugs in Deficiency and Disease**
Volume 15: **Zinc and Its Role in Biology and Nutrition**
Volume 16: **Methods Involving Metal Ions and Complexes in Clinical Chemistry**
Volume 17: **Calcium and Its Role in Biology**
Volume 18: **Circulation of Metals in the Environment**
Volume 19: **Antibiotics and Their Complexes**
Volume 20: **Concepts on Metal Ion Toxicity**
Volume 21: **Applications of Nuclear Magnetic Resonance to Paramagnetic Species**
Volume 22: **ENDOR, EPR, and Electron Spin Echo for Probing Coordination Spheres**
Volume 23: **Nickel and Its Role in Biology**

Contents of Volumes in the *Metal Ions in Life Sciences* Series

edited by the SIGELs

Volumes 1–4

published by John Wiley & Sons, Ltd., Chichester, UK (2006–2008)

<www.Wiley.com/go/mils>

and from Volume 5 on

by the Royal Society of Chemistry, Cambridge, UK (since 2009)

<www.rsc.org/shop/books/series/85.asp?seriesid=85>

Volume 1: Neurodegenerative Diseases and Metal Ions

1. The Role of Metal Ions in Neurology. An Introduction
   Dorothea Strozyk and Ashley I. Bush
2. Protein Folding, Misfolding, and Disease
   Jennifer C. Lee, Judy E. Kim, Ekaterina V. Pletneva, Jasmin Faraone-Mennella, Harry B. Gray, and Jay R. Winkler
3. Metal Ion Binding Properties of Proteins Related to Neurodegeneration
   Henryk Kozlowski, Marek Luczkowski, Daniela Valensin, and Gianni Valensin
4. Metallic Prions: Mining the Core of Transmissible Spongiform Encephalopathies
   David R. Brown
5. The Role of Metal Ions in the Amyloid Precursor Protein and in Alzheimer's Disease
   Thomas A. Bayer and Gerd Multhaup
6. The Role of Iron in the Pathogenesis of Parkinson's Disease
   Manfred Gerlach, Kay L. Double, Mario E. Götz, Moussa B. H. Youdim, and Peter Riederer

Volume 2: Nickel and Its Surprising Impact in Nature

Volume 3: The Ubiquitous Roles of Cytochrome P450 Proteins

Volume 4: Biomineralization. From Nature to Application

Volume 5: Metallothioneins and Related Chelators

Volume 6: Metal-Carbon Bonds in Enzymes and Cofactors

Volume 7: Organometallics in Environment and Toxicology

Volume 8: Metal Ions in Toxicology: Effects, Interactions, Interdependencies

Comments and suggestions with regard to contents, topics, and the like for future volumes of the series are welcome.

Met. Ions Life Sci. **2011**, *9*, 1–35

1

Metal Ion Binding to RNA

Pascal Auffinger, [1,*] ***Neena Grover*** [2] ***and Eric Westhof*** [1]

[1]Architecture et Reactivité de l'ARN, Université de Strasbourg, IBMC, CNRS, 15 rue René Descartes, F-67084, Strasbourg, France
< p.auffinger@ibmc-cnrs.unistra.fr >
< e.westhof@ibmc-cnrs.unistra.fr >

[2] Department of Chemistry and Biochemistry, The Colorado College, Colorado Springs, CO 80903, USA
< ngrover@coloradocollege.edu >

Metal Ions in Life Sciences, Volume 9 Edited by Astrid Sigel, Helmut Sigel, and Roland K. O. Sigel

Published by the Royal Society of Chemistry, www.rsc.org DOI: 10.1039/978184973251200001

ABSTRACT: RNA crystal structures have provided a wealth of information on localized metal ions that are bound to specific sites, such as the RNA deep groove, the Hoogsteen face of guanine nucleotides and anionic phosphate oxygens. With a number of crystal structures being solved with heavy metal derivatives and other "reporter" ions, sufficient information is available to estimate global similarities and differences in ion binding properties and to begin determining the influence of RNA and ions on each other. Here we will discuss the ions that are observed bound to RNA, their coordination properties, and the roles they play in RNA structural studies. Analysis of the crystallographic data reinforces the fact that ion interactions with nucleic acids are not easily interchanged between similarly charged ions. The physiological relevance of RNA-ion interactions, mainly involving K^+ and Mg^{2+} cations, needs to be analyzed with care as different structures are solved under very diverse ionic conditions. The analysis is complicated by the fact that the assignment is not always accurate, often done under sub-optimal conditions, which further limits the generalization about the types of interactions these ions can establish.

KEYWORDS: anions · cations · crystallography · divalent metal ions · hexamines · ions · lanthanides · magnesium · manganese · metal binding · monovalent metal ions · potassium · RNA · sodium · solvation · transition metals · trivalent metal ions

1. INTRODUCTION

The roles of metal ions in the formation of RNA structures and catalysis are inherently intertwined (for reviews, see [1–5]) and have been extensively investigated since the sixties [6]. Only a fraction of the ions that are involved in forming functional RNA molecules show up in RNA crystal structures. In addition, the X-ray structures may not necessarily correspond to active forms [7–9], as crystallography often requires non-physiological ionic conditions. Therefore, a series of biochemical and biophysical studies are needed, in conjunction with structural studies, to fully understand the role of metal ions in RNA structure formation and to clarify their potential role in RNA function [10]. To achieve this goal, knowledge of ion binding motifs is required and can only be gained by a thorough examination of existing crystal structures.

In this chapter, we are examining, primarily from a structural point of view, interactions of various monovalent, divalent, trivalent, and tetravalent ions seen in RNA crystal structures.

2. DETAILS OF ION COORDINATION

For precisely characterizing ion binding sites, it is essential to first gather structural details on ion binding modes [11]. Coordination distances and geometries are two of the criteria that are used for assigning ionic species to electron density spots [12–17]. For example, Li^+ and Mg^{2+}, display average Li^+/Mg^{2+}...Ow coordination distances below 2.2 Å. Other ions, like Rb^+ and Cs^+ display Rb^+/Cs^+...Ow coordination distances around 3.0 Å (even 3.6 Å for iodide). Statistical analysis of the structures deposited in the Cambridge Structural Database (CSD), the world repository of small-molecule crystal structures [18,19], provides estimates of coordination distances with a better accuracy than the larger and generally less precise structures deposited in the Protein Data Bank (PDB).

Several web tools are now available for analyzing metal ion distributions in the PDB and other databases. Programs, such as MeRNA, MINAS, and SwS, were designed for mapping metal ion interactions with nucleic acids. These tools are based on an exploration of sub-ensembles of nucleic acid structures with specific goals in mind.

MeRNA (http://merna.lbl.gov) is a database for metal ion binding sites in RNA (see Chapter 2 of this volume) that currently scans 389 PDB files to characterize RNA interactions with 23 different metal compounds [20].

MINAS (http://www.minas.uzh.ch) contains exact geometric information on the first and second-shell ligands of metal ions present in nucleic acid structures (see Chapter 2 of this volume).

SwS (http://www-ibmc.u-strasbg.fr/arn/sws.html) is a Solvation web Service designed to provide a statistical overview of the structure of the first solvation shell of nucleic acid base pairs that is useful for gaining a better understanding of crystal structures and for validating the results of molecular dynamics simulations [21]. Among all the features of this web service, SwS characterizes cation and anion binding sites located in the first solvation shell of selected base pairs extracted from RNA and DNA structures (Figure 1).

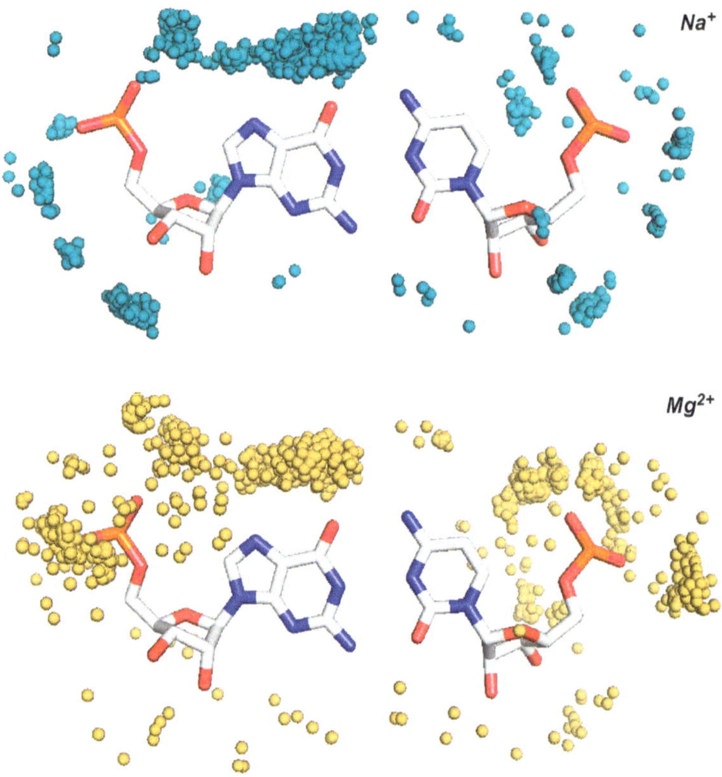

Figure 1. Na^+ (top) and Mg^{2+} (bottom) ion distributions around an rG=C base pair as calculated from PDB structures by the SwS web service [21]. These views depict the ion positions around an average rG=C base pair. For Na^+ and Mg^{2+}, 1227 and 1164 ions are shown, respectively.

3. PHYSIOLOGICAL RELEVANCE OF METAL IONS

A significant number of RNA structures show metal ions interacting with nucleic acids, either directly or through their first solvation shell. Almost all possible ion combinations have been tested in crystallization assays. Many of these metal ions are not biologically relevant and some are even toxic. Others are known cofactors for proteins with specific enzymatic functions, are found only in trace concentrations in cells and are unlikely to interact with nucleic acids *in vivo*. Indeed, the two cations that are the most frequently found in the vicinity of nucleic acids are the divalent Mg^{2+} and monovalent K^{+} ions. Compared to all other cations, Mg^{2+} ($\approx$1 mM) and K^{+} ($\approx$140 mM) display the highest concentrations of divalent and monovalent ions in intracellular fluids (cytosol) of practically all living cells and are consequently the cations with the greatest likelihood to interact with nucleic acid components [22–25]. The most prevalent monovalent cation in extracellular fluids is Na^{+} ($\approx$12 and 150 mM in cytosol and extracellular fluids, respectively). Therefore, it is assumed that Na^{+} is rarely in contact with nucleic acid elements. Indeed, besides Mg^{2+} cations, numerous experiments point to the necessity of having K^{+} cations in buffer solutions for activating RNA systems. For instance, the importance of K^{+} ions for ribosome structure and function became evident when significant unfolding of mammalian ribosomes was observed in their absence [26–28].

Other studies report that ribosomal activity is dependent on the presence of K^{+} or NH_4^{+} cations but is inhibited by Na^{+} [29–31]. A recent study suggests that Na^{+} and K^{+} cations allosterically regulate cooperative binding of human progesterone receptor to DNA [32]. The cooperative binding of the receptor to DNA is activated or repressed by Na^{+} and K^{+}, respectively. Interestingly, the apparent binding affinities of Na^{+} and K^{+} are comparable to their intracellular concentrations. Furthermore, progesterone receptors directly regulate the genes of a number of ion pumps and channels. This type of regulation seems analogous to that of Mg^{2+} riboswitches which are Mg^{2+}-sensing RNAs regulating Mg^{2+} concentration in cytosols [33–37]. A majority of Mg^{2+} in the cell is bound to cellular components and a significant portion of Mg^{2+} is bound to phosphometabolites, such as ATP [38], where its role is to neutralize the charge density of phosphates and to assist in the enzymatic hydrolysis reactions. Magnesium also stabilizes ribosome structures [28,39,40]. The growth of *Escherichia coli* cells under conditions of Mg^{2+} starvation results in ribosome depletion [41] and the *in vitro* association of the small and large ribosomal subunits to form intact ribosomes depend strongly on Mg^{2+} concentration [42–44].

4. MONOVALENT CATIONS

4.1. The Sodium and Potassium Cations

K^+ cations prevail over Na^+ cations in intracellular fluids and, consequently, are more likely to interact with nucleic acids. Despite this knowledge, Na^+ cations were (and still are) used in NMR and crystallographic buffers [40]. An analysis of ion distributions, based on crystallographic structures, around an rG=C pair using SwS web service [21] reveals that 2218 Na^+ and only 88 K^+ cations are detected in the solvation shell of this pair.

The first monovalent cations to be tentatively detected in nucleic acid structures were Na^+ cations in the vicinity of a DNA/drug complex [45]. Analysis of coordination distances indicates that assignment of K^+ cations would also have been appropriate. When a hybrid solvent model was applied to a DNA dodecamer, the solvent sites that were previously associated with water, were defined as hybrid sites that were alternatively occupied by water molecules and cations [46–50]. These findings were rapidly followed by the detection of bound cations through the use of crystallographic difference maps [51] and confirmed by Tl^+ substitution experiments [48].

Use of Na^+ cations in crystallographic experiments might be associated with improved crystallization conditions although this has not been documented. It was noted that substituting K^+ ions for Na^+ slightly decreased RNA hairpin stability [52] as well as the stability of some ribozymes [53]. However, using crystallization mother liquors containing predominantly Na^+ ions seems odd since K^+ cations are much easier to distinguish from water molecules for several reasons: first, because a K^+ cation displays an excess of 8 electrons over Na^+ cations; second, Na^+ ions are almost isoelectronic to water molecules; third, K^+ cations provide anomalous signals whereas Na^+ cations do not. K^+ specific anomalous signals have been successfully used in several studies [48,54]. For structural assignments one can solely rely on the specific Na^+...Ow coordination distance of 2.4 Å to detect sodium cations [46]. For potassium cations, similarity between the average K^+...Ow and O/N...Ow coordination distances (2.8 Å) may, however, slightly complicate the detection of K^+ cations.

The first specific monovalent binding site in RNA was identified crystallographically in the vicinity of an AA platform of a tetraloop receptor [55] (Figure 2). The presence of K^+ cations was inferred from Tl^+ and Cs^+ soaking experiments. The specificity of the AA platform for K^+ was confirmed by the fact that the related *Azoarcus* intron is six times more active in K^+ instead of Na^+ buffer. Another strong and specific K^+ binding site is buried deep inside the structure of a 58-nucleotide ribosomal RNA

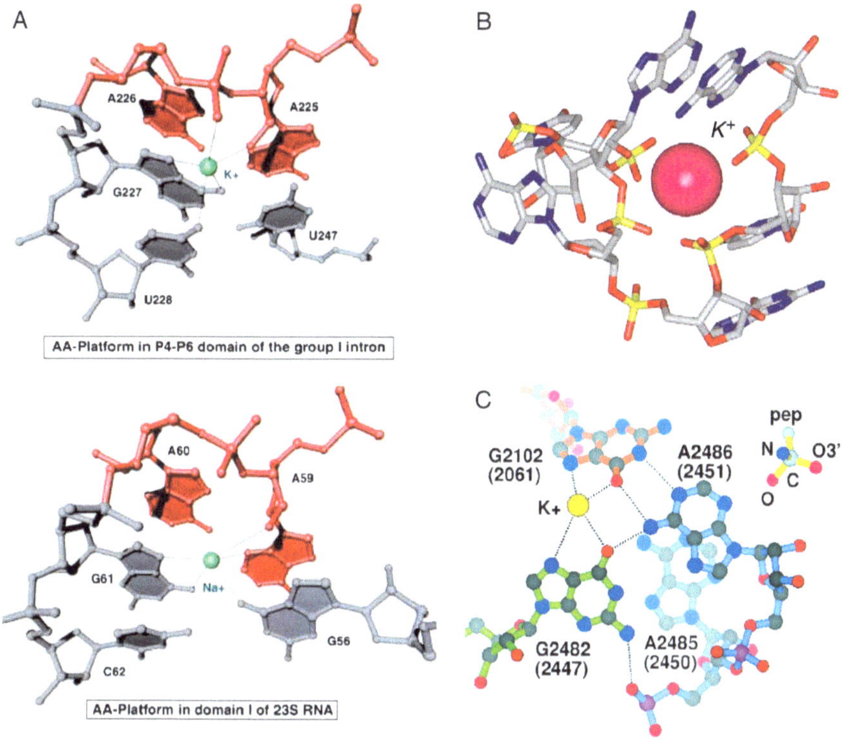

Figure 2. Various K^+ and Na^+ binding sites. (**A** – top panel) The AA platform with its bound K^+ from the P4-P6 domain of the group I intron from *Tetrahymena thermophila* [55] and (**A** – bottom panel) K^+ binding in the large ribosomal subunit of *Haloarcula marismortui* [40]. (Reproduced by permission from Ref. [174]). (**B**) K^+ binding pocket within a 58-nt ribosomal fragment (Reproduced by permission from Ref. [54]). (**C**) A conserved K^+ binding site in the ribosomal peptidyl transferase active site. Reproduced by permission from [57]; copyright (2000).

fragment. In this structure, six oxygen atoms from buried phosphate groups contact a K^+ cation [56] (Figure 2). A large number of monovalent ion binding sites (86 Na^+ and 2 K^+) contacting primarily the rRNA have been identified in the crystal structure of a large ribosomal subunit [40]. Although some interactions with anionic phosphate oxygens were identified, like the AA platform motif binding a Na^+ cation (Figure 2), the most common monovalent cation-binding sites appears to be located in the major groove of G·U wobble pairs. Binding of a Na^+ cation to an AA platform is similar to that observed for a tetraloop receptor [55].

Interestingly, a K^+ binding site has also been identified in close vicinity to the ribosomal peptidyl transferase active site [57] (Figure 2). In the ribosome, it is very likely that under physiological conditions most, if not all, of the monovalent binding sites are occupied by K^+ rather than Na^+ [40]. Indeed, functional specificity for K^+ has been observed in several natural RNAs [55]. For binding of K^+ and Na^+ to RNA see also [34,54,58–60].

Molar concentrations of monovalent cations often stabilize RNA structures [53,61,62]. Although they allow charge neutralization that collapses the polynucleotidic chains into a more compact structure, monovalent ions do not necessarily lead to the active fold of large RNAs in the absence of divalent cations [63]. For ribozymes Na^+ and K^+ cations do not seem to be directly involved in catalytic mechanisms but are necessary for the activity of all ribozymes [64].

4.2. Other Monovalent Cations

4.2.1. Lithium Cations

Lithium ions are often used in molar concentrations in crystallization solutions. Yet their low numbers of electrons (3 e^-) makes detection of Li^+ ions very difficult. There are 35 structures in the PDB database in which Li^+ was assigned to an electron density spot and only one of these is a nucleic acid structure, a DNA quadruplex that contains a Li^+ cation coordinated to an O2' atom and three water molecules. A tetrahedral arrangement with coordination distances under 2.0 Å is likely to be the most precise signature for these cations [15,17,65]. Regular tetrahedral arrangements should be systematically sought to identify Li^+ when this ion is present in the crystallization solution.

Interestingly, Li^+ cations were shown to promote catalytic activity of various ribozymes such as the hammerhead, the hairpin and the VS ribozymes at molar concentrations [62,66]. Li^+ cations are also considered being good at RNA folding [67]. An *in vitro* vitamin B_{12} RNA aptamer selected in 1 M LiCl buffer needed Li^+ cations; these ions could not be substituted by Na^+ or K^+ indicating that this aptamer is adapted to the conditions in which it was selected [53].

Li^+ cations are used to control bipolar disorders. It has been suggested that competition between Li^+ and Mg^{2+} ions for divalent ion binding sites in cellular components is the underlying theme in putative mechanisms of Li^+ action [68].

4.2.2. Rubidium Cations

Rubidium cations could be identified in four different DNA structures. Rubidium is not normally found in living organisms. Due to its similarity with potassium cations, Rb^+ concentrates into intracellular fluids and short-lived radioisotopes can be used as biomarkers. Crystallographers have used rubidium as a probe for detecting, otherwise difficult to characterize, Na^+ cation binding sites with the assumption that both cations would bind at the same locations [40,69,70]. Water coordination distances for Rb^+ are not well defined but are expected to be around 3.0 Å with fluctuating coordination numbers between 6 and 8 [17]. With the exception of Na^+ and Li^+ cations, K^+, Rb^+, Cs^+ and Tl^+ cations exhibit significant anomalous signals, especially the latter three cations, providing an opportunity to use single wavelength anomalous diffraction (SAD) technique to solve nucleic acid structures. The combination of high resolution and anomalous diffraction data can also pinpoint partially occupied binding sites [69]. Yet, Rb^+ and Cs^+ substitutions are, at present, seldomly used for detecting monovalent cation binding sites in biomolecular systems.

4.2.3. *Cesium Cations*

Cesium cations have a slightly larger ionic radius than rubidium cations ($\approx$3.1 Å [17]). These have been characterized in one DNA and three RNA structures deposited in the PDB. Use of Cs^+ in crystallography is similar to that of Rb^+ [55]. Cs^+ displays larger anomalous signals but its coordination properties differ more from those of K^+ than Rb^+ thus limiting its use. Cs^+ cations, like $[Co(NH_3)_6]^{3+}$ and unlike Mg^{2+} cations, bind to the major groove of single and tandem G•U. This specificity led to the development of a strategy to solve the phase problem using the insertion of appropriate G•U motifs [71].

4.2.4. *Thallium Cations*

Tl^+ cations represent a more appropriate K^+ mimic than Rb^+ and Cs^+ cations, given very similar coordination distances ($\approx$2.8 Å) and enthalpies of hydration (≈ -78 kcal/mol) [48]. Tl^+ cations are nearly interchangeable with K^+ in Na^+/K^+ pumps [72] as well as in several other biochemical systems [73,74]. However, one must be careful in interpreting Tl^+ containing structures, as Tl^+ is not a biologically relevant cation. It has *d* electrons that favor covalent bonding. Hence, soft atoms such as sulfur might interact differently with Tl^+ than with K^+ [48].

Despite their high toxicity, Tl^+ cations are used in crystallography since they display significant anomalous scattering of X-rays. Moreover, Tl^+ is a strong scatterer and can be identified simply by its peak height in standard electron density maps for fully occupied sites [59]. Hence, the inclusion of Tl^+ cations in crystallization liquors provides a sensitive monovalent cation detection system. K^+/Tl^+ substitution was used in the interpretation of several nucleic acid structures such as RNA and DNA duplexes [48,69,70,75–77], the *Tetrahymena* ribozyme P4-P6 domain [55], a 58 nucleotide ribosomal fragment [56], the signal recognition particle (SRP) [78], a viral RNA pseudoknot [59] and the HDV ribozyme [64]. With strong scattering and strong anomalous signal, it is not surprising that more monovalent binding sites are observed with Tl^+ than with any other monovalent ion [48]. By using anomalous diffraction techniques, partial occupancies can be easily attributed to Tl^+ [48,55]. In a crystal structure of a DNA duplex, 13 partial Tl^+ binding sites could be identified. It was inferred that the number of monovalent cation binding sites substantially exceeds previous observations using less strongly scattering Rb^+ (one 50% occupied position) or Cs^+ ions (four 20% occupied positions). Indeed, the cumulative occupancy for the 13 sites is close to 2.26 [48]. ^{205}Tl can be detected by NMR techniques and is useful in characterizing bound monovalent cations that exchange slowly [79].

4.2.5. Ammonium Cations

Ammonium cations are non-metal ions that are very difficult to observe by crystallography and are certainly not physiologically relevant given their cellular toxicity. It has been reported that the hammerhead, hairpin and VS ribozymes are all catalytically active in solutions containing ammonium ions and no divalent cations [62]. The hairpin ribozyme, in particular, shows a fourfold increase in cleavage rate when magnesium ions are replaced by ammonium ions at a 1.0 M minimal concentration pointing to the fact that NH_4^+ cations have a great structuring potential. NH_4^+ cations mainly bind to phosphate groups and display coordination distances comparable to those of water molecules (2.8–3.0 Å).

5. DIVALENT CATIONS

5.1. The Magnesium Cation

5.1.1. General Properties

Mg^{2+} cations are physiologically the most relevant divalent ions for the formation of RNA structures and its functions. Mg^{2+} has a smaller ionic radius than Na^+, K^+, and Ca^{2+}. It is a hard ion with only ten electrons. Mg^{2+} predominantly binds six ligands in an octahedral geometry and prefers hard ligands, such as oxygen, over nitrogen and sulfur atoms. The Mg^{2+}...Ow distance is 2.1 Å. Mg^{2+} binding to RNA is difficult to study as this cation is not amenable to direct spectroscopic studies, and is difficult to distinguish from Na^+ and H_2O in crystal structures (given that they have almost equal numbers of electrons). High resolution (1.5 Å or better) and high occupancy are necessary to accurately identify Mg^{2+} ions in crystal structures. The metal-water distance (bond length) is what distinguishes Na^+ from Mg^{2+} and is only interpretable at high resolution. Deprotonated water ligands further complicate unambiguous assignment of Mg^{2+} ions in crystal structures.

5.1.2. Outersphere Binding

Mg^{2+} interacts with RNA in both outersphere and innersphere modes. In the former Mg^{2+} associates with the negatively charged phosphate backbone in a diffuse manner [80,81]. Mg^{2+} is preferred for charge neutralization over the monovalent ions (K^+, physiological) because of its greater charge density and a lower entropic cost for localization of these ions [2,3,82]. A compactly folded RNA has a greater charge density and is likely to accumulate a greater number of Mg^{2+} ions in its surroundings. The assumption is that Mg^{2+} ions that are freely exchanged with bulk Mg^{2+} (diffusely bound) are not likely to show up in the crystal structures due to their short occupancy times. A few Mg^{2+} outersphere interactions that show up in RNA are presumed to be playing a role that may be different from the bulk ions.

Preferred ligands for outersphere interactions with Mg^{2+} are anionic phosphate oxygens and electroneagative atoms of bases. Outersphere interactions are primarily seen in the major groove of A-form helices, that display a pronounced negative electrostatic surface potential. Hydrated Mg^{2+} ions bind to the floor of the deep groove of A-form helices by forming hydrogen bonds to acceptors atoms (mostly G(N7) and G(O6)) especially in GpG and GpU steps. The residence time for these ions is long enough to differentiate them from bulk ions [83,84]. The hexahydrated Mg^{2+} prefers to bind to the Hoogsteen face of guanines, especially when the guanine is part of a non-canonical base pair. This is seen in the HIV-1 dimer initiation site [54], in the NMR structure of a small construct containing $^{5'}GU^{3'}/^{3'}UG^{5'}$ base pairs [85], and in the P4-P6 domain of *Tetrahymena* (at a tandem $^{5'}GG^{3'}/^{3'}UU^{5'}$) [86]. A water-mediated charge transfer from guanine to magnesium is also thought to play a role in stabilization of interactions with guanine and is thought to occur through a cooperative mechanism [87].

Although, magnesium outersphere interactions with guanine, especially at the non-canonical G•U base pairs are often cited, it should be noted that no magnesium binding is seen in $^{5'}GU^{3'}/^{3'}UG^{5'}$ in the crystal structure of the recognition domain [78] or in the NMR structure of the tetraloop receptor complex [88]. Expectation of outerspace ion interactions with the G•U base pairs is due to the presence of G(N7), G(O6), and U(O2) atoms in the major groove, creating an unusual hydrogen bond acceptor environment. A variety of ions interact with this site while magnesium hexahydrates are not systematically located there [58,89]. Several different explanations are possible for this observation, including low-resolution of the structures or use of magnesium analogs, making the assignment of magnesium binding to a particular site problematic. Hence, caution is advised before overanalyzing specific ion binding properties.

In a 2.4 Å resolution structure of the 50 S subunit of *Haloarcula marismortui*, nine out of 116 Mg^{2+} ions are bound via outersphere interactions to the 3045 nucleotides of the RNA (Figure 3). These Mg^{2+} ions were characterized by visually identifying the octahedral arrangement of water molecules located in the major groove of A-form helices [40]. As expected, hexahydrated Mg^{2+} cations are much more mobile than partially dehydrated cations that form direct bonds with RNA atoms [90,91].

NMR and crystallographic data often yield slight differences in metal binding sites [58,85,88]. The data for magnesium binding are reliable when magnesium ions are being used, instead of substitutes, as demonstrated throughout this chapter.

5.1.3. Innersphere Binding

For an RNA fragment to form direct interactions with metal ions, the local electrostatic environment and the energetics of metal dehydration are among the many factors that play a role. Partial dehydration of Mg^{2+} is more likely in the non-helical region where neighboring phosphates are in close proximity [92]. As seen in the P4-P6 group I intron and 58-nucleotide rRNA crystal structure [86,92,93], chelated metal ions would be expected to be buried and relatively inaccessible to solvent due to phosphate charge neutralization. Even though dehydration of hexahydrated magnesium is energetically expensive, crystal structures of RNA show that specific environments allow for the formation of multiple direct bonds between magnesium and RNA. The properties of metal ions in the presence of RNA are likely to be different from the free metal ion; hence, estimates of the energetic requirements of forming innersphere coordination are rather imprecise. Studies on model systems such as nucleotides and small RNAs will lead to better understanding of direct interactions between metal ions and RNA [94,95]. Advances in theoretical and computational modeling, such as non-linear Poisson-Boltzmann calculations, microenvironment analysis, and molecular dynamic simulations (in concert with improved NMR approaches) will play a key role in understanding the dynamic behavior of magnesium-RNA interactions [96,90–92].

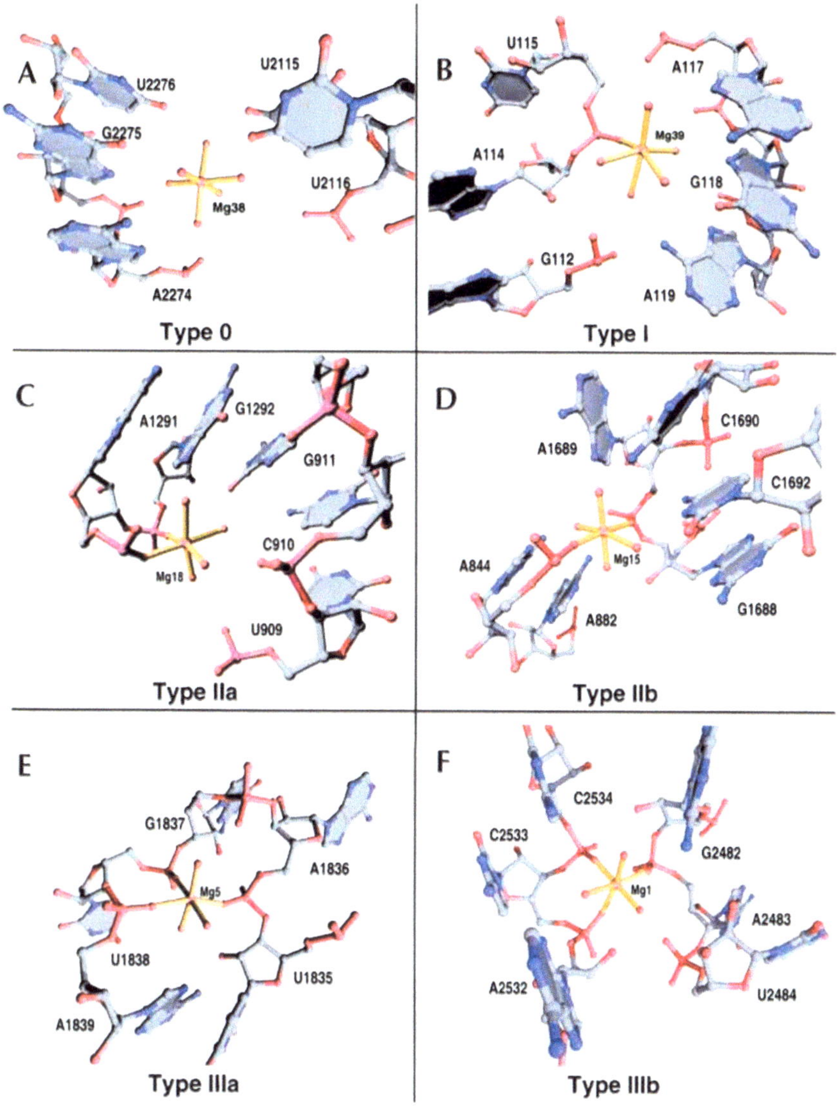

Figure 3. An example of each of the six binding modes of interaction between Mg^{2+} and RNA derived from a crystal structure of the large ribosomal subunit of *Haloarcula marismortui* crystallized at 2.4 Å resolution involving six to three water molecules and zero to three direct contacts with RNA hydrophilic atoms. Reproduced by permission from [174]; copyright (2001).

Several high-resolution small RNA structures show potential magnesium binding sites [54,59,97]. A few comparative analysis of ion binding sites have also been performed [54,88]. Differences in crystallization conditions along with the use of magnesium analogs makes it challenging to generalize magnesium innersphere interactions. Some examples of magnesium binding sites and types of interactions are discussed here to provide insight into the role of innersphere interactions in RNA structures.

The loop E motif of the ribosomal RNA is a classical example of innersphere coordination of Mg^{2+} ions with RNA. This structure has been solved at 1.5 Å and shows Mg^{2+} bound in its octahedral and water in its tetrahedral geometry. Due to the high resolution, Mg^{2+} and water are distinguishable by their bond lengths of 2.1 Å versus 2.7 Å, respectively [97]. Two of the bound Mg^{2+} are hexahydrated and interact with the backbones of both strands. Three of the bound Mg^{2+} show one direct interaction each with RNA. One Mg^{2+} ion is primarily coordinated to G(N7), G(O6), and anionic phosphate oxygens. Three additional waters are proposed to bridge interactions between two Mg^{2+} ions that are seen bound to adjacent phosphate groups [97].

A classification of Mg^{2+} binding sites was derived from the observation of 116 Mg^{2+} binding events to the large ribosomal subunit of *Haloarcula marismortui* [40]. Ten different binding modes were observed: those with zero (type 0), one (type I), two (type II), three (type III), four (type IV), or five (type V), bonds between the Mg^{2+} ion and RNA. Each of these were found 9, 37, 45, 18, 1, and 1 times, respectively (Figure 3). The Type II and III interactions were further subdivided into two groups based on the three-dimensional arrangement of their RNA, protein, and/or water ligands. Complete dehydration of Mg^{2+} cations has not been reported yet. No octahedral coordination geometry was observed in five cases leading to description of a type X. Most innersphere interactions involve phosphate moieties that are in close proximity and require charge neutralization to form a particular structure. Anionic phosphate oxygens serve as ligand for Mg^{2+} ions in a majority of innersphere interactions. The N7 of purines or O6 of guanine are good outersphere ligands for Mg^{2+} and form innersphere interactions with Mg^{2+} in approximately 20% of the cases.

Recently, two different magnesium sensing riboswitches have been reported [33,34,37]. In the M-box RNA, six Mg^{2+} binding sites are seen, three of which are considered crucial for forming the compact structure [37]. One Mg^{2+} is bound via four innersphere interactions in the loop 5 region and is considered to be the most significant. It is likely that additional magnesium binding motifs exist in RNA that have not yet been identified.

Experimental approaches that allow studying RNA under physiologically relevant conditions, such as fluorescence resonance energy transfer (FRET), transient electric birefringence (TEB) or gel shift assays provide crucial insights into changes in RNA structures under varying concentrations of magnesium ions and built a link between RNA structure and function [95,98]. Examining site-specific binding of magnesium is remains a challenging problem as Mg^{2+} cannot be perfectly substituted by any other ions.

5.2. Manganese(II) as a Magnesium Substitute

Divalent manganese (Mn^{2+}) is similar to Mg^{2+} cations in some of its chemical properties. Yet, these cations have different biological roles. An early survey of crystal structures from the CSD and PDB indicated that Mg^{2+} cations bind preferentially to oxygen atoms while Mn^{2+} prefers nitrogen [99]. It has also been reported that the affinity of Mn^{2+} and Mg^{2+} cations for A(N7) is similar, while that of Mn^{2+} for G(N7) is larger than that of Mg^{2+} [94,100]. The fact that Mn^{2+}, as a softer metal, is better able to coordinate to the softer sulfur than Mg^{2+} has been quite successfully used for rescue experiments [10].

Manganese has two major functions in enzymes: (1) as a Lewis acid, for which its properties can be compared with those of magnesium, zinc, and calcium, and (2) as an oxidation catalyst, for which it can be compared with iron and copper. Divalent manganese has a radius of approximately 0.75 Å, somewhat larger than that of magnesium (0.65 Å) leading to Mn^{2+}...O coordination distances larger by 0.1 Å than the Mg^{2+}...Ow coordination distance (≈2.2 and 2.1 Å, respectively [14]). When divalent manganese replaces magnesium in the active site of a magnesium-utilizing enzyme, the catalytic activity of the enzyme is often maintained. This observation and the known paramagnetic properties of manganese have led to the widespread use of Mn^{2+} in NMR [88,101] and EPR studies [10,102–104]. Magnesium, however, is rarely a competent replacement for divalent manganese in enzymes. Given this higher affinity for nitrogens, Mn^{2+} prefers, as a transition metal with *d* electrons, to bind to lysine and histidines [99] and, therefore, probably discriminates between proteins and RNA on this basis.

The effects of the substitution of Mg^{2+} by Mn^{2+} cations or of Mn^{2+} soaking experiments on RNA structure and catalytic activity are diverse. Sometimes, Mn^{2+} ions appear as effective as Mg^{2+} ions for folding and catalytic activity. For the RNase P [105] and the HDV ribozyme [106], Mn^{2+} can substitute for Mg^{2+} with only a slight change in activity. In other instances, the reaction rate can be accelerated by such substitutions as seen for the minimal hammerhead ribozyme [107]. Mn^{2+} cations are also able to inhibit the catalytic reaction of ribozymes such as the hairpin ribozyme. In this case, Mn^{2+} cations are not able to correctly fold this RNA, whereas Sr^{2+} and Ca^{2+} can [108]. Additional non-specific cleavage with Mn^{2+} was also observed with group II introns [109].

Mn^{2+} cations can induce localized conformational changes. For the SRP system [78] and a DIS sub-type [54], localized conformational changes were reported when Mg^{2+} is replaced by Mn^{2+}. For the glycine riboswitch, Mg^{2+}, Mn^{2+}, and Ca^{2+} facilitate glycine binding while other divalent cations did not [110]. For the SRP particle, Cs^{+} and Mn^{2+} significantly increase the protein-RNA binding affinity over that observed in the presence of K^{+} and Mg^{2+} [78] and biochemical studies assessed that Mn^{2+} cations are more efficient than Mg^{2+} for folding an mRNA pseudoknot [67]. Mn^{2+} soak of the lysine riboswitch did not replace Mg^{2+} [60]. Replacement of Mg^{2+} by Mn^{2+} cations cause ribosomal crystals to become twinned [40].

Overall, the data with these substituent metal ions may lead to incorrect assumptions about the role of magnesium ions in RNA structure and functions, since Mn^{2+} and Mg^{2+} do not always bind to the same sites as a result of different affinities for hard and soft atoms.

5.3. Other Alkaline Earth Metal Cations

Calcium is by far the most abundant divalent ion in the human body. Yet, its concentration in the cytosol is below the micromolar range while that of Mg^{2+} is in the millimolar range. Hence, Ca^{2+} cations are not likely to interact with nucleic acid components. The Ca^{2+}...Ow coordination distance (2.4 Å) is close to that of Na^{+}. The intermediate Sr^{2+} cations have no particular biological functions. Non-radioactive strontium is not toxic in low doses. In the human body, Sr^{2+} can substitute for Ca^{2+} cations in bones. The Sr^{2+}...Ow coordination distance is $\approx$2.6 Å. Ba^{2+} is the largest cation in the alkaline earth series with Ba^{2+}...Ow coordination distances around 2.8 Å (close to that of K^{+}) and is poisonous.

Ca^{2+}, Sr^{2+}, and Ba^{2+} display anomalous signals [69] and were reported to behave similarly. DIS crystal structures indicated that Ca^{2+}, Sr^{2+}, and Ba^{2+} cations display a clear preference for outersphere coordination to deep grove Hoogsteen sites of guanines leading to much less localized binding, although, Ba^{2+} establishes in one occurence innersphere coordination to a G(N7) atom [111]. These group II cations induce a distortion of the DIS helix leading to a lack of polymorphism with Mg^{2+} structures [54]. On the other hand, in a high-resolution leadzyme structure crystallized in the presence of a mix of Mg^{2+} and Sr^{2+}, strontium cations did not displace or perturb bound Mg^{2+} cations seen in the absence of Sr^{2+} [7]. A subdomain of the hepatitis C virus displays similar conformations when crystallized in the presence of Sr^{2+} or Mg^{2+} cations [112] (Figure 4).

Like other divalent cations, Ca^{2+}, Sr^{2+}, and Ba^{2+} do not bind to G•U motifs [71]. Ca^{2+} binds directly to phosphate groups [113,114] (Figure 4). In

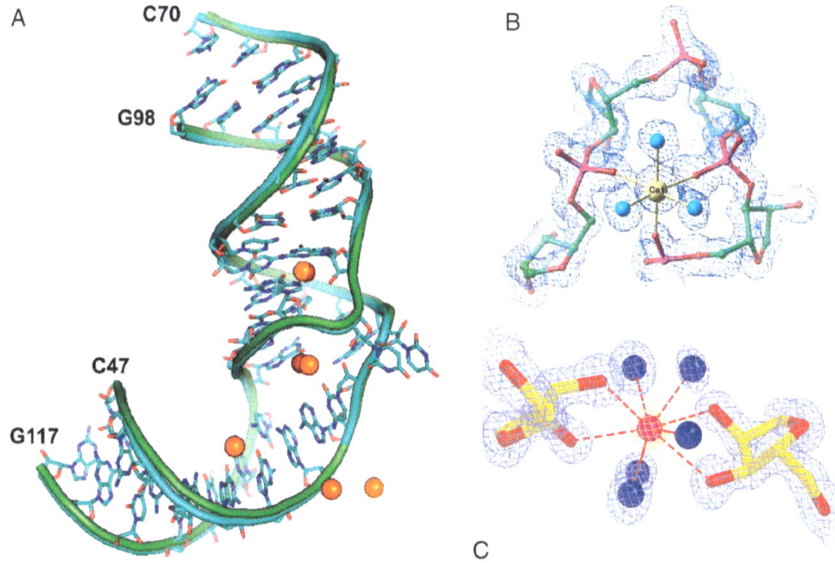

Figure 4. Examples of binding of Ca^{2+} and Sr^{2+} to RNA. (**A**) Superposition of two structures (one with Mg^{2+}, the other with Sr^{2+}) of the subdomain IIa of the hepatitis C virus. Reproduced by permission from [112]; copyright (2008). (**B**) A Ca^{2+} cation stabilizing a sharp turn by interacting with three anionic oxygen atoms. Reproduced by permission from Ref. [113]; copyright (1998). (**C**) A Sr^{2+} ion involved in crystal packing interactions. Reproduced by permission from Ref. [116]; copyright (1999).

the TAR RNA structure (1.3 Å), Ca^{2+} binding hinders the formation of an active structure [113]. The crystallization was performed in the presence of Mg^{2+}, indicating that the structure formed in the presence of Ca^{2+} is prefered over the presumably active structure in the presence of Mg^{2+}. Thermodynamic analysis of TAR RNA constructs shows differences in stability in the two divalent ions [95]. In the crystal structure of the malachite green aptamer Sr^{2+} is required for crystallization (Ca^{2+} and Ba^{2+} did not substitute for it). Indeed, a Sr^{2+} cation bridges two symmetry-related molecules in the crystal lattice through innersphere contacts with an U(O2) and a phosphate group oxygen atom [115]. In another structure, a Sr^{2+} cation interacts with the O3' and O2' atoms of two symmetry-related terminal nucleotides and completes its coordination sphere with five additional water molecules [116] (Figure 4).

Sr^{2+} and Ba^{2+} maintain the catalytic activity of the HDV ribozyme [106]. Yet the affinity of Ba^{2+} and Sr^{2+} for the P4-P6 domain of the *Tetrahymena* group I intron ribozyme was two orders of magnitude smaller than that of other cations [117]. Similarly, these cations are less effective than Mg^{2+} in folding a mRNA pseudoknot, with Ca^{2+} displaying the weakest activity [67].

5.4. Divalent Transition Metal Cations and Lead(II)

5.4.1. *Cobalt(II) Cations*

Co^{2+} is a key component of cobalamin (vitamin B_{12}) and is only found in trace amounts in most organisms. Hence, it is not likely to interact with nucleic acid components. It is used in crystallography and biochemical assays. DIS crystal structures show Co^{2+} binding to the major groove N7/O6 atoms of guanine Hoogsteen sites [54]. However, the site occupancies are weak (between 0.16 and 0.82). Crystal structures of the hammerhead ribozyme display Co^{2+} in both inner- and outersphere coordination [62,118]. Most crystal structures of DNA molecules indicate a strong direct binding to G(N7) atoms (see also an early $tRNA^{Phe}$ structure [119]). Direct contacts with hydrophilic atoms imply average coordination distances around 2.1 Å. Note that a quantum mechanical study emphasizes the preference of transition metals (Co^{2+} and Mn^{2+}) over Mg^{2+} for binding to N7 atoms of guanines [100].

5.4.2. *Nickel(II) Cations*

Nickel plays important roles in the biology of microorganisms and plants. As found in six DNA structures, Ni^{2+} coordinates exclusively to G(N7) atoms (see for example [120]) (Figure 5). In protein/nucleic acid structures, nickel atoms prefer to interact with amino acid residues [121]. Average Ni^{2+}...OW coordination distances are around 2.1 Å.

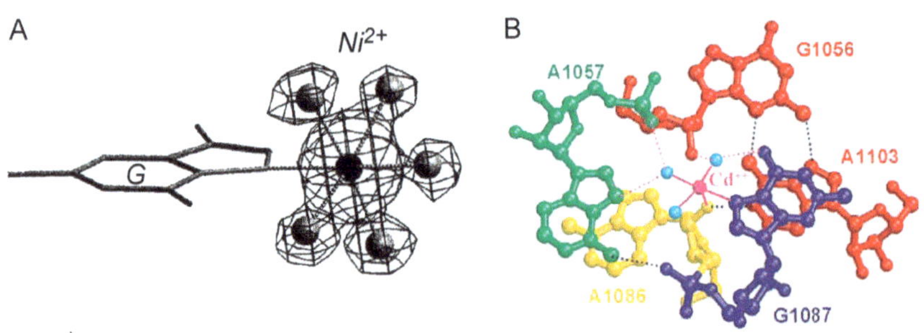

Figure 5. Binding of Cd^{2+} and Ni^{2+} ions to guanine residues. (**A**) Binding of Ni^{2+} and associated waters to a guanine Hoogsteen site as observed in a netropsin bound DNA duplex at 1.58 Å resolution. Reproduced by permission from [120]; copyright (1999). (**B**) A Cd^{2+} cation that stabilizes sharp turns in a four-way junction as observed in a ribosomal RNA/protein structure. Reproduced by permission from [130]; copyright (1999).

5.4.3. *Copper(II) Cations*

Copper is the third most abundant transition metal in the body and in the brain. The major oxidation states for copper ions in biological systems are cuprous Cu^{+} and cupric Cu^{2+}; intriguingly, the former is more common in the reducing intracellular environment and the latter in the oxidizing extracellular environment [122].

No RNA structure associated with Cu^{2+} has been deposited into the PDB. It has been stated that Cu^{2+} displays a higher affinity for G(N7) than Zn^{2+}, Cd^{2+}, and Mg^{2+} [123]. In this respect, one Z-DNA structure shows clear interactions of Cu^{2+} with G(N7) atoms [124]. Average coordination distances are close to 2.0 Å and Cu^{2+} binds, like Hg^{2+} and Au^{3+} to the N3 atom of a modified purine [125].

5.4.4. *Zinc(II) Cations*

Next to iron, zinc is the second most abundant transition metal in the human body. Its concentration is the highest in the brain where it matches that of magnesium [122]. Zinc cations have a different binding profile than Mg^{2+} or Mn^{2+} cations. The coordination number of Zn^{2+} varies between four and six with Zn^{2+}...Ow distances around 2.1 Å. These cations display a strong preference for interacting with nitrogen and sulfur atoms [14] and, in nucleic acids, with G(N7) atoms [123].

Zn^{2+}, like all non-Mg^{2+} divalent ions, is not likely to interact with RNA molecules *in vivo*. Their preference for binding to proteins arises from their increased affinity for nitrogen and sulfur atoms. In DIS structures, Zn^{2+}, like most transition metals, interacts with nucleobases, especially with G(N7) atoms [54]. It has been reported that the minimal hammerhead is catalytically active with Zn^{2+} cations in the presence of spermine [126]. Zn^{2+} binding is similar to Mg^{2+} at tandem G•A steps [127].

5.4.5. *Cadmium(II) Cations*

Cadmium has only been found in marine diatoms living in zinc-depleted environments. In all other organisms cadmium is considered to be toxic. The Cd^{2+} cation shows a clear preference for binding to G(N7) atoms when compared to Mg^{2+} cations [123]. Cadmium (^{113}Cd) is used in NMR spectroscopy [79] and in EPR silencing experiments [128]. Given their thiophilic character, Cd^{2+} cations are used, similarly to Mn^{2+}, in phosphorothioate rescuing experiments [129]. However, it is important to note that addition of a thiophilic metal ion rescues some reaction steps but has deleterious effects on others.

Three Cd^{2+} cations are found in a ribosomal fragment bound to protein L11. One ion is thought to stabilize the association of the two complexes in the asymmetric unit, but the other two are integral to the RNA structure [130] (Figure 5). Cd^{2+} was shown to bind like Mg^{2+} to tandem G•A steps [127]. Its average coordination distance is close to 2.1 Å.

5.4.6. Mercury(II) Cations

Although considered to be toxic, mercury, is an important component of Chinese, Tibetan, and Ayurvedic medicine. Mercury has no known physiological role as a trace element. One weak Hg^{2+} ion binding site is observed in a crystal structure of a 5S rRNA domain [97]. Hg^{2+} intercalates also between a T-T noncanonical base pair [131] that resembles the binding of an Au^{3+} between a guanine and a cytosine of a GxC base pair (see Figure 8 in Section 7.1). In such complexes, mercury induces deprotonation of the N3 atom of each thymine, which yields a highly stable complex. In a ribosomal fragment bound to protein L11, a mercury cation binds to the N3 atom of a U with a coordination distance of 2.4 Å [130]. Interestingly, Hg^{2+} is bound to nucleobases and not to amino acids in this structure. Hg^{2+} was also observed in an early structure of $tRNA^{Phe}$ [132].

5.4.7. Lead(II) Cations

Lead is a poisonous metal with no known medicinal applications [133]. It was recognized very early that Pb^{2+} is involved in RNA-specific position cleavage [134,135]. Hence, it is commonly used as a chemical probe for determining RNA three-dimensional structures.

A low-resolution crystallographic structure reveals three Pb^{2+} tRNA binding sites [136]. In another low-resolution structure of the specificity domain of the RNase P, 23 Pb^{2+} atoms were identified [137]. Finally, in a DIS structure, three partially occupied binding sites were described [54]. Unfortunately, given the cleaving activity of Pb^{2+} on the *in vitro* selected leadzymes, no crystal structures with lead were obtained [7]. Hence, no clear consensus on Pb^{2+} binding sites can be proposed. Pb^{2+} has been used to obtain anomalous diffracting derivatives of an RNase P fragment [138].

6. TRIVALENT CATIONS

6.1. Hexammine Cations ($[Co/Ru/Rh/Os/Ir(NH_3)_6]^{3+}$)

Hexammine cations have no biological functions but are sometimes used to aid crystallization [139] and as heavy atom derivatives that can provide phase information through multiple or single wavelength anomalous diffraction (MAD or SAD) methods [140,141]. They are often observed to bind to nucleic acids through direct contacts involving the ammine groups of the cations. Hexammine cations bind essentially to major groove atoms, to oxygen atoms of phosphate groups and less frequently at other sites. Interestingly, they are primarily used for solving the structure of nucleic acid and nucleic acid analogs (129 structures) and are much less useful in protein crystallography (12 structures). The use of $[Co(NH_3)_6]^{3+}$ cations prevails (109 structures) over the use of the heavier ruthenium, rhodium, osmium, and iridium hexammines (20 structures in total).

For DNA systems, hexammine cations are known inducers of the left-handed Z-form and are claimed to stabilize this form much more effectively than Mg^{2+} cations [142,143]. This justifies their use in the crystallization of a large number of Z-DNA double helices. Yet, for RNA systems, there is no real evidence that these cations drastically alter conformational equilibria. Examination of the SRP in the presence of various cations showed that this RNA structure accommodates a variety of metal ions without structural changes [78].

In *Neurospora* VS ribozyme, $[Co(NH_3)_6]^{3+}$ ions induce a "similar" fold to Mg^{2+} cations at a 33-fold lower concentration but without leading to an active structure [144]. In mixed-metal kinetic experiments of this ribozyme, $[Co(NH_3)_6]^{3+}$ does not inhibit Mg^{2+}-induced self-cleavage. More generally it was reported that hexammine cations are efficient at folding RNA [63]. When included in crystallization solutions, low concentrations of cobalt hexammine chloride (0.1–1.0 mM) dramatically increased the number, size, and growth rate of P4-P6 crystals [139]. In the hairpin ribozyme, cobalt(III) hexammine can replace Mg^{2+} cations for all folding and catalytic functions implying that Mg^{2+} cations are not directly involved in the catalytic reaction [108,145,146]. On the other hand, cobalt hexammine is a potent inhibitor of the HDV ribozyme [64].

The question of whether hexammine cations are appropriate mimics of $[Mg(H_2O)_6]^{2+}$ cations is often debated. It is generally understood that these cations are not perfect analogs of magnesium hexahydrate. Although their ionic radii are close (4.1 Å for Mg^{2+} hexahydrates and 3.9 to 4.5 Å for Co^{3+} and Ir^{3+} hexammines, respectively), their charge (+3/+2) and the number of associated hydrogen atoms (18/12) differ. In several instances, $[Co(NH_3)_6]^{3+}$ was observed to bind at locations also occupied by $[Mg(H_2O)_6]^{2+}$ and $[Mn(H_2O)_6]^{2+}$ ions [78]. In many structures, cobalt hexammine coordinates to GpG steps in the major groove [139]. However, the details of coordination of hexammine and hexahydrate cations are different [78]. Clearly, some sites can discriminate between these ions, as observed in the P4-P6 domain, the SRP [78], and the smaller HIV-1 dimerization initiation site [54]. Tandem G•U pairs were reported to be good hexammine cation binders [71,78] but displayed no specificity for divalent cations [89,147]. However, an NMR investigation found that tandem G • A pairs displayed a good affinity for divalent cations and no affinity for hexammine cations [148]. In at least one instance, a cobalt hexammine cation had all six of its ammine groups engaged in hydrogen bond contacts with RNA atoms [149]. A rare innersphere coordination to the RNA by a $[Co(NH_3)_4]^{3+}$ cation has also been reported [54]. Besides recent Raman microscopic analysis of RNA crystals supported the fact that $[Co(NH_3)_4]^{3+}$ cations can displace innersphere bound magnesium cations [150], indicating that the properties of these coordination complexes is affected by the RNA environment.

The observation that hexammine and monovalent cations bind more efficiently to the minor groove of G • U pair motifs, such as tandem G • U pairs [78,89] led to an efficient general strategy for solving the phase problem in RNA crystallography [71]. In this method, G • U tandems are grafted into the target sequences with the expectation to create a highly specific hexammine binding site. This "rational" soaking strategy is an improvement over earlier "soak and pray" strategies and has been successfully used in the resolution of large RNA structures [151].

Along with cobalt hexammine, four other hexammine compounds, based on ruthenium, rhodium, osmium, and iridium have been used as crystallization agents. In the resolution of an hybrid RNA duplex [152], $[Co(NH_3)_6]^{3+}$, $[Rh(NH_3)_6]^{3+}$ and $[Ir(NH_3)_6]^{3+}$ complexes were isomorphous, although the resolution increases from the lighter to the heavier cation (2.2 Å for cobalt hexammine; 1.8–1.6 Å for rhodium hexammine; 1.5 Å for iridium hexammine). A Z-DNA structure crystallized in the presence of ruthenium hexammine, suggested that the interaction of these cations with adenines induced a tautomeric shift from the amino to the imino form, disrupting the canonical A-T base-pairing scheme [153]. Indeed, metal ion binding to nucleobases can alter their pK_a values [154]. Iridium hexammine is observed in six RNA and one DNA structure (see for example [155]). These heavy and bulky cations were used to obtain the heavy atom derivatives needed for solving the RNA nucleic acid structures (Figure 6). Osmium hexammine was used in several studies [56,139]. Even though osmium hexammine is almost isoelectronic to $[Ir(NH_3)_6]^{3+}$, the observed electron density does not allow detecting with precision the ammine groups (see for example [138]). The only exception are the unpublished structures of the P4-P6 introns in which a specific site could bind a putative osmium pentammine cation [139].

6.2. Lanthanide Cations (La^{3+} to Lu^{3+})

The lanthanide series (from Ce^{3+} to Lu^{3+}) was systematically tested to create heavy atom derivatives in the P4-P6 structure [139], as lanthanides

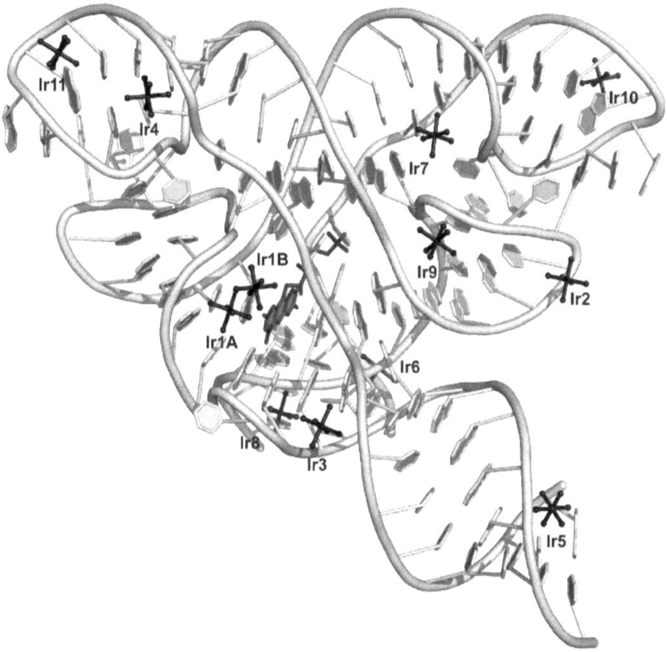

Figure 6. Binding of twelve $[Ir(NH_3)_6]^{3+}$ cations to a flavin mononucleotide riboswitch inferred from a crystal structure at ≈3.2 Å resolution. Reproduced by permission from [155]; copyright (2009).

proved useful in the determination of the first tRNA [119,132] and hammerhead ribozyme crystal structures [156]. Diffraction experiments on lanthanide-soaked crystals focused on Sm^{3+} as a potential heavy-atom derivative for two reasons. First, the unit-cell dimensions changed as a function of decreasing ionic radius for lanthanides in the series from Lu^{3+} to Sm^{3+} after which they remained constant (Sm^{3+} to Ce^{3+}). Second, the mosaic spread of the diffraction pattern increased as a function of increasing ionic radius for all lanthanides except for Sm^{3+}, for which the mosaicity was only slightly worse than that of native crystals. Yet, due to poor isomorphism and weak diffraction, the samarium derivative was abandoned in favor of the osmium hexammine derivative [139].

Lanthanide ions such as samarium bind specifically between phosphate oxygens in an adenosine-rich corkscrew structure at the junction of three helices by displacing a magnesium ion in this region; this ion is critical to the folding of the entire P4-P6 domain [139]. In order to find crystallization conditions that included metal cations that were able to generate anomalous scattering derivatives for a large RNase P fragment, the La^{3+}, Pr^{3+}, Sm^{3+},

Gd^{3+}, Dy^{3+}, Yb^{3+}, and Lu^{3+} lanthanides were tested. From this series, only Sm^{3+}, Gd^{3+}, and Yb^{3+} produced anomalous diffracting derivatives where lanthanide cations bound solely to phosphate oxygen atoms [138]. In a SRP fragment, a Lu^{3+} cation was found to bridge two phosphate groups [157]. It seems that lanthanide cations do not form direct coordination complexes to bases but bind exclusively to backbone phosphate oxygens, in agreement with earlier observations dealing with Pr^{3+}, Eu^{3+}, Gd^{3+}, Tb^{3+}, and Lu^{3+} cations. These cations were found to bind to the same sites as Sm^{3+} [119]. Regarding catalytic reaction rates, the addition of small quantities of Nd^{3+} in the presence of Pb^{2+} increased significantly the yield of the RNA cleavage reaction in a leadzyme, although other rare earth ions or divalent ions did not promote the reaction suggesting that subtle recognition phenomena are at play [158].

It was proposed that terbium inhibits the hammerhead ribozyme's catalytic activity by competing with Mg^{2+} cations leading to the characterization of three Tb^{3+} binding sites that displayed, like almost all other metal ions bound minimal hammerhead ribozyme structures, occupancies below 0.5 [159–161]. These Tb^{3+} ions are located in close proximity to electropositive and are reminiscent of SO_4^{2-} binding sites. They display coordination distances of 3.8–4.0 Å compatible with sulfate binding and suggesting a possible error in the assignment of these electronic densities [162]. Moreover, lanthanides are expected to coordinate to phosphate oxygens than to base atoms. Note that the average "lanthanide . . . Ow" coordination distance is around 2.4 Å.

Unfortunately, most of these lanthanide atom derivatives have not been made public (only three nucleic acid structures containing terbium, ytterbium or lutecium cations are archived in the PDB) probably because of the modest resolution of these derivatives [58] (Figure 7).

7. OTHER TRIVALENT AND TETRAVALENT CATIONS

7.1. Gold Cations

Au^{3+} is sometimes included in soaking experiments in order to produce anomalous diffracting derivatives [138]. Soaking experiments of a DIS subtype revealed the formation of an intriguing base pair: The Au^{3+} cation induced a deprotonation of a G(N1) atom leading to a gold mediated G=C pair [54] (see also Sections 5.4.3 and 5.4.6 devoted to the Cu^{2+} and the Hg^{2+} cations). This very unique base pair is emblematic of the difficultly of predicting cation binding modes, as some can be induced by soaking experiments themselves (Figure 8). Note that in $tRNA^{Asp}$ Au^{3+} was found to interact with more common binding sites [6].

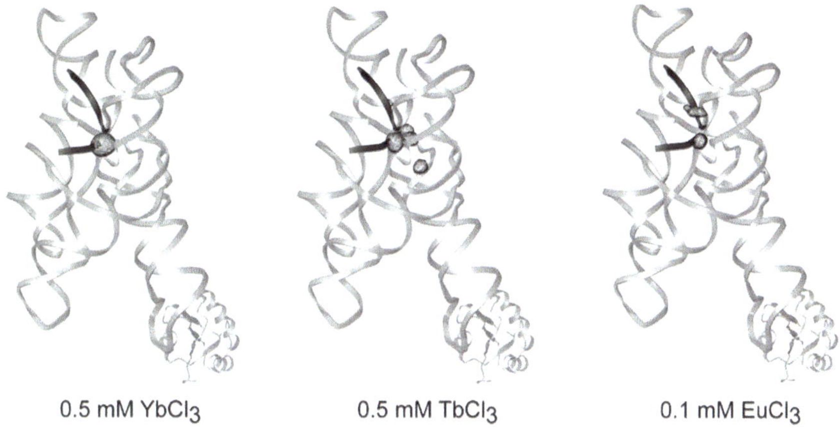

Figure 7. Anomalous difference maps showing the binding of Yb^{3+}, Tb^{3+}, and Eu^{3+} to the core of the *Azoarcus* group I intron in X-ray structures of 4.6, 5.0 and 5.5 Å resolution, respectively. Reproduced by permission from [58]; copyright (2007).

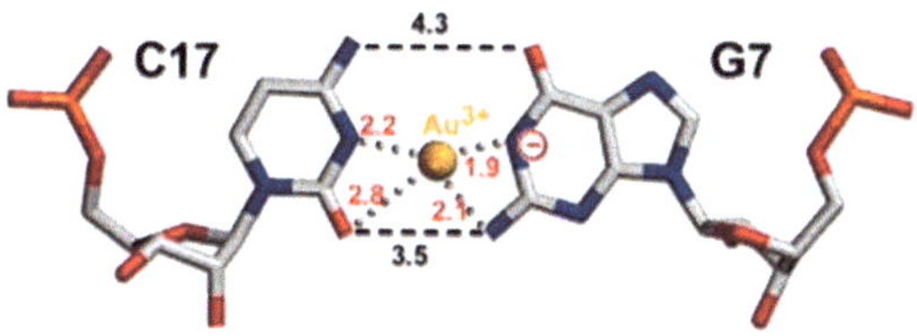

Figure 8. Intercalation of an Au^{3+} cation between the two nucleotides of a rG=C pair observed in a crystal structure of a DIS system after soaking with $AuCl_3$. Reproduced by permission from [54]; copyright (2003).

7.2. Platinum Cations

Pt^{4+} cations provide anomalous signals. They seem to have a good affinity for $^{5'}ApC^{3'}$ sites as revealed by a DIS crystal structure [54] and early structures of $tRNA^{Phe}$ [119,163]. They were also tentatively located close to a G(N7) atom [119]. Hence, they seem to bind exclusively to nucleobases with an octahedral coordination and associated binding distances exceeding 2.5 Å.

8. ANIONS

A statistical survey of nucleic acid structures [162] established that anions can bind directly to nucleotides despite the negative charges carried by the polyanionic RNA and DNA molecules. This survey led to a map of anion binding sites to nucleotides. Among the possible anions, SO_4^{2-}, SeO_4^{2-} [164], and acetate anions ($CH_3CO_2^-$) [165] were found to bind to the same guanine Watson-Crick and Hoogsteen sites as Cl^- anions. It is worth noting that some metals are found in their anionic rather than cationic form. For instance, vanadium and tungsten atoms form vanadate (VO_4^{2-}) and tungstate (WO_4^{2-}) anions that are iso-structural with SO_4^{2-}, SeO_4^{2-}, and PO_4^{2-} anions.

The presence of well localized electron densities around nucleic acid structures has generated some significant assignment errors where metal cations such as Mg^{2+} instead of anions such as Cl^- or SO_4^{2-} were placed incorrectly in the vicinity of nucleic acid electropositive atoms [162].

9. SUBJECTIVITY IN THE STRUCTURE DETERMINATION PROCESS

At the end of this analysis, it is important to recall that a certain level of subjectivity is inherent in the structure determination process and thus it is likely that some errors are present in macromolecular crystal structures. If uncorrected, these errors become a part of databases and statistical surveys. Consequently, reports based on locally deficient crystal structures, integrate some of these errors, like errors in the assignment of electron densities to water, cations or anions. Many of these potential sources of errors have been reviewed in the following articles [162,166–170]. Thus, there is still a need for revising existing structural databases, and for the development of efficient validation techniques. The SwS web service is one such validation tool [21].

Paraphrasing Georg Cristoph Lichtenberg (1742-1799) [171], it is worth remembering that, "crystallographers (scientists in general) are 'creative' and manage to produce new types of errors with regularity".

10. SUMMARY

A significant number of crystal structures that include metal cations have been solved, and deposited into the PDB. These structures provide important data on water and ion binding sites [172]. Many of these metal-RNA interactions have been further investigated through biochemical and biophysical studies that help us to better understand the role played by mono- and multivalent cations in maintaining the structure and the activity of RNA systems.

The specific roles played by the diffuse ionic atmosphere surrounding nucleic acids are still difficult to address despite numerous studies devoted to this topic [92,173]. With the advent of more sophisticated detection techniques, it appears that a significant part of the "diffuse ionic cloud" is rather localized, and that very specific binding sites are associated with RNA motifs such as, for example, the AA platform for K^+ cations. It is quite clear that much of the progress in understanding specific ion binding to RNA lies ahead of us. In the meantime, experiments on model systems such as nucleosides, nucleotides [6,94], and small RNAs [95] are still necessary to understand the effect of metal ions on RNA and vice versa.

While a large variety of ions are reported to associate with nucleic acids (close to 30 ions interacting with RNA and DNA are discussed in this review), only two of them, K^+ and Mg^{2+}, are physiologically relevant and interact directly with RNA *in vivo* (although rare associations with other ions cannot be fully excluded). All the other ions are used as crystallographic probes in order to provide heavy atom derivatives for solving crystal structures or for specific spectroscopic purposes, with the most common use being associated with specific paramagnetic properties.

Monovalent cations are easier to dehydrate and can therefore intrude into the first hydration shell of RNA nucleotides. They interact essentially with the deep groove G(N7/O6) atoms and less with other electronegative atoms. They are also seen close to the anionic oxygen atoms of phosphate groups in turn regions or participating in crystal packing interactions. Monovalent cations that do not directly bind to nucleic acids are difficult to detect. Even though various monovalent ions can be used, they are not interchangeable – for example, Tl^+ is considered to be the best mimic for K^+ (beside Na^+) but it does not always occupy the same binding sites or, if it is the case, coordination details may change. One interesting use of non-physiological monovalent cations is related to the high affinity of Cs^+ ions for G•U containing motifs leading to targeted insertion of these base pairs into large RNAs for solving phase issues.

Probably the most important and best characterized metal-RNA interactions are those involving Mg^{2+} cations. Again, the G(N7/O6) binding sites seem to be preferred along with anionic oxygen atoms. Single dehydration of the hexahydrated Mg^{2+} cations is observed more often than multiple dehydration. Mg^{2+} seems to be a part (along with K^+ cations) of several specific metal binding motifs that have to be more thoroughly characterized. Fully hydrated Mg^{2+} cations are often observed in the electronegative major groove of GpG steps.

Since Mg^{2+} cations are difficult to distinguish from water molecules in crystal structures, soaking experiments with heavier cations are often utilized.

The best mimic for Mg^{2+} is Mn^{2+} that displays anomalous properties. Yet, sufficient evidence exists that magnesium cannot be perfectly substituted by other ions. When Mg^{2+} is substituted by other cations, such as for example Mn^{2+}, Ca^{2+}, Sr^{2+}, Co^{2+}, Ni^{2+}, Zn^{2+}, Cd^{2+}, or $[Co(NH_3)_6]^{3+}$ to study general ion binding properties or to locate a particular ion binding site, the results need to be interpreted carefully due to differences in coordination chemistry and binding properties of these cations. Many of these transition metals bind exclusively to G(N7) atoms. Such substitutions can lead to incorrect conclusions regarding the RNA structure or the role of the associated metal ions [94].

Lanthanides generally bind to phosphate oxygen atoms and not to electronegative nucleobase atoms. Their use is generally limited to obtaining heavy atom derivatives since they quite often generate crystal structures of low resolution and/or deform the crystal cell in soaking experiments.

In conclusion, it is highly desirable that crystallization and other biophysical experiments focus more on the use of the two biologically relevant K^+ and Mg^{2+} ions and limit the use of Na^+ cations in crystallization buffers. The detection of cations in the vicinity of nucleic acids remains a challenge that is complicated by partial occupancy issues. Indeed, anomalous diffraction techniques often provide occupancies below one for the larger monovalent cations (K^+, Rb^+, Cs^+, Tl^+) suggesting that many ion-binding sites are alternatively occupied by monovalent cations, divalent cations and water molecules.

ACKNOWLEDGMENTS

Neena Grover is funded by a grant from the National Science Foundation MCB-0950582.

ABBREVIATIONS AND DEFINITIONS

ATP	adenosine 5'-triphosphate
CSD	Cambridge Structural Database
DIS	HIV dimerization initiation site
EPR	electron paramagnetic resonance
FRET	fluorescence resonance energy transfer
HDV	hepatitis delta virus ribozyme
HIV	human immunodeficiency virus
MAD	multiple anomalous diffraction

MeRNA	metal binding to RNA database
mRNA	messenger RNA
NMR	nuclear magnetic resonance
Ow	oxygen of water
PDB	Protein Data Bank
SAD	single anomalous diffraction
SRP	signal recognition particle
SwS	solvation web service for nucleic acids
TAR	trans-activation region
TEB	transient electric birefringence
tRNA	transfer RNA
VS	virus satellite ribozyme

REFERENCES

1. S. A. Woodson, *Curr. Opin. Chem. Biol.*, 2005, **9**, 104–109.
2. V. K. Misra and D. E. Draper, *J. Mol. Biol.*, 2002, **317**, 507–521.
3. D. E. Draper, *RNA*, 2004, **10**, 335–343.
4. J. Schnabl and R. K. O. Sigel, *Curr. Opin. Chem. Biol.*, 2010, **14**, 269–275.
5. R. K. O. Sigel and A. M. Pyle, *Chem. Rev.*, 2007, **107**, 97–113.
6. V. Swaminathan and M. Sundaralingam, *CRC Crit Rev Biochem*, 1979, **6**, 245–336.
7. J. E. Wedekind and D. B. McKay, *Nat. Struct. Biol.*, 1999, **6**, 261–268.
8. M. Martick and W. G. Scott, *Cell*, 2006, **126**, 309–320.
9. J. A. Ippolito and T. A. Steitz, *J. Mol. Biol.*, 2000, **295**, 711–717.
10. J. K. Frederiksen and J. A. Piccirilli, *Methods*, 2009, **49**, 148–166.
11. M. H. Harding, *Acta Cryst.*, 2004, **D60**, 849–859.
12. M. H. Harding, *Acta Cryst.*, 1999, **D55**, 1432–1443.
13. M. H. Harding, *Acta Cryst.*, 2002, **D58**, 872–874.
14. M. H. Harding, *Acta Cryst.*, 2006, **D62**, 678–682.
15. C. W. Bock, G. D. Markham, A. K. Katz and J. P. Glusker, *Theor. Chem. Acc.*, 2006, **115**, 100–112.
16. Y. Marcus, *Chem. Rev.*, 1988, **88**, 1475–1498.
17. Y. Marcus, *Chem. Rev.*, 2009, **109**, 1346–1370.
18. F. H. Allen, *Acta Cryst.*, 2002, **B58**, 380–388.
19. F. H. Allen and R. Taylor, *Chem. Soc. Rev.*, 2004, **33**, 463–475.
20. L. R. Stefan, R. Zhang, A. G. Levitan, D. K. Hendrix, S. E. Brenner and S. R. Holbrook, *Nucleic Acids Res.*, 2006, **34**, D131–134.
21. P. Auffinger and Y. Hashem, *Bioinformatics*, 2007, **23**, 1035–1037, http://www-ibmc.u-strasbg.fr/arn/sws.html.
22. L. N. Csonka, *Microbiol. Rev.*, 1989, **53**, 121–147.
23. J. H. Christian and J. A. Waltho, *J. Gen. Microbiol.*, 1961, **25**, 97–102.
24. A. Romani, *Arch. Biochem. Biophys.*, 2007, **458**, 90–102.
25. R. D. Grubbs, *Biometals*, 2002, **15**, 251–259.

26. P. H. Naslund and T. Hultin, *Biochim. Biophys. Acta*, 1970, **204**, 237–247.
27. P. H. Naslund and T. Hultin, *Biochim. Biophys. Acta*, 1971, **254**, 104–116.
28. H. L. Ennis and M. Artman, *Biochem. Biophys. Res. Commun.*, 1972, **48**, 161–168.
29. R. Miskin, A. Zamir and D. Elson, *J. Mol. Biol.*, 1970, **54**, 355–378.
30. A. Zamir, R. Miskin and D. Elson, *J. Mol. Biol.*, 1971, **60**, 347–364.
31. F. Cahn and M. Lubin, *J. Biol. Chem.*, 1978, **253**, 7798–7803.
32. K. D. Connaghan, A. F. Heneghan, M. T. Miura and D. L. Bain, *Biochemistry*, 2010, **49**, 422–431.
33. M. J. Cromie, Y. Shi, T. Latifi and E. A. Groisman, *Cell*, 2006, **125**, 71–84.
34. C. E. Dann, 3rd, C. A. Wakeman, C. L. Sieling, S. C. Baker, I. Irnov and W. C. Winkler, *Cell*, 2007, **130**, 878–892.
35. S. Brantl, *Trends Biotechnol.*, 2006, **24**, 383–386.
36. S. Brantl, *ACS Chem. Biol.*, 2007, **2**, 656–660.
37. C. A. Wakeman, A. Ramesh and W. C. Winkler, *J. Mol. Biol.*, 2009, **392**, 723–735.
38. A. M. Romani and M. E. Maguire, *Biometals*, 2002, **15**, 271–283.
39. C. Hsiao and L. D. Williams, *Nucleic Acids Res.*, 2009, **37**, 3134–3142.
40. D. J. Klein, P. B. Moore and T. A. Steitz, *RNA*, 2004, **10**, 1366–1379.
41. B. J. McCarthy, *Biochim. Biophys. Acta*, 1962, **55**, 880–888.
42. F. C. Chao, *Arch. Biochem. Biophys.*, 1957, **70**, 426–431.
43. F. C. Chao and H. K. Schachman, *Arch. Biochem. Biophys.*, 1956, **61**, 220–230.
44. A. Tissieres and J. D. Watson, *Nature*, 1958, **182**, 778–780.
45. A. H. Wang, G. Ughetto, G. J. Quigley and A. Rich, *Biochemistry*, 1987, **26**, 1152–1163.
46. X. Shui, L. McFail-Isom, G. G. Hu and L. D. Williams, *Biochemistry*, 1998, **37**, 8341–8355.
47. C. C. Sines, L. McFail-Isom, S. B. Howerton, D. VanDerveer and L. D. Williams, *J. Am. Chem. Soc.*, 2000, **122**, 11048–11056.
48. S. B. Howerton, C. C. Sines, D. VanDerveer and L. D. Williams, *Biochemistry*, 2001, **40**, 10023–10031.
49. V. Tereshko, G. Minasov and M. Egli, *J. Am. Chem. Soc.*, 1999, **121**, 3590–3595.
50. M. Egli, *Chem. Biol.*, 2002, **9**, 277–286.
51. X. Shui, C. C. Sines, L. McFail-Isom, D. VanDerveer and L. D. Williams, *Biochemistry*, 1998, **37**, 16877–16887.
52. J. Vieregg, W. Cheng, C. Bustamante and I. Tinoco Jr., *J. Am. Chem. Soc.*, 2007, **129**, 14966–14973.
53. D. Lambert, D. Leipply, R. Shiman and D. E. Draper, *J. Mol. Biol.*, 2009, **390**, 791–804.
54. E. Ennifar, P. Walter and P. Dumas, *Nucleic Acids Res.*, 2003, **31**, 2671–2682.
55. S. Basu, R. P. Rambo, J. Strauss-Soukup, J. H. Cate, A. R. Ferré-D'Amaré, S. A. Strobel and J. A. Doudna, *Nat. Struct. Biol.*, 1998, **5**, 986–992.
56. G. L. Conn, A. G. Gittis, E. E. Lattman, V. K. Misra and D. E. Draper, *J. Mol. Biol.*, 2002, **318**, 963–973.

57. P. Nissen, J. Hansen, N. Ban, P. B. Moore and T. A. Steitz, *Science*, 2000, **289**, 920–930.
58. M. R. Stahley, P. L. Adams, J. Wang and S. A. Strobel, *J. Mol. Biol.*, 2007, **372**, 89–102.
59. M. Egli, G. Minasov, L. Su and A. Rich, *Proc. Natl. Acad. Sci. USAS A*, 2002, **99**, 4302–4307.
60. A. Serganov, L. Huang and D. J. Patel, *Nature*, 2008, **455**, 1263–1268.
61. R. Shiman and D. E. Draper, *J. Mol. Biol.*, 2000, **302**, 79–91.
62. J. B. Murray, A. A. Seyhan, N. G. Walter, J. M. Burke and W. G. Scott, *Chem. Biol.*, 1998, **5**, 587–595.
63. S. Moghaddam, G. Caliskan, S. Chauhan, C. Hyeon, R. M. Briber, D. Thirumalai and S. A. Woodson, *J. Mol. Biol.*, 2009, **393**, 753–764.
64. A. Ke, F. Ding, J. D. Batchelor and J. A. Doudna, *Structure*, 2007, **15**, 281–287.
65. C. Creze, B. Rinaldi, R. Haser, P. Bouvet and P. Gouet, *Acta Cryst.*, 2007, **D63**, 682–688.
66. J. L. O'Rear, S. Wang, A. L. Feig, L. Beigelman, O. C. Uhlenbeck and D. Herschlag, *RNA*, 2001, **7**, 537–545.
67. T. C. Gluick, N. M. Wills, R. F. Gesteland and D. E. Draper, *Biochemistry*, 1997, **36**, 16173–16186.
68. D. Mota de Freitas, M. M. Castro and C. F. Geraldes, *Acc. Chem. Res.*, 2006, **39**, 283–291.
69. V. Tereshko, C. J. Wilds, G. Minasov, T. P. Prakash, M. A. Maier, A. Howard, Z. Wawrzak, M. Manoharan and M. Egli, *Nucleic Acids Res.*, 2001, **29**, 1208–1215.
70. D. Watkins, S. Mohan, G. B. Koudelka and L. D. Williams, *J. Mol. Biol.*, 2010, **396**, 1145–1164.
71. A. Y. Keel, R. P. Rambo, R. T. Batey and J. S. Kieft, *Structure*, 2007, **15**, 761–772.
72. P. A. Pedersen, J. M. Nielsen, J. H. Rasmussen and P. L. Jorgensen, *Biochemistry*, 1998, **37**, 17818–17827.
73. V. Villeret, S. Huang, H. J. Fromm and W. N. Lipscomb, *Proc. Natl. Acad. Sci. USAS A*, 1995, **92**, 8916–8920.
74. Z. Tao, A. Gameiro and C. Grewer, *Biochemistry*, 2008, **47**, 12923–12930.
75. S. B. Howerton, A. Nagpal and L. Dean Williams, *Biopolymers*, 2003, **69**, 87–99.
76. V. Kacer, S. A. Scaringe, J. N. Scarsdale and J. P. Rife, *Acta Cryst.*, 2003, **D59**, 423–432.
77. T. Moulaei, T. Maehigashi, G. T. Lountos, S. Komeda, D. Watkins, M. P. Stone, L. A. Marky, J. S. Li, B. Gold and L. D. Williams, *Biochemistry*, 2005, **44**, 7458–7468.
78. R. T. Batey and J. A. Doudna, *Biochemistry*, 2002, **41**, 11703–11710.
79. M. L. Gill, S. A. Strobel and J. P. Loria, *J. Am. Chem. Soc.*, 2005, **127**, 16723–16732.
80. G. S. Manning, *Quart. Rev. Biophys.*, 1978, **11**, 179–246.

81. M. T. Record, Jr., M. L. Lohman and P. De Haseth, *J. Mol. Biol.*, 1976, **107**, 145–158.
82. D. E. Draper, D. Grilley and A. M. Soto, *Annu. Rev. Biophys. Biomol. Struct.*, 2005, **34**, 221–243.
83. E. Ennifar, M. Yusupov, P. Walter, R. Marquet, B. Ehresmann, C. Ehresmann and P. Dumas, *Structure Fold. Des.*, 1999, **7**, 1439–1449.
84. H. Robinson, Y. G. Gao, R. Sanishvili, A. Joachimiak and A. H. Wang, *Nucleic Acids Res.*, 2000, **28**, 1760–1766.
85. S. B. Jang, L. W. Hung, M. S. Jeong, E. L. Holbrook, X. Chen, D. H. Turner and S. R. Holbrook, *Biophys. J.*, 2006, **90**, 4530–4537.
86. J. H. Cate, R. L. Hanna and J. A. Doudna, *Nat. Struct. Biol.*, 1997, **4**, 553–558.
87. A. S. Petrov, G. Lamm and G. R. Pack, *J. Phys. Chem. B*, 2002, **106**, 3294–3300.
88. J. H. Davis, T. R. Foster, M. Tonelli and S. E. Butcher, *RNA*, 2007, **13**, 76–86.
89. Y. Fan, B. L. Gaffney and R. A. Jones, *J. Am. Chem. Soc.*, 2005, **127**, 17588–17589.
90. P. Auffinger, L. Bielecki and E. Westhof, *Chem. Biol.*, 2003, **10**, 551–561.
91. P. Auffinger, L. Bielecki and E. Westhof, *J. Mol. Biol.*, 2004, **335**, 555–571.
92. D. E. Draper, *Biophys. J.*, 2008, **95**, 5489–5495.
93. J. H. Cate, A. R. Gooding, E. Podell, K. H. Zhou, B. L. Golden, C. E. Kundrot, T. R. Cech and J. A. Doudna, *Science*, 1996, **273**, 1678–1685.
94. E. Freisinger and R. K. O. Sigel, *Coord. Chem. Rev.*, 2007, **251**, 1834–1851.
95. I. Carter-O'Connell, D. Booth, B. Eason and N. Grover, *RNA*, 2008, **14**, 2550–2556.
96. D. R. Banatao, R. B. Altman and T. E. Klein, *Nucleic Acids Res.*, 2003, **31**, 4450–4460.
97. C. C. Correll, B. Freeborn, P. B. Moore and T. A. Steitz, *Cell*, 1997, **91**, 705–712.
98. K. T. Schroeder and D. M. Lilley, *Nucleic Acids Res.*, 2009, **37**, 7281–7289.
99. C. W. Bock, A. K. Katz, G. D. Markham and J. P. Glusker, *J. Am. Chem. Soc*, 1999, **121**, 7360–7372.
100. R. Oliva and L. Cavallo, *J. Phys. Chem. B*, 2009, **113**, 15670–15678.
101. J. Noeske, H. Schwalbe and J. Wohnert, *Nucleic Acids Res.*, 2007, **35**, 5262–5273.
102. S. R. Morrissey, T. E. Horton and V. J. DeRose, *J. Am. Chem. Soc.*, 2000, **122**, 3473–3481.
103. O. Schiemann, J. Fritscher, N. Kisseleva, S. T. Sigurdsson and T. F. Prisner, *Chembiochem*, 2003, **4**, 1057–1065.
104. P. Z. Qin, J. Feigon and W. L. Hubbell, *J. Mol. Biol.*, 2005, **351**, 1–8.
105. B. K. Oh, D. N. Frank and N. R. Pace, *Biochemistry*, 1998, **37**, 7277–7283.
106. A. Ke, K. Zhou, F. Ding, J. H. Cate and J. A. Doudna, *Nature*, 2004, **429**, 201–205.
107. M. Roychowdhury-Saha and D. H. Burke, *RNA*, 2006, **12**, 1846–1852.
108. K. J. Young, F. Gill and J. A. Grasby, *Nucleic Acids Res.*, 1997, **25**, 3760–3766.
109. S. Basu and S. A. Strobel, *Methods*, 2001, **23**, 264–275.
110. J. Lipfert, A. Y. Sim, D. Herschlag and S. Doniach, *RNA*, 2010, **16**, 708–719.

111. M. Perbandt, A. Nolte, S. Lorentz, V. A. Erdmann and C. Betzel, *FEBS Lett.*, 1998, **429**, 211–215.
112. Q. Zhao, Q. Han, C. R. Kissinger, T. Hermann and P. A. Thompson, *Acta Cryst.*, 2008, **D64**, 436–443.
113. J. A. Ippolito and T. A. Steitz, *Proc. Natl. Acad. Sci. USA*, 1998, **95**, 9819–9824.
114. P. B. Rupert and A. R. Ferré-D'Amaré, *Nature*, 2001, **410**, 780–786.
115. C. Baugh, D. Grate and C. Wilson, *J. Mol. Biol.*, 2000, **301**, 117–128.
116. U. Mueller, H. Schübel, M. Sprinzl and U. Heinemann, *RNA*, 1999, **5**, 670–677.
117. K. J. Travers, N. Boyd and D. Herschlag, *RNA*, 2007, **13**, 1205–1213.
118. C. M. Dunham, J. B. Murray and W. G. Scott, *J. Mol. Biol.*, 2003, **332**, 327–336.
119. S. H. Kim, W. C. Shin and R. W. Warrant, *Methods Enzymol.*, 1985, **114**, 156–167.
120. N. G. Abrescia, L. Malinina and J. A. Subirana, *J. Mol. Biol.*, 1999, **294**, 657–666.
121. R. K. O. Sigel and H. Sigel, in *Nickel and Its Surprising Impact in Nature*, Volume 2 of *Metal Ions in Life Sciences*, eds. A. Sigel, H. Sigel, R. K. O. Sigel, John Wiley & Sons, Chichester, UK, 2007, pp. 109–180.
122. E. L. Que, D. W. Domaille and C. J. Chang, *Chem. Rev.*, 2008, **108**, 1517–1549.
123. L. Rulisek and J. Sponer, *J. Phys. Chem. B*, 2003, **107**, 1913–1923.
124. B. H. Geierstanger, T. F. Kagawa, Shun-Le-Chen, G. J. Quigley and P. S. Ho, *J. Biol. Chem.*, 1991, **266**, 20185–10191.
125. S. Atwell, E. Meggers, G. Spraggon and P. G. Schultz, *J. Am. Chem. Soc.*, 2001, **123**, 12364–12367.
126. S. C. Dahm and O. C. Uhlenbeck, *Biochemistry*, 1991, **30**, 9464–9469.
127. G. Wang, B. L. Gaffney and R. A. Jones, *J. Am. Chem. Soc.*, 2004, **126**, 8908–8909.
128. N. Kisseleva, S. Kraut, A. Jaschke and O. Schiemann, *HFSP J.* (http://hfspj.aip.org/), 2007, **1**, 127–136.
129. M. Forconi, J. A. Piccirilli and D. Herschlag, *RNA*, 2007, **13**, 1656–1667.
130. B. T. Wimberly, R. Guymon, J. P. McCutcheon, S. W. White and V. Ramakrishnan, *Cell*, 1999, **97**, 491–502.
131. L. D. Kosturko, C. Folzer and R. F. Stewart, *Biochemistry*, 1974, **13**, 3949–3952.
132. A. Jack, J. E. Ladner, D. Rhodes, R. S. Brown and A. Klug, *J. Mol. Biol.*, 1977, **111**, 315–328.
133. H. Sigel, C. P. Da Costa and R. B. Martin, *Coord. Chem. Rev.*, 2001, **219–221**, 435–461.
134. C. Werner, B. Krebs, G. Keith and G. Dirheimer, *Biochim. Biophys. Acta*, 1976, **32**, 161–175.
135. R. S. Brown, B. E. Hingerty, J. C. Dewan and A. Klug, *Nature*, 1983, **303**, 543–546.
136. R. S. Brown, J. C. Dewan and A. Klug, *Biochemistry*, 1985, **24**, 4785–4801.
137. A. S. Krasilnikov, X. Yang, T. Pan and A. Mondragon, *Nature*, 2003, **421**, 760–764.

138. A. V. Kazantsev, A. A. Krivenko and N. R. Pace, *RNA*, 2009, **15**, 266–276.
139. J. H. Cate and J. A. Doudna, *Structure*, 1996, **4**, 1221–1229.
140. B. L. Golden, *Methods Enzymol.*, 2000, **317**, 124–132.
141. B. L. Golden, *Methods Mol. Biol.*, 2007, **317**, 239–257.
142. R. V. Gessner, G. J. Quigley, A. H. J. Wang, G. A. van der Marel, J. H. van Boom and A. Rich, *Biochemistry*, 1985, **24**, 237–240.
143. M. Gueron, J. Demaret and M. Filoche, *Biophys. J.*, 2000, **78**, 1070–1083.
144. J. L. Maguire and R. A. Collins, *J. Mol. Biol.*, 2001, **309**, 45–56.
145. A. Hampel and J. A. Cowan, *Chem. Biol.*, 1997, **4**, 513–517.
146. S. Nesbitt, L. A. Hegg and M. J. Fedor, *Chem. Biol.*, 1997, **4**, 619–630.
147. M. J. Serra, J. D. Baird, T. Dale, B. L. Fey, K. Retatagos and E. Westhof, *RNA*, 2002, **8**, 307–323.
148. A. H. J. Wang, T. Hakoshima, G. van der Marel, J. H. Boom and A. Rich, *Cell*, 1984, **37**, 321–331.
149. C. MacElrevey, J. D. Salter, J. Krucinska and J. E. Wedekind, *RNA*, 2008, **14**, 1600–1616.
150. B. Gong, J. H. Chen, P. C. Bevilacqua, B. L. Golden and P. R. Carey, *Biochemistry*, 2009, **48**, 11961–11970.
151. D. A. Costantino, J. S. Pfingsten, R. P. Rambo and J. S. Kieft, *Nat. Struct. Mol. Biol.*, 2008, **15**, 57–64.
152. W. Cruse, P. Saludjian, A. Neuman and T. Prange, *Acta Cryst.*, 2001, **D57**, 1609–1613.
153. D. Bharanidharan, S. Thiyagarajan and N. Gautham, *Acta Cryst.*, 2007, **F63**, 1008–1013.
154. B. Lippert, *Chem. Biodivers.*, 2008, **5**, 1455–1474.
155. A. Serganov, L. Huang and D. J. Patel, *Nature*, 2009, **458**, 233–237.
156. H. M. Pley, K. M. Flaherty and D. B. McKay, *Nature*, 1994, **372**, 68–74.
157. L. Jovine, T. Hainzl, C. Oubridge, W. G. Scott, J. Li, T. K. Sixma, A. Wonacott, T. Sharzynski and K. Nagai, *Structure*, 2000, **8**, 527–540.
158. T. Ohmichi and N. Sugimoto, *Biochemistry*, 1997, **36**, 3514–3521.
159. W. G. Scott, J. T. Finch and A. Klug, *Cell*, 1995, **81**, 991–1002.
160. W. G. Scott, J. B. Murray, J. R. P. Arnold, B. L. Stoddard and A. Klug, *Science*, 1996, **274**, 2065–2069.
161. A. Feig, W. G. Scott and O. C. Uhlenbeck, *Science*, 1998, **279**, 81–84.
162. P. Auffinger, L. Bielecki and E. Westhof, *Structure*, 2004, **12**, 379–388.
163. D. Rhodes, P. W. Piper and B. F. C. Clark, *J. Mol. Biol.*, 1974, **89**, 469–475.
164. J. S. Kieft, E. Chase, D. A. Costantino and B. L. Golden, *RNA*, 2010, **16**, 1118–1123.
165. S. D. Gilbert, F. E. Reyes, A. L. Edwards and R. T. Batey, *Structure*, 2009, **17**, 857–868.
166. U. Das, S. Chen, M. Fuxreiter, A. A. Vaguine, J. Richelle, H. M. Berman and S. J. Wodak, *Acta Cryst.*, 2001, **D57**, 813–828.
167. L. D. Williams, *Top. Curr. Chem.*, 2005, **253**, 77–88.
168. G. J. Kleywegt, *Acta Cryst.*, 2009, **D65**, 134–139.
169. Y. Hashem and P. Auffinger, *Methods*, 2009, **47**, 187–197.

170. A. M. Davis, S. A. St-Gallay and G. J. Kleywegt, *Drug Discovery Today*, 2008, **13**, 831–841.
171. G. C. Lichtenberg, *Georg Cristoph Lichtenberg. Aphorismen und andere Sudeleien*, Reclam Verlag, Ditzingen, 2003.
172. P. Auffinger and Y. Hashem, *Curr. Op. Struct. Biol.*, 2007, **17**, 325–333.
173. A. M. Soto, V. Misra and D. E. Draper, *Biochemistry*, 2007, **46**, 2973–2983.
174. D. J. Klein, T. M. Schmeing, P. B. Moore and T. A. Steitz, *EMBO J.*, 2001, **20**, 4214–4221.

Met. Ions Life Sci. **2011**, *9*, 37–100

2

Methods to Detect and Characterize Metal Ion Binding Sites in RNA

***Michèle C. Erat*[1] *and Roland K. O. Sigel*[2]**

[1]Department of Biochemistry, University of Oxford, South Parks Road, Oxford OX1 3QU, United Kingdom
<michele.erat@bioch.ox.ac.uk>

[2]Institute of Inorganic Chemistry, University of Zürich, Winterthurerstrasse 190, CH-8057 Zürich, Switzerland
<roland.sigel@aci.uzh.ch>

Metal Ions in Life Sciences, Volume 9 Edited by Astrid Sigel, Helmut Sigel, and Roland K. O. Sigel

Published by the Royal Society of Chemistry, www.rsc.org DOI: 10.1039/978184973251200037

ABSTRACT: Metal ions are inextricably associated with RNAs of any size and control their folding and activity to a large part. In order to understand RNA mechanisms, also the positioning, affinities and kinetics of metal ion binding must be known. Due to the spectroscopic silence and relatively fast exchange rates of the metal ions usually associated with RNAs, this task is extremely challenging and thus numerous methods have been developed and applied in the past. Here we provide an overview on the different metal ions and methods applied in RNA (bio)chemistry: The physical-chemical properties of important metal ions are presented and briefly discussed with respect to their application together with RNA. Each method ranging from spectroscopic over biochemical to computational approaches is briefly described also mentioning caveats that might occur during the experiment and/or interpretation of the results.

KEYWORDS: equilibrium constants · metal ions · ribozymes · RNA · methods · spectroscopy

1. INTRODUCTION

The detection of metal ions within complex RNA structures is highly challenging. Nevertheless, a profound characterization of this crucial interaction is the basis to understand folding, structure, and function of any RNA, especially catalytically active ribozymes. Three major problems are to be overcome in every method that aims at investigating this relationship: (i) Only a few metal ions are specifically bound, whereas most are just diffusely associated with the negatively charged phosphate-sugar backbone. The differentiation between these two groups is difficult, also because a permanent exchange takes place and RNA structures are dynamic. (ii) Metal ion binding to RNAs is by orders of magnitudes weaker than in metalloproteins, thus often prohibiting a clear picture. In addition, fast ligand exchange rates and thus also varying coordination spheres over time seem to be rather the rule than the exception. (iii) The usually assumed natural cofactors Na^+, K^+, Mg^{2+}, and possibly Ca^{2+} are mostly spectroscopically silent. Hence, only indirect observation or the replacement with another (spectroscopically active) ion is possible. In the latter case though one has to be aware of the introduced deviations from the wild-type system.

In this review, we try to give a comprehensive overview on the methods that have been applied in the past as well as on those that might become more important in the future to investigate and characterize metal ion-RNA (or nucleic acids in general) interactions. Obviously it is impossible to cover every method to the same extent, and hence we will concentrate in more detail on those that we have applied ourselves in the past, whereas others shall be discussed rather briefly and the reader will be referred to the corresponding literature. One aim of this review is not only to highlight the advantages and results, but also to list the caveats and possible difficulties associated with each method. In this sense, many of the aspects discussed in more detail in the context of one method are directly applicable to others.

2. GENERAL CONSIDERATIONS

2.1. Metal Ions to Be Considered for RNA Binding

Neither RNA nor DNA can exist without cations compensating for their accumulated negative charge of the phosphate sugar backbone at physiological pH. In principle, also organic cations, e.g., polyamines, can be used for this purpose but in reality monovalent and divalent metal ions will carry the major burden of charge compensation. For life in general, numerous metal ions are essential and are thus found in living cells in a large range of concentrations, which are tightly regulated (Figure 1). However, most of these metal ions are strongly bound within metalloproteins and are thus not freely available. In fact, only Na^+, K^+, Mg^{2+}, and Ca^{2+} are available in their solvated form in larger amounts. Nevertheless, Na^+ is primarily found outside of the cell compartment and millimolar Ca^{2+} concentrations are found only in specialized compartments/cells acting as neurotransmitters. Consequently, K^+ and Mg^{2+} are usually considered the natural metal ion cofactors for nucleic acids *in vivo*.

This view might actually be oversimplified. For example, two different *Hammerhead* ribozymes were examined towards their catalytic activity in the presence of various divalent metal ions [1,2]. In both cases, strong increases in cleavage rate were observed when substituting Mg^{2+} with a transition metal ion, e.g., a 400 times accelerated catalysis of the RzB *Hammerhead* in the presence of Mn^{2+} [2]. However, this change in activity does not hint *per*

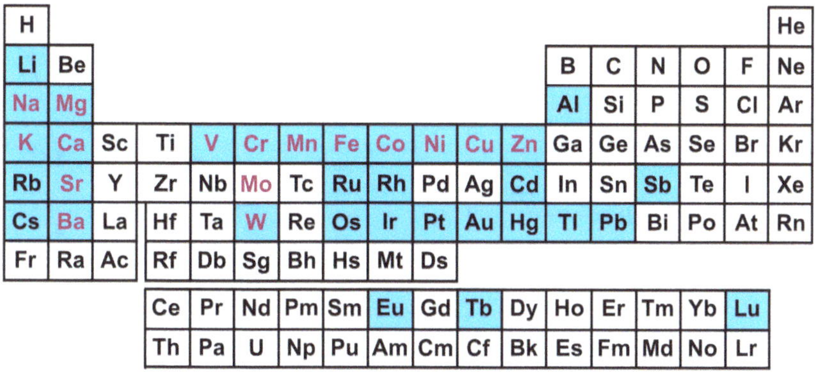

Figure 1. Periodic Table of the elements with the metal ions essential for most organisms colored in magenta [252]. Those elements commonly used in the different methods described here and/or found in the MINAS database are underlaid in light blue.

se towards the usage of other metal ions in nature but just exemplifies that the metal ion-RNA interaction and the basic coordination properties of each metal ion are the key cofactor for RNA structure and activity. The increase in activity of the *Hammerhead* ribozyme in the presence of transition metal ions thereby does not follow the Irving-Williams series, but rather the intrinsic affinities of the metal ions towards phosphate groups [3] (Figure 2 and Table 1). Consequently, the extent of the thermodynamic interaction of the divalent metal ion with the bridging phosphate groups (possibly even the cleavable phosphodiester) is the determining factor for cleavage activity of this small ribozyme. This notion is supported by the fact that the phosphate groups are the major primary binding sites for most metal ions [4,5], whereas further interactions incrementally increase the overall affinity [5].

The situation immediately becomes more complicated in cases where the change in catalytic activity deviates from such basic metal ion properties (Table 1). For example, a ribozyme derived from the yeast mitochondrial group II intron *Sc*.ai5γ is strongly inhibited in the presence of small amounts of Ca^{2+} [6] (see also Chapter 7). By single molecule FRET studies it has been shown that this inhibition possibly results from the formation of a misfolded species [7]. As the phosphate affinity of Mg^{2+} is larger only by a factor of about 1.3 (see also Section 2.2), but Ca^{2+} already strongly inhibits at a 20-fold excess of Mg^{2+}, this means that specific binding pockets for Ca^{2+} must exist, possibly resulting in a cooperative accumulation of Ca^{2+} within the misfolded RNA. Considering that Ca^{2+} in mitochondria can reach millimolar levels and thus almost equal Mg^{2+} concentrations, a biological role of the Mg^{2+}/Ca^{2+} switch is well feasible, although possible roles of metal ions

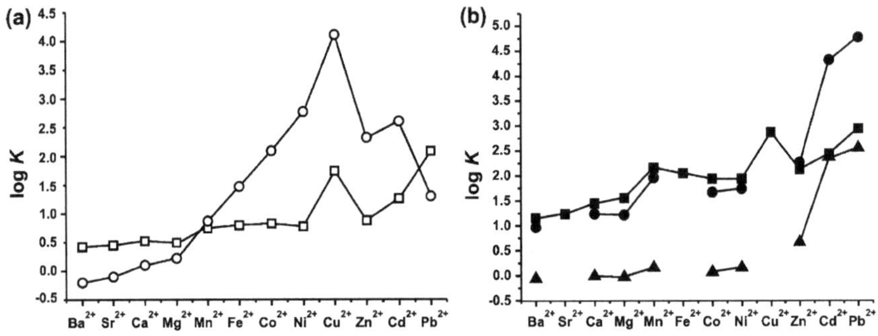

Figure 2. (**a**) Binding of divalent metal ions towards acetate (□, log K_O) and NH_3 (○, log K_N) follows the Irving-Williams series. The log K values as given in Table 1 are depicted and are valid for 25°C and $I = 0.5$ M; regarding Fe^{2+} and Cu^{2+}, see [14]. (**b**) Affinities of divalent metal ions towards the 5′-terminal phosphate group of AMP^{2-} (■, log $K_{AMP,op}$) do not follow the Irving-Williams series. The same trend is valid for binding towards the thiophosphate group of $AMPS^{2-}$ (●, log $K_{AMPS,calc}$) except that Cd^{2+} and Pb^{2+} show extraordinarily high affinities. The intrinsic increase in affinity upon substitution of a terminal oxygen by a sulfur atom (Δ log $\Delta_{M(AMPS)}$) is also given (▲).

other than Mg^{2+} *in vivo* remain largely unexplored in the context of ribozymes.

In contrast to this obvious lack in knowledge of the situation *in vivo*, such experiments with various metal ions to elucidate structural and functional aspects of metal ions in nucleic acids are commonly performed *in vitro*. The specific chemical and physical properties of specific metal ions are thereby used to investigate for example catalytic mechanisms as described in the following section.

2.2. Chemical and Physical Properties of Commonly Used Metal Ions

Aside from Mg^{2+}, a multitude of different metal ions is applied in combination with nucleic acids in daily research experiments (Figure 1). These can be either divalent ions (Table 1) or for many purposes also members of the lanthanide series (Table 2) are used. Especially for X-ray crystallography, a large part of the Periodic Table of metal ions has been explored, as is obvious from the deposited structures in the PDB (Protein Database) (Figure 1). One thereby takes advantage of the better diffracting properties of heavier metal ions (see also Section 3.1.3), often enabling to solve the

Table 1. Comparison of some physico-chemical properties of Mg^{2+}, Ca^{2+}, Sr^{2+}, Ba^{2+}, Mn^{2+}, Co^{2+}, Ni^{2+}, Zn^{2+}, Cd^{2+}, and Pb^{2+} in aqueous solution. Given are the ionic radii and the preferred coordination number (CN) of the M^{2+} ions, the enthalpy of hydration (ΔH_{Hydr}), the acidity constant ($pK^{H}_{M(H_2O)_n}$) of a water molecule in the aqua complex, the ligand exchange rate from the first coordination sphere of the metal ion (k_{exch}), the logarithm of the affinity towards acetate ($\log K_O$), the affinity towards NH_3 ($\log K_N$), the affinity towards the phosphate group in AMP^{2-} ($\log K_{AMP,op}$) and the thiophosphate group in $AMPS^{2-}$ ($\log K_{AMPS,calc}$) as well as the change in affinity by going from AMP^{2-} to $AMPS^{2-}$ ($\Delta \log \Delta_{M(AMPS)}$) (see also Figure 2).

	Ca^{2+}	Sr^{2+}	Ba^{2+}	Mg^{2+}	Mn^{2+}	Co^{2+}	Ni^{2+}	Zn^{2+}	Cd^{2+}	Pb^{2+}	Ref.
Ionic radius (Å)	1.00 (1.12)[a]	1.18 (1.26)[a]	1.35 (1.42)[a]	0.72	0.83[b]	(0.58)[c] 0.75[b]	(0.55)[c] 0.69	(0.60)[c] 0.74	0.95	1.19	[242]
CN	6 (8)[a]	6 (8)[a]	6 (8)[a]	6	6	4/6	4/5/6	4/6	6	6 (8)[a]	
Preferred ligands	O	O	O	O	O/N	O/N	O/N	O	N/O	O	
ΔH_{Hydr} ($kJmol^{-1}$)	1570 (1657)[a]	1386 (1461)[a]	1290 (1361)[a]	1858	1762	(1708)[c] 1888	(1758)[c] 1951	(1691)[c] 1871	1640	1432 (1524)[a]	[243]
$pK^{H}_{M(H_2O)_n}$ (25°C)	12.85±0.1	13.29±0.1	13.47±0.1	11.44±0.1	10.59	9.8	9.86	8.96	10.2	7.8	[113]
k_{exch} (s^{-1})	$\approx 5\times10^{8}$	$\approx 8\times10^{8}$	$\approx 3\times10^{9}$	6.7×10^{5}	2.1×10^{7}	3.2×10^{6}	3.2×10^{4}	4.1×10^{8}	6.8×10^{8}		[244–246]
$\log K_O$	0.52±0.05	0.45±0.07	0.42±0.06	0.49±0.04	0.75±0.11	0.83±0.05	0.78±0.06	0.88±0.19	1.26±0.07	2.08±0.04	[187]
$\log K_N$	0.1	−0.1	−0.2	0.22±0.02	0.87	2.09	2.77±0.03	2.32±0.1	2.60±0.08	1.3	[187]
$\log K_{AMP,op}$	1.45±0.05		1.16±0.04	1.56±0.03	2.16±0.05	1.94±0.06	1.94±0.05	2.13±0.06	2.44±0.05	2.94±0.08	[161,247]
$\log K_{AMPS,calc}$	1.24±0.08		0.97±0.07	1.22±0.06	1.96±0.11	1.67±0.12	1.74±0.09	2.27±0.20	4.32±0.14	4.79±0.15	[161,248]
$\Delta \log \Delta_{M(AMPS)}$	−0.01±0.10		−0.06±0.08	−0.03±0.07	0.16±0.12	0.07±0.14	0.16±0.10	0.66±0.21	2.37±0.15	2.55±0.13	[161,248]

[a]The numbers in brackets refer to CN = 8.
[b]High-spin electron configuration.
[c]The numbers in brackets refer to CN = 4 (tetrahedral).

Table 2. Comparison of some physico-chemical properties of the lanthanide(III) ions including La^{3+}. Given are the ionic radii ($R_{M^{3+}}$, Å) [242] for coordination 8 (top line) and 9 (2nd line), the preferred coordination number (CN) of the aquated M^{3+} ions [249], the enthalpy of hydration (ΔH_{Hydr}, kJ mol^{-1}) [250], the mean distance between the metal ion (Å) and a coordinated water molecule in crystal structures of $[Ln(H_2O)_9]^{3+}$ ions [251], the acidity constant ($pK^{H}_{M(H_2O)_n}$, 25°C) of a coordinated water molecule [113], the ligand exchange rate from the first coordination sphere of the metal ion (k_{exch}, s^{-1}) [249], as well as the oxidation states in aqueous solution [250].

	La^{3+}	Ce^{3+}	Pr^{3+}	Nd^{3+}	Pm^{3+}	Sm^{3+}	Eu^{3+}	Gd^{3+}	Tb^{3+}	Dy^{3+}	Ho^{3+}	Er^{3+}	Tm^{3+}	Yb^{3+}	Lu^{3+}
$R_{M^{3+}}$	1.160	1.143	1.126	1.109	1.093	1.079	1.066	1.053	1.040	1.3027	1.015	1.004	0.994	0.985	0.977
	1.216	1.196	1.179	1.163	1.144	1.132	1.120								
CN	9	9	9	9	8, 9	8, 9	8, 9	8	8	8	8	8	8	8	8
ΔH_{Hydr}	3238	3370	3413	3442	3478	3515	3547	3571	3605	3637	3667	3691	3717	3739	3760
$M\cdots OH_2$	2.539	2.587	2.500	2.494	–	2.435	2.444	2.439	2.371	2.434	2.404	2.340	2.315	2.391	2.323
$pK^{H}_{M(H_2O)_n}$	8.5	8.3	8.1	8.0		7.9	7.8	8.0	7.9	8.0	8.0	7.9	7.7	7.7	7.6
k_{exch}			$>5\times10^8$	$>4\times10^8$				8.3×10^8	5.6×10^8	4.3×10^8	2.1×10^8	1.3×10^8	0.9×10^8	0.5×10^8	
ox. states	3	3, 4	3	3	3	3	3	3	3	3	3	3	3	3	3

phase problem by Patterson methods. Other common examples where different metal ions are employed, include thio-rescue and related experiments (see Sections 4.1 and 4.2), hydrolytic and oxidative cleavage experiments (see Sections 4.3 and 4.4) to probe metal ion binding sites or even specific liganding atoms on the nucleic acid.

Often also the intrinsic properties of the (solvated) metal ion itself are used: The correlation of the pK_a of coordinated water molecules with the pH profile of the reaction allows conclusions on the (metal ion-assisted) acid-base mechanism of catalysis [2]. EPR-active metal ions can be applied to investigate directly the coordination sphere or elucidate the number of tightly bound ions in a folded RNA (see also Section 3.3). Lanthanide ions have interesting spectroscopic properties allowing conclusions on their binding (see Section 3.4), and NMR-active nuclei also allow a view on the metal ion itself (see Section 3.2.3).

Some chemical and physical properties of often applied divalent and lanthanide metal ions are summarized in Tables 1 and 2. The comparison of these accumulated data immediately illustrates that there are clearly two sides to the coin. On the one hand, the usage of different metal ions provides the opportunity to observe the metal ion directly, shift the acid base profile of catalysis, or selectively probe a coordination site, thus yielding crucial information on the metal ion binding. On the other side, no two metal ions will behave identically, and thus results obtained with metal ions other than the natural cofactor will always be burdened with the caveat that one cannot be sure if the system applied is meaningful. Nicely illustrated is this fact by crystal structures of identical RNAs, but containing different metal ions, which often do not occupy identical positions (see also Section 3.1). Nevertheless, if one is aware of the specific coordinating properties of each metal ion, the advantages and opportunities of varying the kind of metal ion are clearly manifold.

2.3. Prerequisits to the RNA and the Experimental Conditions

For all the methods described in this review, the RNA should be as uniform as possible. The RNA sequences, depending on their length and possible modifications, can be obtained by chemical synthesis [8,9] or transcription by T7 RNA polymerase. Transcription and purification of RNA to obtain a uniform sample has been described before [10–12]. To quantify metal ion binding to RNA under situations where the metal ion is not in large excess, it is crucial to be aware that the 5′-end of the transcribed RNA carries a triphosphate chain. This site is likely to be the primary binding site for metal ions [13] and thus needs to be accounted for, if it is not removed by kinases [12].

Further factors can influence the metal ion affinity considerably: Both, ionic strength [14] and solvent polarity [15–18] are well known factors. In addition, it is important to use buffers that are not good ligands for metal ions at the same time. For example, citrate, phosphate, or the very commonly used Tris-HCl bind metal ions rather well [19]. As most buffer substances are applied in much higher concentration than the metal ion itself, these ligands can interfere with binding of metal ions to nucleic acids [20,21].

3. SPECTROSCOPIC METHODS

3.1. X-ray Crystallography

X-ray crystallography is a magnificent method to obtain insight into structural relationships between RNA and its coordinating metal ions. As of May 2010, 21799 Mg^{2+} ions bound to RNA that have been identified by X-ray crystallography are deposited in the PDB. If the resolution is sufficient, the bound alkaline and alkaline earth metal ions can be directly observed. However, more commonly, metal ion binding sites are obtained by soaking a crystal with solutions of heavier metal ions to identify binding sites. The following section gives a brief overview of the major advantages and pitfalls of the method and describes a few examples, where X-ray crystallography has been useful in deciphering metal ion binding sites in RNA.

3.1.1. Setup of RNA Crystal Screens

Prior to setting up crystallization screens, the concentrated RNA needs to be refolded into its native state. The RNA is heated to 90°C for a few minutes and left to cool slowly to room temperature. To avoid self-cleavage of the RNA, it is advisable to keep the pH slightly acidic and to add divalent cations only around 35°C [11].

Once the sample is ready, one can proceed to set up crystallization screens. The main crystallization method used in macromolecular crystallography today is the vapor diffusion technique in a hanging or sitting drop setup. Crystallization is initiated by concentrating the RNA-containing droplet against a comparatively large volume of reservoir solution. Crystallization robots enable the screening of many thousands conditions in a relatively short time. For initial broad screens, specific crystallization sparse matrices have been developed and many of them are commercially available, e.g., from Hampton Research, Emerald Biosystems or Qiagen. General considerations for crystallization of nucleic acids are similar to those for protein

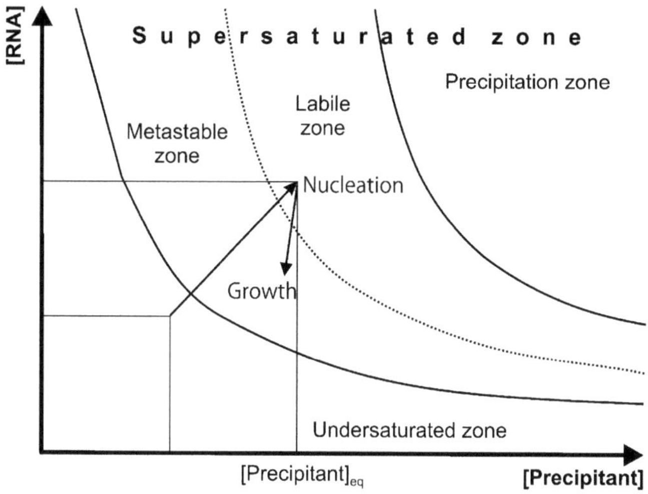

Figure 3. Nucleation of crystal formation occurs in the labile zone, when the precipitant concentration reaches equilibrium. Due to subsequent depletion of the RNA concentration in solution, the system moves to the metastable zone where crystals can grow.

crystallization, but it has been noted that higher temperatures (37°C) seem to favor RNA crystal formation [11]. Further factors that can be varied in order to find the optimal crystallization conditions include buffer, pH, precipitant and salt identity, and concentration (Figure 3). Although RNA is less sensitive to small pH changes than proteins, this factor should not be entirely discarded, as tertiary contacts in larger RNAs can shift the pK_a of titratable groups closer to neutrality [22].

If no crystals appear in any of the conditions tried, it may be worthwhile to consider a change in construct. Sometimes minor variations in length or sequence can result in a shape that has a higher propensity to pack in a regular manner. Once a first hit has been obtained, the long process of optimization starts. The following rules of thumb might be useful when analyzing initial results: Precipitates usually call for a decrease in RNA and/or precipitant concentration. Note, that microcrystals can be mistaken for small precipitates. It is therefore advisable to carefully inspect those conditions under the microscope. If microcrystals or clusters of homogenous crystals are found, it can be helpful to slow down the diffusion process by either lowering the temperature, or covering the reservoir solution with oil. If phase separation is a problem, the monovalent salt concentration can be increased and/or the precipitant concentration lowered. If the droplets remain clear, it may be that the construct has a low propensity to crystallize (see above). Alternatively, the RNA concentration simply may lie outside the nucleation zone (Figure 3), in which case a repetition of the screen with higher RNA and/or salt concentrations may solve the problem. At the end of the day, macromolecular crystallography remains to some extent a "black art"; but the more conditions are tried, the higher the probability of success. A helpful link in this respect is provided by Hampton Research (http://hamptonresearch.com/tips.aspx).

3.1.2. *Direct Observation*

Due to the equal number of electrons of Na^+, Mg^{2+}, and a water molecule, it is difficult to unambiguously distinguish the three. This problem as well as strategies to solve it have been excellently described and reviewed before [23,24]. Most importantly, a high enough resolution is crucial to directly observe metal ions in crystal structures. One of the most prominent examples is the structure of the large ribosomal subunit of *Haloarcula marismortui*, in which 116 Mg^{2+} and 88 M^+ ions have been unambiguously positioned [25]. This was done by an iterative method: First, a list of candidates was generated in an automated fashion by searching F_{obs}–F_{calc} difference electron density maps for peaks that were 1.9–2.1 Å from ribosomal nitrogen or oxygen atoms. If the surrounding electron density features agreed with an octahedral conformation, these electron density peaks were modeled as Mg^{2+} (Figure 4). The modeling of the first hydration shell was also based on F_{obs}–F_{calc} difference maps. The identity of the complex was confirmed if a refined distance of 2 Å between Mg^{2+} and H_2O was found. At last, fully hydrated Mg^{2+}_{aq} ions were identified by visual analysis of the electron density maps.

3.1.3. *Identification of Metal Ion Sites by Heavy Metal Soaking*

For lower resolution structures, metal ion binding sites can potentially be identified by soaking the crystals in solutions of heavier metals, such as Tl^+, Rb^+, Mn^{2+}, Pb^{2+}, Co^{3+}, Os^{3+}, Eu^{3+}, or Ir^{3+} [26]. This method has already been applied to identify metal ions in the first crystal structure of the hammerhead ribozyme in 1994 [27]. Traditionally this strategy is used to solve the phase problem by Patterson methods. As outlined above in Section 2.2, metal ions are very distinct with respect to their binding properties. It is thus important to use metal ions that are as similar as possible in their chemical and structural features to the ion that is suspected to bind under native conditions, i.e., mostly Mg^{2+}. If possible, it is useful to validate the analogous behavior of the light and heavy metal ions by biochemical methods. Unfortunately this approach is not always successful. In case of

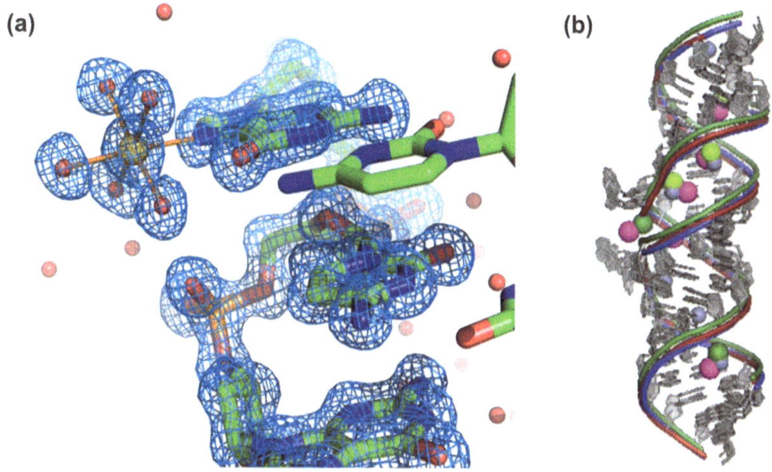

Figure 4. (**a**) Octahedral Mg^{2+} placed into the electron density map illustrating how Mg^{2+} can be identified in crystal structures of nucleic acids and distinguished from, e.g., H_2O carrying the same number of electrons (Figure by courtesy of PD Dr. Bernhard Spingler, University of Zürich, Switzerland). (**b**) Overlay of three crystal structures showing the coordination of Mg^{2+}, Ba^{2+}, and Zn^{2+} in the HIV dimerization initiation site. The three metal ions do not overlap completely, but exhibit slightly different coordination spheres. This panel has been prepared by PYMOL (W. L. DeLano, 2002, http://www.pymol.org) based on the PDB IDs 462D, 1Y6S, and 1NCL.

the large ribosomal subunit for example, replacement of $MgCl_2$ by $MnCl_2$ invariably caused crystal twinning [25].

An interesting application of heavy metal soaking of RNA crystals was reported by Wedekind and McKay [28], in a study of metal ion binding sites in the leadzyme. This small ribozyme, that has been discovered through *in vitro* selection [29,30], depends on Pb^{2+} for activity [31,32] (see also Section 4.3.2). It can be inhibited by addition of Mg^{2+} whereas ions that displace Mg^{2+} restore activity. Because of the disorder induced by cleavage at the scissile phosphate bond in the presence of Pb^{2+}, it is not possible to determine the exact differences between Mg^{2+} (ground state) and Pb^{2+} (catalytic state) binding by soaking the leadzyme extensively with Pb^{2+} salts. Here, instead of Pb^{2+}, Sr^{2+} was used, which has a similar ionic radius and mimics Pb^{2+} in binding, but not in catalysis. After crystallization, the crystals were soaked with either Mg^{2+} only or Mg^{2+} and Sr^{2+}. The metal-soaked samples were flash-frozen and data were collected to a resolution of 1.8 Å, at which the coordination geometry of the metal ions could be observed directly. Sr^{2+} ions were identified in difference Fourier maps. On the basis of this experiment, the participation of a metal ion binding to a remote site as a catalytic regulator of the leadzyme was deciphered [28].

3.1.4. Time-Resolved Crystallography

If homologous structural intermediates exist that can be either rapidly visualized or trapped, time-resolved crystallography allows the direct observation of structural changes that occur during the reaction. Trapping of intermediates is achieved by either adjusting a controlling variable (pH, metal ion concentration, etc.) or by flash-freezing the crystal. Scott et al. [33] captured a structure of the hammerhead ribozyme in the ground state with bound metal ions by lowering the pH < 5.0 and subsequent flash-freezing. In a second experiment, Mg^{2+} was allowed to induce cleavage at pH 8.5 prior to flash-freezing. In this way, conformational rearrangements of the ribozyme to reach the active transition state could be elucidated.

Time-resolved crystallography is a very elegant method to decipher differences in metal ion coordination between the ground- and an intermediate state. In general, it is of course possible that the trapped conformations would not significantly accumulate in solution and further biochemical experiments are necessary to verify the reaction pathway proposed by this technique.

3.1.5. Limitations

Crystal structures will only provide snapshots at a specific point in time under strictly controlled conditions that may not necessarily coincide with the most highly populated conformation in solution [26]. Especially for ribozymes it is important to know that in order to capture enzyme-substrate complexes, one needs to perturb the active site or a parameter that is important for activity (see Section 3.1.4: time-resolved crystallography), which may lead to a change in metal ion positioning [34].

Due to the equal number of electrons in Na^+, Mg^{2+}, and H_2O, these ions can only be distinguished on the basis of geometric criteria requiring high resolution data [35]. To unambiguously identify a particular metal ion, it may be necessary to carefully analyze the anomalous signal or solve the structure in the presence of heavier metal ions that prefer a similar coordination environment [24]. However, Ennifar et al. [36] have shown in a detailed crystallographic study of two HIV-1 dimerization initiation site RNA duplexes containing various different metal ions, that Mg^{2+}-containing crystals were not isomorphic to crystals with other ions and that no two kinds of ions bind identically to the RNA [3,4,36], as is also illustrated in

Figure 4b. This is clear evidence that different ions may stabilize slightly different RNA conformations and care needs to be taken when extrapolating results obtained with one ion to another.

In summary, crystal structures, albeit a great tool to gain insight into the coordination of metal ions to nucleic acids, need to be treated with a healthy skepticism. If structures from the PDB database are used to interpret results of biochemical or spectroscopic experiments, it is essential to have a close look at the experimental data and the conditions used for crystallization.

3.2. Nuclear Magnetic Resonance (NMR)

Besides X-ray crystallography, NMR has become the second power horse for the determination of biomolecular structures at atomic resolution. NMR is particularly suited to study interactions between biomolecules and their ligands in solution. Especially for nucleic acids, NMR has often proven to be the method of choice. 42.4% of all nucleic acid structures deposited in the PDB database by June 2010 were solved by NMR. And because metal ions are always intricately involved in RNA folding, it comes as no surprise that already the first ribozyme structure to be determined in its entirety by NMR revealed details about the importance of metal ion binding [37].

Naturally abundant isotopes in nucleic acids that are suitable for detection by NMR are ^{1}H and ^{31}P. Furthermore, it has become standard practice to produce isotopically enriched RNA by *in vitro* transcription with commercially available ^{13}C,^{15}N-NTPs [10].

NMR is a very powerful method for the observation of metal ion-induced changes of RNA structure in solution. Metal ion binding alters the chemical shifts of nearby NMR active nuclei. These chemical shift changes can be plotted against the metal ion concentration and fitted to a 1:1 binding model, to calculate affinity constants [38,39] (see also Section 6). The diamagnetic Mg^{2+} ion, that is, the most common divalent ion to bind nucleic acids *in vivo*, is well suited for this purpose. It is important to note that chemical shift changes only report the change in the electronic environment of the observed nucleus, which can be caused by the electron-withdrawing effect of the metal ion binding nearby or by structural changes in the local geometry and/or the global fold of the molecule. Chemical shift maps alone do not give any information as to what exactly caused the observed change, and thus have to be interpreted with caution. Fortunately, NMR itself lends a hand in finding alternative ways for the detection of metal ion binding sites.

In the following section, we try to outline the most commonly used NMR methods for the detection and characterization of metal ion binding sites in RNA.

3.2.1. *One-dimensional NMR Methods*

3.2.1.1. ^{1}H-NMR. The quickest way to detect interaction of metal ions with RNA by NMR is to record simple ^{1}H-spectra and observe changes in the imino proton resonances of uracil NH3 and guanine NH1 upon addition of increasing amounts of metal ions (Figure 5). This approach does not require any labeling of the RNA and is sensitive enough to be carried out with relatively low concentrations of the nucleic acid (minimum 0.2 mM). It is thus often used to screen for optimal experimental conditions. Because the imino protons can exchange with the protons of the solvent, these experiments are recorded in 90% H_2O/10% D_2O and water suppression is essential to obtain valuable spectra. Furthermore, low temperatures slow the exchange with the solvent. However, we observed surprisingly sharp signals at 293 K [40]. Considering the general line-broadening effect due to slower molecular tumbling at low temperatures, it may not only be unnecessary, but detrimental to cool the sample too much.

Depending on the metal ion and whether an interaction with the RNA is taking place, changes in chemical shift or line width broadening is observed upon metal ion addition. Whereas paramagnetic ions, such as Mn^{2+}, induce broader lines in a distance-dependent fashion [41], the diamagnetic Mg^{2+} ion is commonly used to observe chemical shift changes. Unfortunately, the exchange rate with Mg^{2+} often falls in the µs to ms range, corresponding to an intermediate exchange on the NMR timescale, thus also leading to line broadening [42,43] (see also Section 3.2.2.1). Although this is commonly regarded as a drawback of the method, it can be used to verify the significance of the observed chemical shift changes: If the proton resonances that experience the largest chemical shift change also broaden during the titration, this implies that the observed chemical shift changes are due to a direct effect of the metal ion and not due to long range structural changes.

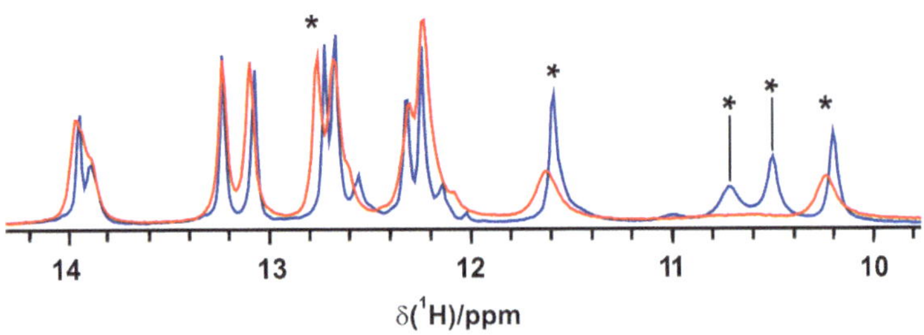

Figure 5. ^{1}H-NMR spectra of a group II intron 27 nucleotide hairpin (D6-27) without (blue) and in the presence of 30 mM $MgCl_2$ (red). Shown are the imino proton resonances of uracil NH3 and guanine NH1. Upon addition of Mg^{2+}, chemical shift changes and line width broadening are observed. Resonances with the strongest effects are marked with asterisks.

However, imino protons are usually located far away from the metal ion coordination site and their chemical shift is additionally dependent on the local structure and the exchange rate with the solvent [38]. Even more importantly, adenine and cytosine moieties cannot be observed and metal ion binding to bulge regions can often not be followed, as such imino protons become broadened beyond detection due to the exchange with the solvent [38]. Thus, observing changes in the imino proton resonances in ^{1}H-NMR spectra is solely a quick and easy way to determine whether a metal ion has any influence on the studied RNA. However, it is a highly insufficient and also misleading method to yield any more details on the M^{n+}-RNA interaction. Hence, once an interaction is detected, it is advisable to use one or several of the methods described below to gain more insight.

3.2.1.2. ^{31}P-NMR. The phosphate group that links sequential nucleotides is negatively charged and thus presents a primary cation binding site in RNA [3–5,44–46]. ^{31}P-NMR seems the obvious choice to detect metal ion-phosphate oxygen interactions in nucleic acids: Already in 1999, Pardi and coworkers [47] have identified a novel metal ion binding site in the hammerhead ribozyme by ^{31}P-NMR. They were able to observe a single resonance that broadened before any other resolved resonance upon addition of the paramagnetic Mn^{2+}. This resonance was assigned by the incorporation of a non-bridging ^{17}O using enriched $H_2{}^{17}O$, because ^{17}O efficiently broadens the attached ^{31}P by quadrupolar relaxation [48].

Unfortunately, apart from the 5′-end of a nucleic acid chain and some very special cases, most resonances are ill resolved in the ^{31}P spectrum, clustering around 0 ppm (Figure 6). One way to overcome this problem is to exchange a non-bridging phosphate oxygen at the position of interest with sulfur. Phosphorothioates are shifted far downfield to around 60 ppm in the ^{31}P NMR (Figure 6). Potential interactions with thiophilic metal ions, such as Cd^{2+}, can then be followed by observation of chemical shift changes of these well resolved residues [49–51]. Care has to be taken when interpreting these data: Large downfield shifts upon addition of M^{n+} ions do not necessarily mean a direct coordination [52], as it was erroneously claimed in studies with nucleoside 5′-triphosphates and 5′-diphosphates in the 1970s [53,54].

The two non-bridging phosphate oxygens are pro-chiral, i.e., the R_P phosphate oxygen points towards the major and the S_P towards the minor groove. The R_P oxygens can be replaced easily and stereoselectively, using α-thio NTPs in the *in vitro* transcription reaction: T7 RNA polymerase only incorporates S_P-α-thio-NTPs in an S_N2-type reaction by inversion of configuration without effect on the transcription yield (see also Section 4.1.1). Naturally, *in vitro* transcription will introduce a phosphorothioate 5′ of each nucleotide class that has been added in the form of α-thio-NTP. For the site-specific incorporation of phosphorothioates, solid-phase RNA synthesis is the

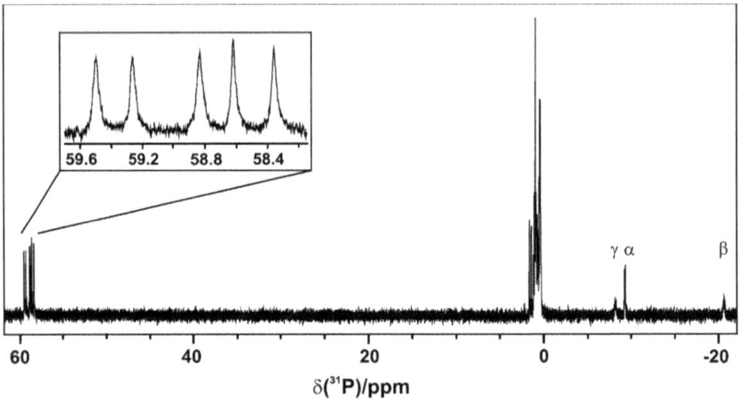

Figure 6. ^{31}P-NMR spectra of a group II intron 27 nucleotide hairpin (D6-27) where the non-bridging R_P oxygens of all five uracil residues present were replaced by sulfur. ^{31}P resonances of bridging phosphodiester groups are clustered around 0 ppm whereas those of thiophosphate groups appear well resolved at around 60 ppm (inset). The ^{31}P resonances of the 5′-terminal triphosphate chain (α-, β-, and γ-phosphates) appear in the negative chemical shift range [253].

only option. Unfortunately, solid-phase synthesis cannot distinguish between R_P and S_P phosphorothioate introduction and therefore this approach will always lead to a mixture of stereoisomers. Although separation of the two isomers should technically be achievable by HPLC [55], we did not succeed with a 27mer oligonucleotide [56]. Nonetheless, such site-specifically incorporated phosphorothioates can be useful to help in resonance assignments.

The replacement of a phosphate by a thiophosphate group is not necessarily an innocent mutation [57]. Hence, it needs to be checked whether the incorporation of sulfur has an effect on the stability of the RNA, e.g., by UV melting curves. In addition, ^{1}H,^{1}H-NOESY and ^{1}H,^{13}C-HSQC fingerprint spectra can indicate structural differences by changes in the chemical shift pattern between the wild-type and thioated RNA.

3.2.1.3. *^{15}N-NMR.* Another one-dimensional method that has been put forward mainly by Tanaka and Taira [58,59], is the use of direct ^{15}N-NMR spectroscopy to probe metal ion binding to nucleobase nitrogen atoms. For this approach, the RNA needs to be isotopically enriched with ^{15}N (spin 1/2). This can be achieved by using ^{15}N-enriched NTPs in the *in vitro* transcription reaction. Tanaka et al. identified a metal ion binding site on an RNA motif derived from the hammerhead ribozyme [58] and analyzed its binding properties in detail [59].

In contrast to ^{1}H-NMR, ^{15}N-NMR provides a tool to directly observe a coordinating atom. Tanaka and Ono have investigated the effect of Hg^{2+} coordination to a DNA duplex in detail and conclude that innersphere coordination is accompanied by an upfield shift of the affected ^{15}N resonances, whereas proton–metal cation exchange leads to a downfield shift [60]. In the case of an imidazole-modified DNA duplex, Ag^{+} coordination caused an upfield shift of the liganding N3 by almost 20 ppm [61]. Again, broadening of the line width can be an issue, if the exchange regime proves unfavorable [58].

The same authors have also tried to directly detect metal ion binding through 1-bond ^{1}J couplings between the metal ion and the nucleobase ^{15}N in 1D ^{15}N-NMR spectra, as it has been done for metalloproteins [62,63]. To our knowledge this approach has not been successful so far, probably due to exchange processes between the metal-bound and the metal-free states of the RNA [58–60].

At least in theory, ^{15}N-NMR is an ideal method to probe the direct interaction between a metal cation and a nucleobase nitrogen in solution because the ^{15}N chemical shift is very sensitive to changes in the local environment [64]. For example, a 20 ppm upfield shift has been measured for the N7 of guanosine in DMSO upon addition of one equivalent of Zn^{2+} or Hg^{2+} and the protonation of an adenosine N1 causes a 70 ppm upfield shift [65,66]. Unfortunately, the low gyromagnetic ratio of nitrogen leads to a rather low signal of directly observed ^{15}N nuclei, and thus, extensive recording times (10–20 h) are needed to obtain a simple one-dimensional spectrum. Also spectral assignments of larger RNAs are complicated. To overcome these issues, it may in fact be easier to record two-dimensional proton-nitrogen correlated HSQCs as described below, to probe direct interactions between metal ions and nucleobase nitrogens.

Using indirect detection in an HSQC-type experiment, Johannsen et al. measured a 86 Hz splitting of the N3 resonances of three consecutive ^{15}N-labeled imidazole moieties in a palindromic modified duplex DNA structure, coordinating three Ag(I) ions [61]. This splitting was interpreted as corresponding to the average coupling of $^{1}J(^{15}N,^{107}Ag)$ and $^{1}J(^{15}N,^{109}Ag)$, as both silver isotopes have an almost identical natural abundance. The observed coupling constant reports directly on the coordination of Ag(I) to the N3 position of the imidazole moieties and represents the largest of this type published to date [67,68].

3.2.2. *Two-dimensional NMR methods*

Although a wealth of information can already be obtained by analysis of one-dimensional NMR spectra, there are limits, which can be overcome by applying multidimensional spectra: First and foremost, the spectral overlap is significantly reduced and often spectral assignment of any RNA would not be possible otherwise. Also, by recording heteronuclear experiments, chemical shift changes affecting multiple nuclei can be traced simultaneously. And last but not least, insensitive nuclei, such as ^{15}N (see also Section 3.2.1.3) can be probed indirectly via heteronuclear quantum coherence experiments, with the signal intensity of the much more sensitive, directly attached protons. The diversity of multidimensional NMR experiments for the assignment of peaks and thus structure determination of nucleic acids is enormous: Many of them could in theory be used to analyze the effect of metal ion binding to RNA. This section does not aspire to be complete in any way, but we want to outline the most commonly used methods, their advantages and their drawbacks and hopefully inspire scientists to try one or the other in their own research.

3.2.2.1. Chemical Shift Change Analysis with Diamagnetic Metal Ions. The caveats of using exchangeable protons to detect metal ion binding have been indicated in Section 3.2.1.1. Much more sensible is thus the observation of non-exchangeable protons. The classic experiment for resonance assignment of protons in nucleic acids, the $^{1}H,^{1}H$-NOESY [69] can also be used to monitor the effects of M^{n+} coordination to RNA at both the sugar and the nucleobase moieties. These observed protons, e.g. the H8, H2, H6, H5 and H1′, are located close to potential binding sites, such as N7 of purines, carbonyl oxygens atoms of guanines, cytosines, and uracils, or 2′-OH of the ribose sugar. Chemical shift changes that are observed during titration of increasing amounts of $MgCl_2$ can be fitted to a 1:1 binding model to calculate affinity constants [38,39] (for a more detailed description thereof see also Section 6). These chemical shift changes occur either due to direct metal ion binding or because of structural changes induced by coordination of the metal nearby, i.e., increased or reduced stacking interactions. From the information content of the chemical shift data alone, it is impossible to distinguish between these two effects. As described above, Mg^{2+} addition can lead to a substantial line broadening of some RNA proton resonances, an effect that has been allocated to an intermediate exchange of the Mg^{2+}_{aq} ion at its binding site on the NMR time scale [42,43]. The line broadening can thus be used to detect specific Mg^{2+} sites, as it should only be observed at sites, where Mg^{2+} binds in close proximity. By including this information into the evaluation, an accurate picture of metal ion binding can be obtained. Indeed, in the study of metal ion binding to the shortened domain 6 of the group II intron *Sc*.ai5γ (D6-27), resonances that experienced line-broadening were the same or close in space to resonances that experienced a large chemical shift change upon the addition of Mg^{2+} [13].

The exchange phenomena observed with Mg^{2+} become severe above a concentration of about 10 mM, impairing further chemical shift analyses. Especially for weak binding sites that are not approaching saturation at this metal ion concentration, it would therefore be useful to have an alternative metal ion at hand that mimics Mg^{2+} binding, but has faster solvent exchange kinetics than Mg^{2+}. Metal ions with this property include Cd^{2+}, which exhibits a markedly faster solvent exchange rate (4×10^8 versus $5.3\times10^5\,s^{-1}$, see also Section 2.2). In our studies with *Sc*.ai5γ D6-27, we have titrated the RNA with Cd^{2+} and observed the chemical shift changes of the RNA proton resonances in the $^1H,^1H$-NOESY [40]. Like Mg^{2+}, Cd^{2+} preferentially forms non-distorted octahedral complexes. Although Cd^{2+} has a larger ionic radius than Mg^{2+} (0.97 versus 0.72 Å), it has been widely applied to mimic Mg^{2+} in biochemical studies [50,51,70–72]. Furthermore, Cd^{2+} selectively replaces Mg^{2+} bound to N7 via an innersphere coordination mode [4,36]. In our studies, Cd^{2+} concentrations above 10 mM led to increased line width, suggesting that even with cadmium(II), the exchange regime was not entirely on the fast NMR timescale. But compared to Mg^{2+}, saturation was reached in most cases, indicated by the ceased move of chemical shifts. In general, Cd^{2+} thus binds with a slightly higher affinity to the RNA than Mg^{2+}, and insofar Cd^{2+} titrations prooved useful to elucidate weaker binding sites. In the light of the inherent ambiguity concerning the physical origin of a particular chemical shift change, it may be revealing to compare the chemical shift pattern obtained from titrations with different metal ions.

3.2.2.2. Line Broadening Analysis Using Paramagnetic Metal Ions. The paramagnetic Mn^{2+} has widely been used to probe for the spatial proximity of an RNA residue to the metal ion [73–75]. Line broadening is caused by the distance-dependent ($1/r^6$) relaxation of NMR active nuclei near the binding site of the paramagnetic metal ion [41]. Therefore, Mn^{2+} broadening studies can be used to distinguish residues that bind directly to divalent metal ions from residues that exhibit a structural change upon metal ion binding. For example, in a group II intron branch domain D6-27 the crosspeaks associated with the GU wobble pairs flanking the branch-adenosine A20 became distinctly broadened upon addition of Mn^{2+} [13]. In contrast, the branch adenosine H1′-H8 and H1′-H2 as well as the interstrand A20H2-G8H1′ crosspeaks were hardly affected, indicating that rather the flanking GU wobble pairs than A20 N7 are involved in binding of a divalent metal ion. Line broadening is observed already at micromolar metal ion concentration, whereas millimolar concentrations of Mg^{2+} are needed to induce chemical shift changes. If the Mn^{2+} concentration exceeds 10–20% of a 1 mM RNA concentration, non-selective broadening of all peaks in parallel has been observed for D6-27.

3.2.2.3. Cobalt(III) Hexammine as Probe for Outersphere Interactions. When talking of Mg^{2+} binding sites in RNA, a question that commonly occurs is, whether the coordination mode is innersphere, e.g., the metal ion is in direct contact with the RNA, or outersphere, in which case the interaction is mediated by water molecules. Because of their similar geometry $[Co(NH_3)_6]^{3+}$ is commonly used as a probe for binding sites of fully solvated magnesium ions $[Mg(H_2O)_6]^{2+}$ [76–79]. If the cobalt(III) hexammine is in fast exchange between the bound and unbound state, a single resonance at 3.65 ppm represents all 18 ammine protons [77]. Fortunately, no protons from the RNA resonate in this region and therefore NOEs between the $[Co(NH_3)_6]^{3+}$ protons and uracil NH3 and guanine NH1 protons are well resolved. By recording a $^{1}H,^{1}H$-NOESY of the RNA in the presence of $[Co(NH_3)_6]^{3+}$ in 90% H_2O/10% D_2O, strong NOEs can be detected between the RNA and $[Co(NH_3)_6]^{3+}$ (Figure 7). This experiment naturally can only be successful, if $[Co(NH_3)_6]^{3+}$ is relatively tightly bound and no rapid dynamics occur.

Rüdisser and Tinoco have used this approach to position a $[Mg(H_2O)_6]^{2+}$ in their structure of a GAAA tetraloop [76]. They used distance constraints based on NOEs in the restricted molecular dynamics calculation. In order to calculate the geometry of the GAAA tetraloop-ligand complex, $[Co(NH_3)_6]^{3+}$ was placed randomly into the RNA starting structures with randomized torsion angles. Out of 17 converged structures, ten folded into geometries with $[Co(NH_3)_6]^{3+}$ in the major groove and seven structures had $[Co(NH_3)_6]^{3+}$ bound in the minor groove to the GAAA tetraloop. The predicted NOEs for the major groove binding mode were consistent with all experimental data, but the minor groove binding mode predicted NOEs that could not be seen. Therefore, the minor groove binding mode was excluded.

$[Co(NH_3)_6]^{3+}$ is considered a rather good mimic for fully hydrated $[Mg(H_2O)_6]^{2+}$ ions. However, it has to be kept in mind that $[Co(NH_3)_6]^{3+}$ is an exchange inert complex. Hence, any Mg^{2+} binding site with more common partially dehydrated Mg^{2+}-aqua complexes will be missed.

3.2.2.4. Innersphere Coordination Detected by ^{2}J-[$^{1}H,^{15}N$]-HSQC. The N7 of purine nucleotides is a well known metal ion binding site in nucleic acids [4,44]. One possibility to detect metal ion binding to this atom is the use of direct ^{15}N detection (see Section 3.2.1.3). A big caveat of this method is the low ^{15}N gyromagnetic ratio leading to very weak signal and concomitantly high concentration requirements and prolonged recording times. To overcome these problems, we have recently used ^{2}J-[$^{1}H,^{15}N$]-HSQC experiments [80,81] that were optimized for the 11.4 Hz coupling between H8 and N7 of purine bases, to identify Mg^{2+} coordination to N7 of the 27 nucleotide long RNA hairpin D6-27 [82]. In this approach, either all adenosine or guanosine nucleotides are $^{13}C,^{15}N$ isotope-labeled by *in vitro* transcription. Spectra are recorded upon addition of increasing concentrations of $MgCl_2$ and line-broadening effects as well as chemical shift changes are observed. In the case of D6-27, all monitored resonances show chemical shift changes in either the ^{1}H, the ^{15}N, or both dimensions.

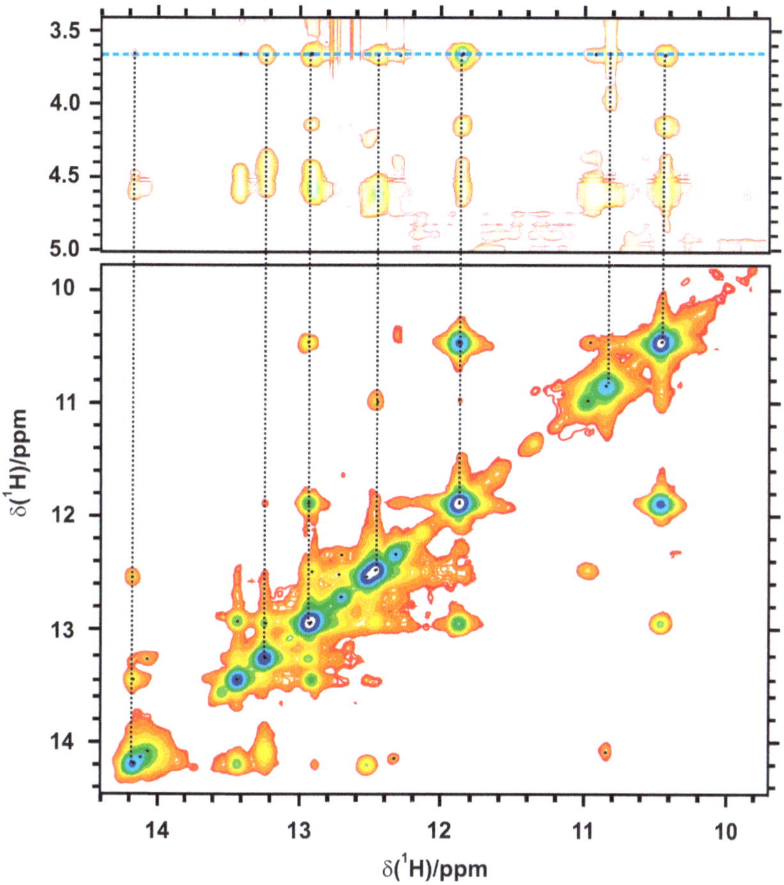

Figure 7. $^1H,^1H$-NOESY of D6-27 in the presence of 2.5 mM $[Co(NH_3)_6]^{3+}$ in 90% H_2O/10% D_2O. The resonance of the ammine protons of $[Co(NH_3)_6]^{3+}$ at 3.65 ppm is indicated by a blue dotted line. The imino protons of D6-27 that give the largest NOEs to $[Co(NH_3)_6]^{3+}$are also indicated.

Some N7 chemical shifts experience line broadening and large upfield shifts, indicating direct coordination of a Mg^{2+} ion (Figure 8). Other resonances are "only" affected in their ^1H shift, being indicative of Mg^{2+}-induced local structural changes, but no direct binding. Due to the indirect detection of ^{15}N in the HSQC experiment, the signal intensity is comparable to the one of the directly observed ^1H. This renders the experiment much more sensitive than direct ^{15}N detection. Thus, direct metal ion innersphere coordination to N7 can be detected rapidly and extensive titration series can be performed without fear of sample degradation.

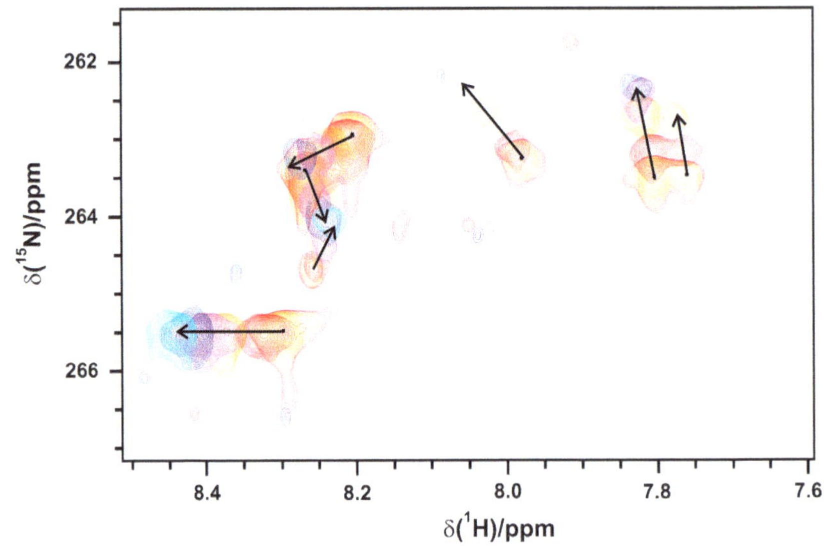

Figure 8. Section of a $^{1}H,^{15}N$-HSQC of D6-27, optimized for the ^{2}J-N7H8 coupling of purines (11.1 Hz for guanine and 11.4 Hz for adenine), showing crosspeaks between H8 and N7 of adenines. Chemical shift changes (as highlighted by arrows) in the nitrogen dimension upon addition of 0–2.5 mM $MgCl_2$ indicate direct metal ion coordination to N7.

The corresponding experiment has recently been used to investigate the binding of Ag^+ to an imidazole-modified DNA duplex, in which Ag^+ bridges two opposite imidazoles [61]. In this case, the ^{15}N resonances of the coordinating N3 atoms are not only strongly upfield shifted but also split by 86 Hz due to the $^{1}J_{15N,107/109Ag}$ coupling with the kinetically inert Ag^+ of spin 1/2 (see Section 3.2.1.3). It is well feasible that corresponding couplings are observable with ^{113}Cd in RNA, if the Cd^{2+} is tightly bound and experiences no exchange. Similar experiments have been performed with proteins in the past: M^{2+} binding to either N1 or N3 of histidine can be resolved by corresponding $^{2}J_{1H15N}$ experiments and in the case of cysteine coordination, ^{3}J couplings can be observed in $^{1}H,^{113}Cd$-HMBC experiments [83–85].

3.2.3. Direct Observation of Metal Ions

Although the indirect methods of identifying metal ion binding sites via the classical "NMR-nuclei" ^{1}H, ^{15}N, ^{13}C or ^{31}P are generally robust, it would of course be preferable to obtain a NMR signature directly from the bound metal ions. Most metal ions that bind to nucleic acids possess significantly abundant, NMR-active isotopes [86]. Unfortunately, most of them also have nuclear spin numbers of I = 3/2, 5/2 and 7/2 and thus suffer from fast quadrupole relaxation that broadens the NMR spectral lines.

In solid state NMR, line-width broadening due to quadrupolar relaxation is less of a problem because of the absence of rapid reorientation [87]. This potential has been exploited in a number of studies using ^{23}Na, ^{25}Mg or ^{59}Co as NMR active nuclei [87–89].

On the other hand, also some successes have been reported using liquid state NMR. ^{25}Mg-NMR was used in a line shape analysis together with atomic absorption spectroscopy to characterize weak magnesium binding sites on yeast $tRNA^{Phe}$ [90]. Magnetic relaxation dispersion experiments were employed to detect ^{23}Na in the spine of hydration in the minor groove of B-DNA [91] and ^{205}Tl has been used as a K^+ mimic in the characterization of monovalent cation binding to a DNA G-quadruplex [92,93].

3.3. Electron Paramagnetic Resonance (EPR) and Related Methods

EPR spectroscopy is based on the analogous physical concept as NMR, but employs the electron spin (instead of nuclear spin) of one or more unpaired electrons. Unpaired electrons are commonly introduced into RNA by attaching a spin label, usually a stable nitroxide radical, at a specific location by site-directed spin labeling (SDSL) [94]. Alternatively, also paramagnetic metal ions can be used, providing a direct handle to observe the environment of these ions within large biomolecular structures. EPR and related methods have been used in the past to investigate global folding and the dynamics, as well as to monitor metal ion binding to RNA. To look at global folding, two spin labels need to be placed at specific locations in an oligonucleotide by chemical methods [94]. Pulsed EPR techniques like pulsed electron-electron double resonance (PELDOR or DEER) [95,96] or more specialized ones like double quantum coherence (DQC-EPR) [97] are then applied to obtain accurate distances in the range from 1.5–8 nm. This application including a detailed methodical description has been excellently reviewed recently [98]. Dynamic features and interactions of biomolecules in the range from picoseconds to milliseconds can be measured by continuous-wave EPR (cw-EPR) and has been applied for example with the *Hammerhead* ribozyme [99]. This method has been thoroughly reviewed lately [94] and we encourage the reader to refer to this literature for more information and practical details.

In the following we will concentrate on the usage of EPR to monitor metal ion binding to RNA and provide a brief overview on the different EPR techniques applied. Again, this topic has recently been excellently reviewed including a detailed experimental description [100] to which we refer the more interested reader.

3.3.1. *Mn^{2+} as an EPR Probe for Metal Ions in RNA*

EPR has developed into a major method to investigate the interaction of paramagnetic ions to RNA. Mostly Mn^{2+} is used as a paramagnetic mimic for Mg^{2+}, but also other ions like Cu^{2+} [101] or the vanadyl ion [102] have been applied in the past. The physico-chemical properties of Mn^{2+} and its differences compared to Mg^{2+} have been summarized above in Section 2.2. Mn^{2+} has five unpaired electrons, is nearly always high spin, and usually displays an octahedral coordination sphere. A single isolated unpaired electron with spin $S = 1/2$ gives rise to a single resonance in the EPR spectrum. High-spin Mn^{2+} with $S = 5/2$ shows six major lines that are separated by about 90 Gauss. However, these lines can be further split due to interactions with other electrons or nuclear spins with $I > 0$. This spin-spin interaction is called hyperfine interaction and can occur either through-space (dipolar) or through-bond. Excluding additional second-order effects, the energy levels of the unpaired electrons are all split by the hyperfine-interaction with the nucleus of ^{55}Mn ($I = 5/2$) giving rise to theoretically 36 transitions. However, a significant degeneracy of the energy levels is true for most Mn^{2+} binding sites and thus only the six major lines are detected.

Upon coordination of Mn^{2+} to, e.g., RNA, the EPR properties change to allow examination of the binding site with respect to (i) a quantification of bound versus free Mn^{2+}, (ii) gain information on the binding site of the biomolecule, and (iii) the identification of the ligating atoms by more advanced methods [100]. Methods to investigate these three characteristics shall be briefly summarized below, whereas a detailed description of the methodical work is reviewed in [100].

3.3.2. *Quantification of Bound Mn^{2+} by Room-Temperature EPR*

Mn^{2+} concentrations in solution can be quantified based on the integral of the peaks or estimated based on the signal height. If recorded at room temperature, binding of Mn^{2+} to a large biomolecule leads to a significant broadening of the EPR lines. This effect is due to the change from a perfectly symmetrical $Mn(OH_2)_6^{2+}$ to a Mn^{2+} coordinated to one or more atoms of the RNA. This effect increases with the size of the biomolecule because of slow molecular tumbling and the consequently longer rotational correlation times of the bound ion. By this method, Mn^{2+} binding to tRNA [103], the P5abc subdomain of a group I intron [100] as well as the *Hammerhead* ribozyme [104,105] have been quantified.

Lowering the temperature below the freezing point after addition of a cryoprotectant allows to detect the isolated EPR signal of the bound Mn^{2+}, if conditions have been chosen such that all Mn^{2+} ions are bound to the RNA. Although these changes are very subtle, they allow for a more detailed examination of the coordination sphere as has been shown in the case of the *Hammerhead* ribozyme [106].

3.3.3. Identification of Mn^{2+} Ligands by ENDOR Spectroscopy

It is possible to characterize the Mn^{2+} binding site in RNA by the additional detection of a signal from the nucleic acid. Relevant nuclei on the RNA are ^{31}P (I = 1/2), ^{14}N (I = 1), ^{1}H (I = 1/2), or specifically labeled sites of ^{15}N (I = 1/2) or ^{13}C (I = 1/2). The hyperfine interaction between unpaired Mn^{2+} electrons and the nuclear spin of the surrounding atoms is based on distance, orbital overlap, and the nuclear properties and are generally too weak to be seen in standard EPR. However, by using advanced techniques like electron-nuclear double resonance spectroscopy (ENDOR), such hyperfine interactions with nuclei up to 7 Å away from the paramagnetic ion can be revealed. In ENDOR generally a single position in the EPR spectrum is monitored [100]. The application of ENDOR has for example shown that Mn^{2+} coordinates in an innersphere fashion to the *Hammerhead* ribozyme [106]. Again, the application of ENDOR, both in a cw- or pulsed-EPR mode, is described in detail in [100].

3.3.4. ESEEM Spectroscopy

Electron-spin echo envelope modulation (ESEEM) is the method of choice to detect weak anisotropic hyperfine couplings. The amplitude of modulation for the hyperfine coupling depends on the number of nuclei and the relationship between the nuclear dependent frequencies [100]. In X-band spectrometers (microwave frequency of $\nu \approx 9.75$ GHz), ^{14}N coordinated to Mn^{2+} can be well observed, making ESEEM very useful to detect Mn^{2+}-purine coordination: In the *Hammerhead* ribozyme, Mn^{2+} coordination to adenine and guanine N7 sites has been detected by ESEEM [107–109]. If different types of nuclei are coordinated to Mn^{2+}, the resulting spectrum corresponds to the product of the individual spectra. Consequently, the hydration level of Mn^{2+} bound to RNAs can be quantified [110]. For more detailed informations on ESEEM please refer to [100].

3.4. Lanthanide(III) Luminescence

Aside from their usage as hydrolytic cleaving agents (see Section 4.3.1) and chemical shift or line-broadening reagents (see Section 3.2.2.2), lanthanide(III) ion complexes are very good luminescence sensors for biological applications. These ions have long luminescence life times, are resistant to photobleaching and offer the possibility of using multiple emission bands for ratiometric sensing [111,112]. Luminescence lifetimes are strongly dependent on the solvation state of the Ln(III) ion including also the formation of hydroxo species, as has been shown in the pioneering work compiled by Baes and Messmer [113]. Therefore, direct excitation photoluminescence studies can yield important information on the coordination sphere of these metal ions, mostly Tb(III) and Eu(III). This has been used by Uhlenbeck and coworkers to study the binding of these two ions to the *Hammerhead* ribozyme [114]. However, the obtained results of the recovered luminescence lifetimes are often not easily interpreted and thus in recent years strong efforts have been made by Morrow and others to establish the basics before trying to understand larger nucleic acid systems [115,116].

Only a few studies have thus dealt so far with nucleotides and/or smaller RNAs. Measurements in H_2O and D_2O have shown that at least one water molecule is lost upon binding of the lanthanide(III) ion to either GTP or RNA at slightly acidic or neutral pH [117]. Along the same line, Eu(III) binding to small RNA hairpins and a GAAA tetraloop has been investigated: The results indicate that Eu(III) loses more water molecules from its inner shell of hydration than Mg^{2+} as comparison with ITC experiments show [118–120]. Recently, Greenbaum and colleagues have used Förster resonance energy transfer (FRET) between Eu(III) ions and Cy3 to identify three site-bound metal ions in the U2-U6 snRNA [121,122].

Taken together, lanthanide(III) luminescence offers a great potential to yield detailed pictures about the stoichiometry of metal ion binding to RNA as well as to gain information on the inner coordination sphere. However, first experiments have shown that the interpretation of the experimental data is often rather complex and one is thus only at the very beginning to really exploit this method.

3.5. Vibrational Spectroscopies

Vibrational spectroscopies have been commonly used as a standard technique to structurally characterize small molecules. The bond stretching, bending and deformation frequencies are strongly dependent on the nature of the atoms involved, the respective binding geometries, as well as their further bonded atoms. These vibrational modes become excited by electromagnetic radiation in the infrared range as long as a change in dipole moment is associated. To obtain an optimal signal-to-noise, mostly Fourier-transform infrared spectroscopy (FT-IR) is applied nowadays.

FT-IR is a well established method to investigate structural changes of proteins, DNA, and RNA, i.e. conformational transitions upon ligand, drug, or protein/nucleic acid binding [123–125]. A more detailed description on experimental aspects has recently been published by Tajmir-Riahi, N′Soukpoe-Kossi and Joly [123]. Metal ion coordination is also directly observable by a change in the energy of the associated vibrational modes. Nevertheless, only very few examples exist in the literature where metal ion binding is investigated by FT-IR, reasons being the dynamic nature of metal ion binding, the rather weak binding, as well as the small number of nucleic acid building blocks, making it very difficult to observe specific atoms within a large RNA. Some of these problems can be overcome by isotopic labeling, or by monitoring H/D exchange [124] but so far, mostly the coordination of kinetically inert ions like Pt^{2+} [126,127] or structural changes induced by metal ion (complexes), e.g., B- to Z-transitions of DNA, have been investigated [126,128]. Nevertheless, there are examples, where FT-IR was used to investigate the interaction of DNA and RNA with Ag^+ [129] and As_2O_3 [130].

Most of the vibrational bands that are not observable by IR, i.e., those that are not accompanied by a change in dipole moment, are Raman-active. The Raman effect is based on the inelastic scattering of monochromatic light and used to study vibrational, rotational, and other low-frequency modes in a system. Raman has the advantage that a much smaller number of bands are observed and that water itself is Raman-silent. With respect to RNA, Raman has been used as an interesting alternative to study catalysis of the hairpin ribozyme [131]: Surface-enhanced Raman spectroscopy (SERS) is a powerful tool to detect trace amounts of RNA that are adsorbed on a metal surface. Upon cleavage the single stranded RNA product dissociates from the hairpin ribozyme and interacts with the surface of silver colloids and can be detected at biological concentration in real time [131].

Raman spectroscopy has been repeatedly used to investigate metal ion binding to the nucleic acid backbone phosphate oxygens. Metal ion interactions thereby lead to a change in energy of the bonds vibrations. In this way, the interaction of various metals like Mg^{2+}, Ca^{2+}, Sr^{2+}, Ba^{2+}, Mn^{2+}, Co^{2+}, Ni^{2+}, Cu^{2+}, Zn^{2+}, Pd^{2+}, and Cd^{2+} with DNA have been examined in a semiquantitative way [132–135]. More recently, innersphere coordination between Mg^{2+} and the backbone of the HDV ribozyme was detected using Raman crystallography [136]. In another study, Harris and coworkers used this method to distinguish between innersphere and outersphere interactions as well as electrostatic effects of metal ion binding. Whereas all three modes contribute to an attenuation of the Raman signal, only innersphere coordination leads to a frequency shift to higher wave numbers of the symmetric vibration of the non-bridging oxygen atoms $\nu_S PO_2^-$ [137]. The authors could show that the shift in $\nu_S PO_2^-$ is dependent on the metal ion identity and can also be used as a quantitative mean to characterize specific metal ion binding events.

Both, FT-IR and Raman spectroscopy have so far been applied rather rarely to solve problems associated with metal ion binding to nucleic acids. However, with the rapid technological advancements in the past years, it becomes clear that the power and possibilities of both methods have only been exploited marginally and one can thus expect exciting results in the future.

3.6. Further Spectroscopic Methods

Aside from the methods described in the above Sections, other methods usually more associated with proteins find application in the characterization of metal ion-nucleic acid interactions. X-rays for example can also be used for small angle X-ray scattering (SAXS), extended X-ray absorption fine structure (EXAFS) or X-ray absorption near-edge structure (XANES) spectroscopy as briefly summarized in the following paragraphs.

Small angle X-ray scattering yields information about folding/unfolding, aggregation, ligand binding, extended conformations, shape and assembly states of macromolecules [138–141]. The resolution limit is with 10 Å rather low, but SAXS can be applied to macromolecules of literally any size. Often in combination with other techniques, a rather accurate picture of shape, flexibility, and conformational changes dependent on cofactors can be obtained [138–141]. For example, SAXS has been used to investigate complete solution structures of the VS ribozyme in solution [142], to look at riboswitch conformations [143], to characterize conformational states of ribozymes [144], to follow folding in a time-resolved manner [145], or to investigate the effect of metal ions on the order of transition states of two group I intron ribozymes [146]. Woodson et al. could show that the *Azoarcus* group I intron folding is described by two components, whereas the *Tetrahymena* folding has at least one intermediate [146] (see also Chapter 7 of this volume). They conclude that in the first case, tertiary contact formation occurs with charge neutralization of the backbone, whereas in the second case, additional counterions are needed.

XANES and EXAFS are two variations of X-ray absorption spectroscopy (XAS) that differ in the incident energy of the synchrotron radiation applied. A prerequisite for any of these methods is that a metal ion is tightly and specifically bound. Accordingly, XAS is regularly applied to metalloproteins or small complexes [147,148]. XAS has recently been used to investigate a highly negatively charged polyoxometalate $[Mo_7O_{24}]^{6-}$ (POM) cluster hydrolyzing 2-hydroxypropyl-4-nitrophenyl phosphate as a RNA model substrate [149]. The investigators could show that the Mo(IV) in $[Mo_7O_{24}]^{6-}$ is actually the catalytically active species. Similarly, it has been shown earlier that a copper(III) hydroxysalen complex leads to site specific cleavage of the

HIV-1 TAR RNA [150]. Very recently Fierke and colleagues have used XANES and EXAFS to look at metal ion binding in the P4 helix of RNase P [151]. The collected XAFS data show that a Zn^{2+} coordinates (at least) partly innersphere to the P4 helix in an octahedral coordination geometry with an average Zn(II)-O/N distance of 2.08 Å [151].

Another method yielding the corresponding information on local geometry is perturbed angular correlation (PAC) of gamma rays [152]. Certain radioactive nuclei decay by sending two gamma rays at a very short (10 nsec to 10 μsec) but defined interval at a specific angle to each other. The hyperfine interactions of the probe nucleus with the surrounding electric and/or magnetic fields allow to draw conclusions on the closer neighborhood. Originally a method to investigate solids, it has recently been applied to look at metal ion binding in proteins [147,148,153–156], using ^{111m}Cd and ^{199m}Hg. With respect to RNA, PAC is currently applied to investigate bridging Hg^{2+} ions between two uracil units forming U-Hg(II)-U base pairs in a RNA duplex [157].

To summarize, SAXS, XAS, and PAC are new methods that have rarely been applied yet, but with increasing understanding of metal ion interactions, will provide new ways to characterize in detail the coordination spheres and geometries of metal ions in nucleic acids.

4. CHEMICAL AND BIOCHEMICAL METHODS

Aside from spectroscopic methods described above in Section 3, (bio)chemical methods are frequently applied to gather information on metal ion binding, as well as elucidating their possible roles in ribozyme catalysis. All of these methods require either modification of the nucleic acid, the exchange of the *in vivo* cofactor by another metal ion and/or RNA cleavage. Nevertheless, crucial information is gained that could not be obtained by other methods.

4.1. Sulfur Rescue Experiments

In such assays, one or more specific oxygen atoms are probed for their importance in structure and/or catalysis of a given RNA by replacing them with the larger and softer sulfur. Most often a non-bridging phosphate oxygen is probed [158,159], but also other positions like the 3′ ribose oxygen or even a nucleobase oxygen have been replaced in the past [160]. Here we concentrate on the application of such experiments to detect metal ion binding: For example, if Mg^{2+} binding to a specific phosphate oxygen is essential for catalysis, the activity of the ribozyme is reduced upon replacement with sulfur, due to the about two-fold lower affinity of Mg^{2+} towards the softer sulfur ligand [161]. Upon addition of a more thiophilic metal ion like Cd^{2+} or Zn^{2+} [42,161,162] ribozyme activity may be restored [163]. Such a "rescue" is usually interpreted as evidence for innersphere metal ion coordination to the phosphate oxygen.

In the following we will briefly summarize the incorporation of sulfur, the experimental setup, analysis, and caveats of these experiments. For a detailed example the reader should also consider reference [26].

4.1.1. *Incorporating the Phosphorothioate*

Phosphorothioate nucleoside 5′-triphosphates (or phosphoamidites) containing a sulfur substitution at the position of one of the non-bridging oxygens of the α-phosphate are commercially available (e.g., from Glen Research). They can be incorporated into the RNA either by *in vitro* transcription or by solid state synthesis. However, T7 RNA polymerase only accepts the S_P diastereomer of the phosphorothioate analog and incorporates it by inversion of configuration. This leads to RNA transcripts that carry only R_P phosphorothioates (see also Section 3.2.1.2) [164]. In addition, the phosphorotioate nucleotide will be incorporated at every position complementary to the coding template strand.

In the case of solid state synthesis, both isomers are accepted and incorporated, leading to a mixture of R_P and S_P products. These can be separated by reverse-phase HPLC, with the R_P derivative usually emerging prior to the S_P [55]. The identity of the stereoisomer can subsequently be tested by digestion with snake venom phosphodiesterase, which cleaves the S_P isomer more slowly than its R_P counterpart [165]. Unfortunately, chemical synthesis of RNA is restricted to about 70 nucleotides in length and the separation works well only for very small RNAs (up to 13 nt) [166]. Thus, one often depends on *in vitro* transcribed phosphorothioate RNA with the above mentioned caveats. Alternatively, if the modification is required near the end of a long strand, ligating a synthetically produced, modified oligonucleotide to a larger *in vitro* transcript may be an option [166]. In double stranded regions, simple annealing of a modified oligonucleotide may also work as it has been shown with other single atom modifications within the 23s ribosomal subunit [167].

4.1.2. *Experimental Conditions*

Sulfur rescue experiments have been widely used to identify metal ion binding sites in ribozymes and spliceosomal RNA [168–171]. A comprehensive review of the method is given in [26].

Most importantly, to determine the contribution of a specific metal ion to catalysis of a given ribozyme, it is imperative that the catalytic step is rate-limiting. Therefore, one generally uses single turnover conditions, in which the ribozyme is in large excess over the substrate. The 5′ ^{32}P-labelled substrate is cleaved by addition of either the wild-type or the sulfur-substituted ribozyme in the presence of Mg^{2+} or a thiophilic metal ion (e.g., Cd^{2+} or Zn^{2+}), respectively. Aliquots of the reaction solution are taken at different time points, quenched with EDTA, and all samples are then loaded on a denaturing polyacrylamide gel and analyzed by autoradiography.

Thiophilic metal ions often form insoluble metal oxide or hydroxide species at physiological pH. Hence, stock solutions of such metal ions are usually kept at acidic pH. Dilutions are best prepared freshly immediately prior to use. The pH of the final solution needs to be adjusted to fit the reaction conditions [166]. Elevated levels of monovalent salts are often used, both to saturate diffuse metal ion binding sites as well as to dampen indirect effects due to changes in ionic strength upon addition of thiophilic metal ions [166].

4.1.3. *Data Analysis*

The intensity of the cleavage bands can be analyzed using, e.g., the ImageQuant software. For each experiment conducted either in the presence of Mg^{2+} and/or a thiophilic metal ion, the changing intensities are plotted versus time and fit to the appropriate kinetic equation [166]. To obtain a reliable fit, enough time points are required to give a non-linear least squares fit with $R^2 > 0.98$. As a rule of thumb, this requires usually about 10 data points, but optimization is required for each individual experiment.

To distinguish true thiophilic metal ion rescue from other effects, it is necessary to investigate both the modified and the unmodified RNA in the presence and absence of the thiophilic metal ion. The relative rate of rescue (k_{rel}) can then be determined applying equation (1). This equation describes the rate enhancement of the phosphorothioate-modified RNA with the thiophilic metal ion (k_S^M) over the reaction in the presence of Mg^{2+} only (k_S^{Mg}) normalized to the rate enhancement observed for the unmodified RNA (k_O^M/k_O^{Mg}) [26,166]:

$$k_{rel} = \frac{k_S^M / k_S^{Mg}}{k_O^M / k_O^{Mg}} \qquad (1)$$

Metal ion rescue experiments have also been used to determine the apparent affinity of a specific metal ion interaction by measuring the reaction rate over a range of metal ion concentrations [26].

4.1.4. Limitations

Although sulfur rescue experiments have been widely used to identify metal ion binding sites in ribozymes, the approach contains several pitfalls:

(i) First, it is usually assumed that the single atomic sulfur substitution does not perturb the structural integrity of the RNA. For many cases this is true, but there are examples known where the presence of the larger sulfur disturbs the local structure [57,172]. Also from NAIM experiments (see Section 4.2.4) it is well known, that numerous phosphorothioate substitutions cannot be rescued [173]. Hence, the structural integrity has to be carefully confirmed before a sulfur rescue experiment is attempted.

(ii) Sulfur is simply different from oxygen: Aside from the ionic radius and electronegativity, these two elements also lead to different charge distributions (sulfur tends to make sigma bonds and hence forms no double bonds) in phosphorothioates and phosphates, and abilities to accept hydrogen bonds [26]. The larger diameter of the sulfur atom can also potentially impose steric constraints that impair the thiophilic metal ion from binding to the active site.

(iii) In case a rescue is observed, metal ion thiophilicity alone cannot explain the efficiency of this process [174,175]. Whereas the Irving-Williams series of stability constants for metal-sulfur complexes follows the order $Cd^{2+} > Zn^{2+} > Ni^{2+} > Co^{2+} > Mn^{2+}$ [161,176,177], manganese is often the most efficient metal ion in thio rescue experiments [26,173,174]. However, Mn^{2+} is not significantly more thiophilic than Mg^{2+}, but just has a slightly higher intrinsic affinity to both phosphates and thiophosphates [161,178]. A comparison of the basic affinities of various metal ions to phosphates and thiophosphate is given in Table 1 and Figure 2. Consequently other factors, such as the relative size, the preferred coordination geometry, and the hydration energy seem to be critical in determining the ability of a certain metal ion to restore ribozyme activity. Especially the ionic radius of Cd^{2+} (1.03 Å) as compared to Mg^{2+} (0.86 Å) could be detrimental to phosphorothioate rescue because the original metal ion binding site has already been reduced in size by the bulky sulfur substituent [174]. At the same time it is also possible that the thiophilic metal ion binds to allosteric sites within the ribozyme and alters the rate limiting step of the reaction. It is therefore important to perform a rigorous thermodynamic and kinetic analysis of the reaction for all conditions before conclusions are drawn from a sulfur rescue experiment [26].

Hence, although rescue of catalytic activity upon addition of a thiophilic metal ion to a sulfur-substituted RNA is indicative for metal ion binding to this site, the opposite is not true: Lack of sulfur rescue does not prove that there is no metal ion binding site present. Rather, the experimental setup may not have been suitable to detect it.

4.2. Nucleotide Analogue Interference Mapping (NAIM)

In nucleotide analog interference mapping (NAIM), modified nucleotides are incorporated into the RNA by *in vitro* transcription as phosphorothioate nucleotide analogs. The pool of modified RNAs is then subjected to a stringent selection step (folding or catalysis), where active and inactive RNAs are separated. By iodine cleavage of the phosphorothioate linkage, modifications in each pool can be mapped. NAIM is a very elegant technique that allows the simultaneous characterization of chemical groups that are important for RNA function. For comprehensive protocols, we refer to refs [179] and [180]. Here, we just give a brief overview on the general method of NAIM and its applications for the identification of metal ion binding sites in RNA.

4.2.1. Methodology

NAIM consists of three distinct steps (Figure 9):

1. First, a pool of RNAs that contains phosphorothioate nucleotide analogs at random positions throughout the molecule is obtained by *in vitro* transcription. Subsequently, the RNAs are 5′- or 3′-end labeled with radioactive ^{32}P to allow detection by autoradiography.
2. In a second step, the RNA pool is subjected to a stringent selection, to identify chemical groups that are essential for RNA function. The active and inactive populations are separated by chromatography or native gel electrophoresis.
3. In the last step the positions of interference have to be mapped. This is done by iodine cleaveage of the phosphorothioate bond and subsequent resolution of the cleaved positions on a denaturing polyacrylamide gel and visualization by phosphor imaging.

Some key points about the three steps are explained in more detail below:

Step 1 – Transcription and labeling: As discussed above in Sections 3.2.1.2 and 4.1.1, *in vitro* transcription by T7 RNA polymerase [10,164] leads to oligonucleotides that contain only R_P phosphorothioates. The ratio of unmodified and modified NTPs during transcription determines the incorporation level of the phosphorothioate tag and should be adjusted to about 5% [179]. Commercially available nucleotide analogs are usually sold at the required concentration to yield the optimal incorporation level for NAIM. Otherwise, this ratio has to be carefully re-established for each new analog. To incorporate NTPs with modifications at the 2′-hydroxyl position of the sugar moiety, a T7 RNA polymerase Y639F mutant has been developed [181].

Step 2 – Selection: It is of utmost importance to carefully optimize this step to ensure the successful outcome of the experiment [179]. Therefore, a basic characterization of the RNA and the processes that are subject of the NAIM analysis are a prerequisite for the successful optimization of the selection step. Two types of selection systems have been used in the past: In the case of ribozymes, active and inactive molecules are separated based on their length difference upon either self-cleavage or ligation with a substrate [180]. If folding is the required trait, native gel electrophoresis, which separates molecules according to their overall shape, is the method of choice. It is important that the gel solution does not contain denaturants such as urea, and should be performed at low voltage at 4°C. The gel solution should contain all cations that are needed for folding in the appropriate concentrations [180]. Reaction conditions are chosen, such as to yield 20–40% of folded relative to non-folded RNA [179].

Step 3 – Interference mapping: Positions of interference are mapped by subjecting the two separated RNA pools to iodine cleavage and subsequent resolution of the pieces on a denaturing polyacrylamide gel (Figure 9a). Iodine cleaves the phosphodiester linkage of phosphorothioates, without affecting any other bonds in the RNA (Figure 9b). It is important to load all samples on the same gel and include controls for background degradation. For large RNAs (>250 nt), it is advisable to run long gels (90 cm) in addition to the standard sequencing gels (45 cm), to increase the resolution [179]. Also, certain regions of the RNA may be better resolved using a higher or lower percentage of acrylamide, and for optimal results, these parameters need to be adjusted for each new RNA.

The sites of interference are identified by quantification of the band intensities on the gel. The interference value κ (equation (2)) is defined according to Strobel and Cochrane [180] as

$$\kappa = \mathrm{NF}\,\frac{\mathrm{N\alpha S_{experiment}}/\delta\alpha \mathrm{S_{experiment}}}{\mathrm{N\alpha S_{unselected}}/\delta\alpha \mathrm{S_{unselected}}} \tag{2}$$

with NαS being the parent phosphorothioate nucleotide, δαS the phosphorothioate-tagged nucleotide analog, and NF the normalization factor used to account for differences between loading and extent of reaction between lanes (typically between 0.8 and 1.2). A κ value >2 usually defines a site of nucleotide analog interference, whereas a value <0.5 one of enhancement.

4.2.2. *Controls*

In order to distinguish NAIM effects from direct interference of the phosphorothioate tag, it is crucial to transcribe the same RNA also in the presence of αS-NTPs without any further modification and subject these

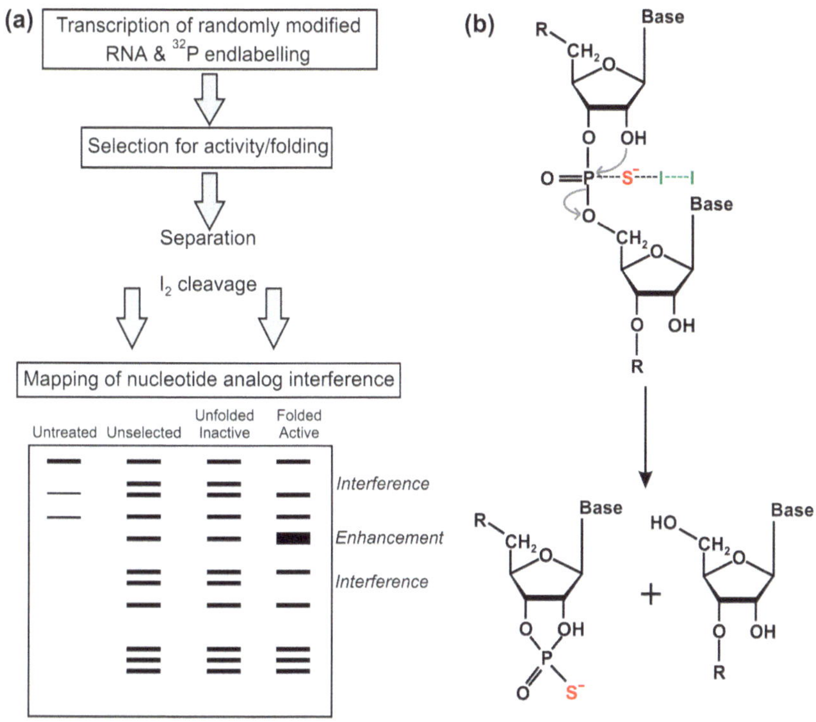

Figure 9. (**a**) The 3 steps of NAIM: First, the phosphorothioate analogue is introduced at random positions in the RNA. In the second step, a stringent selection is applied to be able to separate active from inactive (or folded from unfolded) molecules. After iodine cleavage of the diester bond at phosphorothioate positions, the effect of the substitution is mapped by polyacrylamide gel electrophoresis. Untreated RNA and a pool of RNA that has not been subjected to the selection process are included as controls. A missing band in the folded/active species signifies that incorporation of the nucleotide analog at this position interfered with function/folding. In contrast, stronger bands in the active/folded species mark positions where nucleotide analog incorporation is beneficial. (**b**) Putative mechanism of iodine cleavage of thioated phosphodiester bonds. Alternatively, sulfur could also act as the leaving group.

molecules to the same procedure (NαS in the above equation). Also, RNAs that have not undergone a selection step should be added as controls to analyze the overall level of analog incorporation into the RNA. If the incorporation level is too high, the bands at the bottom of the gel will be much stronger than those at the top of the gel, as multiple iodine-induced cleavages occur in each RNA [180]. Unselected and uncleaved control samples are added to check for background degradation.

4.2.3. *Choice of the Nucleotide Analog*

A wide variety of nucleotide analogs is available these days, many of which commercially, or they can be synthesized using published protocols ([179] and references herein). Three groups of analogs exist: Those deleting a functional group, those adding a steric bulk to the functional group, and those altering the chemical nature of a functional group. It is important to define, which particular biochemical property is going to be investigated and to take into account the available structural and biochemical information on the system, before choosing a particular analog.

4.2.4. *NAIM to Identify Metal Ion Binding Sites*

NAIM is a viable method to identify metal ion binding sites and has been exploited in a number of studies [182–185]. Metal ion binding to the N7 position has been probed by 7-deaza-αS analogs, to the N6 amino group of adenosine by N6-methyl-A-αS, and to probe the amino group in guanosine, inosine was used [184]. Identification of guanosine-O6-carbonyl ligands for metal ions has been achieved with 6-thioguanosine phosphorothioate (S6GαS) [180]. A main problem with this method to identify metal ion binding atoms, is that each of the above modifications can lead to an interference even in the absence of any interaction with a metal ion as the steric and hydrogen bonding properties of the functional groups are severely altered. Hence, again great care has to be taken in interpreting these results.

4.2.5. *Limitations*

Obviously, NAIM only works if the analog can be incorporated successfully into the RNA. Some analogs are very poor substrates for the RNA polymerase and thus incorporation efficiency is very low. It is important to optimize the ratio between non-modified and modified nucleotides for each analog. Furthermore, incorporation of certain analogs can result in severe inhibition of RNA function, making it impossible to perform the selection step. Another drawback is the possibility of interference by the phosphorothioate substitution itself (see also Section 4.1). Apparently, sometimes the inclusion of Mn^{2+} or a thiophilic metal ion in the reaction buffer can help to decrease the phosphorothioate effect [180].

However, one of the most severe limitations with respect to metal ion binding remains the difficulty to unambiguously assign an interference to the disruption of a RNA-metal ion interaction, and not the primary effect of the atomic modification (see also Section 4.1). A rescue by a metal ion might not be due to a direct coordination to the introduced sulfur atom, but rather close-by, e.g., thus reinstalling the active local structure that has been disturbed by the sulfur. Generally, RNAs are very flexible and adaptable molecules that are active under many conditions (i.e., high M^{+} concentrations replacing Mg^{2+}, etc.). It is unclear if under such varying conditions different catalytic strategies are employed, hence the metal ion binding sites found in thiorescue and NAIM experiments may not necessarily be equal to the ones present under physiological conditions.

4.3. Hydrolytic Cleavage Experiments

The detection of metal ion binding sites within large RNAs poses an enormous challenge due to the sheer size of the system, but also because one needs to distinguish between the specifically and diffusely bound ions. The inherent instability of RNA provides a good tool for specific metal ion binding site detection: Solvated, higher valent metal ions are partially deprotonated and can as such act either directly by attacking the phosphodiester bridge or by mediating hydrolysis of the backbone through 2′-OH activation. The resulting fragments can then be detected by gel electrophoresis and the cleavage sites located. As the Mg^{2+}-driven hydrolysis of the backbone is usually rather slow, other ions like lanthanide(III) ions or Pb^{2+} are applied, as discussed in the following sections.

4.3.1. Lanthanide Cleavage

Lanthanide(III) ions are rather good mimics of alkaline earth metal ions, e.g., Tb^{3+} for Ca^{2+}, as has been shown for metal-proteins interactions [186–188]. Also in RNA, lanthanide(III) ions are believed to bind mostly to the same sites as Mg^{2+}, albeit with higher affinity, due to their higher charge. Some physico-chemical properties of lanthanide(III) ions are summarized in Table 2 in Section 2.2. The pK_a values of the Ln^{3+}-coordinated water molecules are much closer to the physiological value than those coordinated to Mg^{2+} [113,189], and hence, hydrolysis of the RNA backbone mediated by the $M(H_2O)_n(OH)^{2+}$ is accelerated around neutral pH [38,190–192]. Especially Tb^{3+}, as well as Eu^{3+} have been successfully used in the past to determine metal ion binding sites in complex three-dimensional structures of RNAs [192,193]. A detailed protocol on this method using terbium(III) can be found in the *Handbook of RNA Biochemistry* [194]. Below, we thus only state the most important points.

In the case of lanthanide(III)-mediated cleavage of the RNA, it is crucial to work with homogenous samples and carefully optimized reaction conditions for every new construct. Parameters that need to be optimized include

(i) reaction temperature (0–45°C), (ii) incubation time (0.5–4h), (iii) Ln^{3+} concentration, (iv) lanthanide(III) ion identity, and (v) Ln^{3+}:Mg^{2+} ratio. Furthermore, the pH needs to be kept below 7.5 (6.5–7.0 is ideal), as more basic conditions lead to the formation of insoluble lanthanide(III) hydroxo species. The total RNA cleavage should be smaller than 10% to ensure that each RNA undergoes only a single cleavage event. The cleavage products can then be separated by denaturing PAGE and the location of the cleavage identified by a concurrently run alkaline hydrolysis cleavage ladder and a RNase T1 digest.

To identify high affinity metal ion binding sites, a competition experiment with increasing Mg^{2+} concentrations is carried out. The intensity of a particular band should decrease if Mg^{2+} is binding at that position. It is possible that not all Mg^{2+} binding sites are detected in a particular RNA, because Tb^{3+} needs to bind close in space to the 2′-OH that acts as a nucleophile in the cleavage reaction. Also sites that are very specific for a certain metal type may escape detection. Nonetheless, lanthanide(III) ion cleavage is a very powerful method to simultaneously identify several specific metal ion binding sites within a RNA.

4.3.2. Pb^{2+} Cleavage as a Probe for Single Stranded RNA Regions

Historically, Pb^{2+} has been one of the first mimics to investigate Mg^{2+} in RNA [190,195,196]. Nowadays, Pb^{2+}-induced RNA cleavage is best known for the leadzyme, an *in vitro* selected ribozyme [29,30] (see also Section 3.1.3). The leadzyme is surprisingly selective for Pb^{2+}, cleaving at the 5′-end of an internal bulge structure, and cannot be replaced by any other metal ion as a cofactor for the reaction. The logarithm of the rate of cleavage increases linearly between pH 5.5 and 7.0, implicating a Pb^{2+}-coordinated hydroxide as a general base [31].

Otherwise, highly specific Pb^{2+} cleavage is rarely observed. More often, cleavage by Pb^{2+} is relatively weak and usually comprises several consecutive phosphodiester bonds [197]. Predominant cleavage sites involve bulges, loops, and other single stranded RNA regions and thus Pb^{2+} is commonly used as a probe for flexible parts within a given RNA [198]. Nevertheless, some sequence specific bias has been found [199]: In general, single stranded regions that undergo extensive stacking interactions, such as tetraloops, are resistant to cleavage by Pb^{2+}. Single nucleotide bulges are more susceptible if they are flanked by pyrimidines rather than purines, whereas multiple nucleotide bulges are relatively little protected by their flanking regions. These findings have promoted the use of Pb^{2+} cleavage to detect conformational changes in RNAs, e.g., upon increasing concentration of other divalent metal ions [197]. This method has been used even *in vivo* with *E. coli* cultures growing under different conditions [200].

The experimental conditions of Pb^{2+} cleavage assays are very similar to the ones carried out with lanthanide(III) ions (Section 4.3.1): ^{32}P-endlabeled RNA is first properly refolded by heating to 90°C in a suitable buffer and subsequent slow cooling to room temperature (see also Section 3.1.1). Cleavage is initiated by the addition of $Pb(acetate)_2$ (10–100 mM) and incubation at room temperature for 5 min. The reaction is stopped with EDTA and the samples are analyzed by denaturing PAGE alongside an RNase T1 and an alkaline cleavage ladder and subsequent autoradiography [197,198]. Lower concentrations of $Pb(acetate)_2$ (μM to low mM range) can be used to probe high affinity metal ion binding sites [197]. As an indication that such strong and specific cleavage sites observed at low Pb^{2+} concentrations co-localize with a Mg^{2+} binding site, competition experiments can be performed with increasing concentrations of Mg^{2+}. If Mg^{2+} competes for binding, cleavage should be suppressed. This approach has been used to determine K_D values for Mg^{2+} binding in RNA [201,202] (see also Section 6.2). Nevertheless, results should be interpreted with care, as just the increasing ionic strength might be the reason for a reduced Pb^{2+}-induced cleavage.

4.3.3. In-Line Probing

In-line probing is a classical method to monitor changes in tertiary structure, relying on the natural instability of nonstructured RNA at elevated pH [203]: Positioning of the nucleophilic 2′-OH "in-line" with the 5′-O-leaving group (i.e., in a linear 180° angle typical for S_N2 reactions) promotes transesterification of the respective bridging phosphodiester [204]. In a typical A-form helix, such an orientation is precluded, whereas more flexible single stranded RNAs are free to explore the full conformational space and invariably sample the reactive "in-line" positioning of the relevant atoms. In-line probing exploits this tendency for RNA to differentially degrade depending on its structure to probe for motifs within a larger RNA. Any structural change, be it induced by a metal ion or any ligand can thus be elucidated by in-line probing.

A good protocol for in-line probing is given in [203]. In short, ^{32}P-endlabeled RNA is incubated together with a suitable ligand at room temperature for 40 h in buffered medium at pH 8.3 under general conditions that allow for proper folding to the native state. The reaction is quenched with EDTA and the sample is loaded immediately on a denaturing polyacrylamide gel, together with a variety of controls: RNAse T1, alkaline hydroxy ladder, a sample of undigested precursor RNA to check for sample integrity at the beginning, and a control in-line probing reaction where no ligand was added. As is obvious from this protocol, in-line probing corresponds to the Ln^{3+}- and Pb^{2+}-induced cleavage described above in Sections 4.3.1 and 4.3.2, but using Mg^{2+} instead (hence the higher pH and much longer reaction time): generally Mg^{2+} is the cofactor that activates the 2′-OH, however, also experiments have been performed where, e.g., amine functional groups of ligands activate the 2′-OH [205].

The cleavage bands are examined for increasing or decreasing intensity as a function of ligand concentration. To calculate the apparent K_D, at which the cleavage reaction is half maximal, the *fraction of RNA cleaved* according to equation (3) is plotted versus the logarithm of the ligand concentration:

$$\textit{fraction RNA cleaved} = \frac{\text{sample } \textit{value} - \text{min.} \textit{value}}{\text{max.} \textit{value} - \text{min.} \textit{value}} \tag{3}$$

Max. and min. refer to the highest and lowest cleavage values measured for each analyzed band. Alternatively, if the binding stoichiometry is known, e.g., 1:1, a corresponding non-linear least-squares fit can be applied (see also Section 6).

It is generally rather difficult to draw any conclusions on the local structure: While a metal ion may stabilize a given motif and thus protect the RNA from degradation, it can also lock the RNA in an "in line" position and thus promote cleavage. It is therefore important to carefully analyze the results obtained from an in-line probing experiment and compare them to complementary structural informations of the system before drawing a definite conclusion.

4.3.4. Limitations and Caveats of Hydrolytic Cleavage Experiments

The hydrolytic cleavage experiments described above in Sections 4.3.1 to 4.3.3 have the great advantages of (i) being widely applicable to all different kinds of RNA, (ii) making even very large RNAs amenable to structural probing, (iii) yielding a large number of metal ion binding sites in a single experiment, and (iv) they can be performed rather easily and fast by standard biochemical techniques [26,188]. In addition, by stepwise increasing the concentration of the hydrolytically active metal ion and plotting the intensity of cleavage versus the concentration, binding constants can be calculated [188]. If applied in higher concentrations, i.e., millimolar, such experiments can also be used to probe secondary structures, as single stranded regions and non-helical regions are more prone to cleavage: For example, the binding of the 5′-exon to the exon binding sites 1 and 2 (EBS 1 and 2) of a group II intron can be followed by the disappearance of the Tb^{3+}-induced cleavage in these regions [188,193].

However, this kind of experiments are by far not "fast-selling" and care has to be taken in several respects: (a) These assays will not yield all metal ion binding sites, as only those are detected that are in close proximity to the 2′-OH, or position a H_2O ideally for hydrolysis. Consequently, metal ions located in the major groove will not be detected. (b) Cleavage intensities do not necessarily correlate with binding affinities of the metal ion at a given site [188] because cleavage efficiency is strongly dependent on the correct geometry for an in-line attack. (c) Competing the hydrolytic metal ion M^{n+} by increasing Mg^{2+} concentrations might not yield a correct picture of Mg^{2+} binding and affinity because in general a higher ionic strength will reduce the binding affinity of a given metal ion. (d) Every metal ion has slightly different properties, like size, charge, pK_a, coordination number, and preference for N/O ligands. Consequently, these ions will also bind not only slightly differently, but possibly also at different places. It has been noted with different RNAs, i.e., in tRNAPhe [190] and in two different group II introns [188,193,206], that depending on the metal ion applied, different cleavage sites occur. Hence, the picture obtained for, e.g., Tb^{3+}, might not reflect the true binding sites of Mg^{2+}.

Taken together, hydrolytic cleavage assays can yield highly valuable information on metal ion binding and tertiary structure, but again they have to be interpreted with care.

4.4. Fenton Cleavage to Probe Metal Ion Binding Sites

Cleavage of the RNA backbone can also be achieved by redox chemistry, making use of the well-known Fenton reaction [207]: Fe^{2+} reacts with H_2O_2 to yield short-lived, highly reactive hydroxyl radicals that cleave the RNA chain in close spatial proximity to the bound metal ion. In combination with RNA, the reaction is often catalytically applied, i.e., ascorbate is added to reduce the Fe^{3+} back to Fe^{2+}:

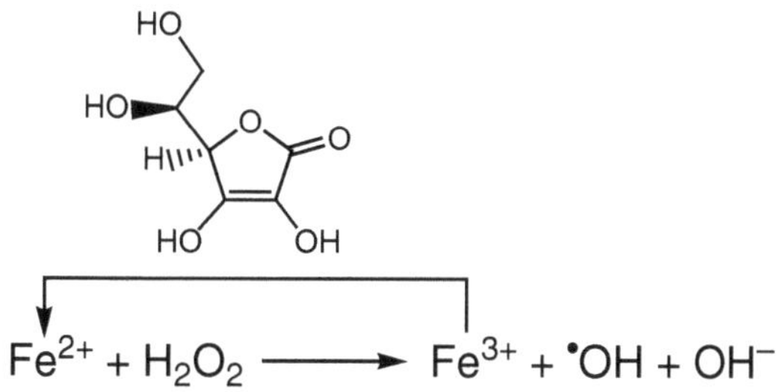

Classically, Fenton chemistry has been applied to probe for tertiary structures and mapping of binding sites of DNAs, proteins, RNA, and interactions thereof [208–214]: The relatively large $Fe(EDTA)^{2-}$ cannot penetrate a folded biomolecule and thus cleaves only the peripheral residues. Using Fenton chemistry it could be shown for the first time that RNAs other than tRNA and the ribosome are distinctly folded [215,216].

The Fenton reaction has also been exploited to characterize Mg^{2+} ion binding sites in RNA by using just the solvated Fe^{2+}. Like Mg^{2+}, Fe^{2+} (and

Fe^{3+}) possesses a predominantly octahedral coordination geometry and can therefore replace the former in its binding pocket [217,218]. If Fe^{2+} is only applied in low micromolar concentrations in the presence of sodium ascorbate [219], addition of H_2O_2 leads to the catalytic formation of hydroxyl radicals in close proximity to the Fe^{2+}/Fe^{3+} binding site. These radicals instantly cleave the RNA backbone and reveal the sites occupied by Fe^{2+}/Fe^{3+}.

Fenton cleavage is induced after proper folding of the ^{32}P-endlabeled RNA into its native structure by incubation in a buffer containing 0.2–20 μM $Fe(NH_4)_2SO_4$ [220]. Incubation time and temperature need to be optimized for each RNA and vary between 5 and 30 min and 0 and 37°C, respectively. The reaction is quenched by addition of loading buffer and the reaction products are separated by denaturing PAGE and analyzed by autoradiography. Care has to be taken to properly adjust the RNA:Fe^{2+} ratio for each individual case because excessive free Fe^{2+} leads to unspecific RNA degradation [219]. Ideally, the Fe^{2+} concentration is adjusted as to yield a single cleavage per molecule.

In general, the same limitations as for the other metal ion-induced cleavage assays also apply here. If Fe^{2+} fails to substitute for a specific binding site of the native metal ion, this site will not be detected. On the other hand, false positives may occur due to the presence of alternative binding pockets for Fe^{2+}. A distinct difference to the hydrolytic cleavage described above in Section 4.3 is that hydroxyl radical cleavage is not limited to any geometry of the RNA binding site. Hence, $Fe^{2+/3+}$ specifically bound anywhere in the RNA, i.e., minor or major groove as well as loops, can be detected. In combination with, e.g., lanthanide(III) cleavage, an attribution to minor or major groove binding is feasible, but has to the best of our knowledge not yet been done.

5. COMPUTATIONAL METHODS

Especially for very large RNAs, the experimental methods are very limited if non-existent to identify metal ion binding sites or specify even single coordinating atoms. Hence, several *in silico* methods and databases have been developed in the past years to help to overcome these deficiencies and to predict binding pockets for metal ions.

5.1. Theoretical Calculations

The negatively charged backbone of RNA molecules presents a large energetic barrier to fold to the complex structure. Thus cations, i.e., mostly metal ions, need to be involved, ranging in their binding behavior from a small number of specifically bound ions to a diffuse cloud of non-specifically bound ions. The latter is usually called an ion atmosphere that surrounds any charged molecule in solution. Classically, the role of metal ions in folding has been described by a simple Hill equation. However, physically, the Hill model can possibly not be correct: The overall charge of the system must equal zero, but a Hill fit to a folding transition yields a much lower number of metal ions than realistically needed. To overcome this obvious problem, Draper and colleagues have employed the non-linear Poisson Boltzman theory (NL-PB) to describe such an ion atmosphere [35,221,222]. In the past years this theory has become widely used and found diverse applications in nucleic acids research (e.g., [35,223,224]). A typical outcome of such calculations is the electrostatic surface potential map that reveals regions of high negative charge potential that are likely to harbor metal ions (Figure 10) [225]. However, as with any model, there are limitations to the

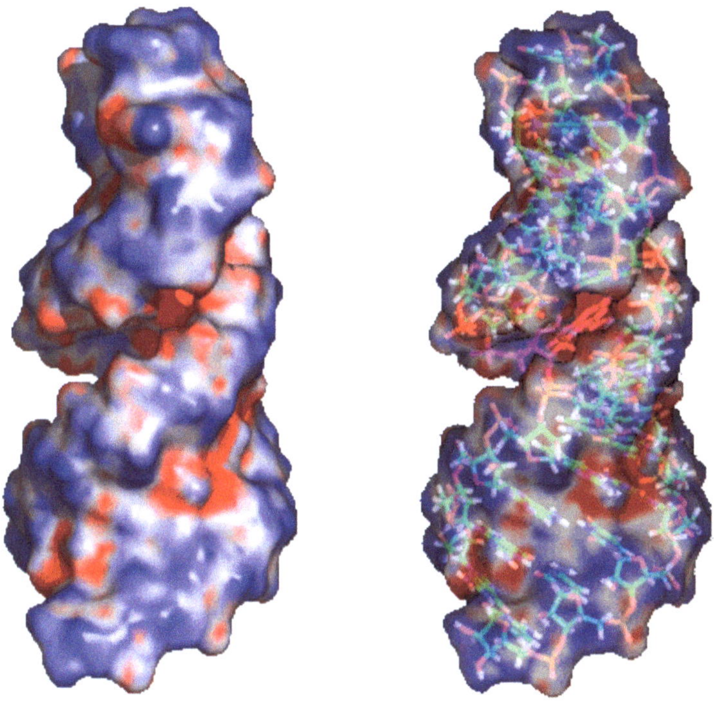

Figure 10. Electrostatic surface potential map of D6-27 with (right) and without (left) the residues shown. The negatively charged regions are shown in red (–40 kTe), whereas the positively charged ones in blue (10 kTe). The surface potential was calculated with QNIFFT [225] and visualized with PYMOL (W. L. DeLano, 2002, http://www.pymol.org) based on the PDB ID 2AHT [39].

NL-PB theory, e.g., the solvent behaves like a simple continuum dielectric medium or solvated ions are point-like with no volume. Lately, one has begun to learn from experiments and to quantify the deficiencies in order to make better model systems. Such efforts have been excellently reviewed by Herschlag, Doniach et al. [226] and shall thus not be explained in more detail here.

An interesting possibility to scan for metal ion binding sites provides the web-based program FEATURE [227,228]: Originally developed to study the microenvironment of functional groups within proteins, it has now also been adapted to scan for metal ion binding sites within RNAs. Based on a set of known metal ion binding pockets and their microenvironment together with a PB-based electrostatic surface potential map, FEATURE scans RNA structures (i.e., the pdb file) of any size and predicts possible metal ion binding sites. In the case of a group I intron, known metal ion binding sites have thus been confirmed, whereas others were newly discovered, and can now be probed for by experimental techniques [228].

Highly sophisticated molecular modeling or quantum mechanics based methods to investigate the dynamics of RNAs and possibly catalytic mechanisms are still very rare. The pitfalls and advantages of these methods, especially QM/MM, which allows to investigate also larger systems, have recently been excellently reviewed [229,230] and shall thus not be described in more detail here.

In the following we will concentrate on available databases that allow structural characterization of metal ion binding pockets in nucleic acids.

5.2. Databases

To the best of our knowledge, three databases are available (all based on the PDB) that list RNAs and are specialized to investigate metal ion binding sites within. Each of these three databases has some special characteristics, which will be briefly described below. In addition to these three databases, also a *Solvation Web Service* (SwS) for nucleic acids has been created by Auffinger and Hashem to provide an exhaustive overview of the solvation of nucleic acid [231].

5.2.1. Metalloprotein Database and Browser (MDB)

The Metalloprotein Database and Browser (MDB; http://metallo.scripps.edu) contains all metal ion-ligand interactions deposited within the PDB and NDB (Nucleic Acid Database) [232]. Its aim is to offer quantitative information on geometrical parameters of the primary metal ion coordination sphere. The database is searchable by different parameters like metal ion, number of ligands, R-value, resolution, author, or experimental technique. A Java applet allows then to visualize the metal ion together with its coordinating atoms.

5.2.2. *MEtals in RNA (MeRNA)*

The MeRNA database (http://merna.lbl.gov) compiles all metal ion binding sites identified in three dimensional structures of RNA available from the PDB and NDB [233]. The metal ion binding pockets are identified according to distances between the metal and the nearest atoms, be they from nucleotides, amino acids, water, or other molecules. Again, the database is searchable by specific parameters like PDB identifier, metal ion, method or distance from metal to any ligand. A speciality of the database is certainly the incorporated classification and listing of eight known metal ion binding motifs, namely the YG pair, sheared GA pairs, the Mg^{2+} clamp, the metal ion zipper, the loop E motif, the AA platform, tetrads, and the G-phosphate Mg^{2+}-binding motif.

5.2.3. *Metal Ions in Nucleic AcidS (MINAS)*

The MINAS database (http://www.minas.uzh.ch) is not restricted to RNA, but also includes all DNA structures within the PDB and NDB. As of June 2010, 2079 structures are deposited and compiled in MINAS containing a total of 43757 metal ions. Out of these, 715 structures contain RNA that bind 26893 Mg^{2+}. Like the MDB and MeRNA databases, also MINAS is searchable with regard to metal ion, PDB ID, experimental method and containing macromolecule (Figure 11).

In addition, several other special features are part of MINAS: MINAS allows to search for any specific combination of coordinating atoms, e.g., O1P, O2P, N7, O6 or H_2O, or just any atom of a nucleotide. One can thereby exploit the first as well as the second coordination sphere, i.e., innersphere and outersphere binding, and define inter- as well as intranucleotide binding patterns. The output file then lists all according occurrences in the databank. Searches can be personalized and be either downloaded or saved online to retrieve at a later time. The output file contains all relevant information, like metal ion, coordinating atoms, PDB ID, as well as a link to a small pdb file that contains the coordinates of the full binding pocket including the nearer surroundings (Figure 11). This extract of the original pdb file contains again all relevant headers and can be downloaded, allowing for a detailed characterization of each binding site.

(a)

1 2 3 4 » - sort results - go Page: 1/9 | Records: 874

Nr	Metal Ion	PDB ID	Classification	Exp. Method	Journal Title	Authors	miniPDB	Ligands
1	MG (9010)	2NZ4	STRUCTURAL PROTEIN/RNA	X-Ray	STRUCTURAL INV..	J.C.COCHRANE,S..	view download	show export
2	MG (9014)	2NZ4	STRUCTURAL PROTEIN/RNA	X-Ray	STRUCTURAL INV..	J.C.COCHRANE,S..	view download	show export
3	MG (9016)	2NZ4	STRUCTURAL PROTEIN/RNA	X-Ray	STRUCTURAL INV..	J.C.COCHRANE,S..	view download	show export
4	MG (8001)	1YJ9	RIBOSOME	X-Ray	STRUCTURES OF ..	D.TU,G.BLAHA,P..	view download	show export
5	MG (8003)	1YJ9	RIBOSOME	X-Ray	STRUCTURES OF ..	D.TU,G.BLAHA,P..	view download	show export
6	MG (8005)	1YJ9	RIBOSOME	X-Ray	STRUCTURES OF ..	D.TU,G.BLAHA,P..	view download	show export
7	MG (8006)	1YJ9	RIBOSOME	X-Ray	STRUCTURES OF ..	D.TU,G.BLAHA,P..	view download	show export
8	MG (8007)	1YJ9	RIBOSOME	X-Ray	STRUCTURES OF ..	D.TU,G.BLAHA,P..	view download	show export
9	MG (8008)	1YJ9	RIBOSOME	X-Ray	STRUCTURES OF ..	D.TU,G.BLAHA,P..	view download	show export
10	MG (8016)	1YJ9	RIBOSOME	X-Ray	STRUCTURES OF ..	D.TU,G.BLAHA,P..	view download	show export

(b)

Ligands for metal MG (9016) from PDB File: 2NZ4

Nr	Type	Name	Element	Residue Name	Residue Number	Distance to Metal Ion
1	outersphere	O2*	O	G	1	4.5771
2	outersphere	O3*	O	C	2	5.6821
3	outersphere	O3*	O	A	28	3.6688
4	outersphere	O2*	O	A	28	4.1191
5	outersphere	OP1	O	C	29	4.7556
6	innersphere	OP2	O	C	29	2.3113
7	outersphere	O5*	O	C	29	4.4017
8	outersphere	O3*	O	C	29	4.5342
9	outersphere	OP1	O	G	30	4.1716
10	innersphere	OP2	O	G	30	2.094
11	outersphere	N7	N	G	30	5.3405
12	innersphere	O	O	HOH	62	1.8601
13	innersphere	O	O	HOH	101	1.8864
14	innersphere	O	O	HOH	102	2.0012
15	outersphere	O3*	O	G	1	3.4884
16	outersphere	OP1	O	C	2	4.2845
17	innersphere	OP2	O	C	2	1.8038
18	outersphere	O5*	O	C	2	4.0232

(c)

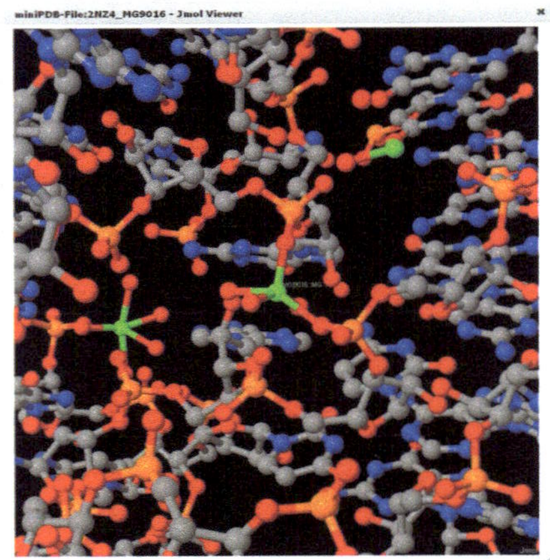

Figure 11. Databank of Metal Ions in Nucleic AcidS (MINAS; J. Schnabl, P. Suter, R. K. O. Sigel, 2010, http://www.minas.uzh.ch). **(a)** Top section of a typical output of MINAS after a search for Mg^{2+} ions bound to three phosphate oxygens in an innersphere fashion. **(b)** Detailed neighborhood of Mg(9010) in the PDB file 2NZ4 as provided by MINAS. **(c)** View of Mg(9016) in 2NZ4 as given in the mini PDB file provided by MINAS.

To summarize, each of the three databases has a different focus, allowing researchers to choose according to their needs. Whereas the MDB concentrates on the geometry of the coordinating atoms, MeRNA allows to search for specific binding motifs. MINAS distinguishes between inner- and outersphere coordination and thus provides the most detailed, but also most complex picture of each binding site.

6. CALCULATION OF BINDING CONSTANTS

Compared to proteins, metal ion binding to RNA is often weak and dynamic in nature. Yet, metal ions play an important role in fine tuning the active fold and the catalytic activity of ribozymes [3,4,44,45,234]. Specific metal ion binding sites within one RNA molecule often display similar affinities (mostly with K_D values in the lower millimolar range), meaning that these sites are in direct competition with each other [13]. Consequently, not all metal ions present in the solution are available at each binding site. This notion has direct consequences on the procedures to calculate intrinsic affinity constants for each site. Developing reliable tools for the calculation of affinity constants for such simultaneous binding events is important in order to understand the exact roles that metal ions play in RNA structure and function.

6.1. ISTAR – Intrinsic STAbilities of RNA Metal Ion Complexes

^{1}H-NMR chemical shifts are strongly affected by metal ion binding, i.e., by the change in electrochemical environment, and have been used previously to calculate affinity constants. Usually this is done by plotting the chemical shift change in ppm versus the added [Mg^{2+}] and subsequently fitting these curves to a non-linear least-squares regression for a 1:1 bimolecular association scheme [13,38,51,235]. Although yielding rough estimates of binding affinities, this method has two major caveats: (i) Chemical shift change data of exchangeable protons should not be used for the calculation, as they are affected by several effects and thus do not follow a 1:1 binding behavior (Figure 12a). (ii) The Mg^{2+} ions bind simultaneously to the various sites, having the consequence that the total amount of Mg^{2+} in solution is not equal to the one available for each site.

The first problem can be solved by using the changes in chemical shifts of non-exchangeable protons or other nuclei, e.g., ^{15}N or ^{13}C. To tackle the second issue, we have recently developed a procedure that simultaneously takes into account all binding sites within one RNA during the calculation procedure [13], as briefly discussed below:

First, one needs to define the number of specific binding sites within the RNA under investigation. To do this, chemical shift change data of as many protons as possible are evaluated and preliminary affinity constants calculated. As one metal ion will affect many protons in close neighborhood, these constants are grouped according to their strengths, i.e., usually up to 10 protons "define" one binding pocket. These binding pockets can be verified by analyzing the line-broadening induced by Mg^{2+} and Mn^{2+} coordination (see also Sections 3.2.2.1 and 3.2.2.2). The individual log K_A values of the evaluated protons are then averaged (av) to obtain one affinity constant log $K_{A,av}$ per binding site.

As explained above, the available Mg^{2+} concentration for each site does not equal the total $[Mg^{2+}]$ in solution, if numerous binding sites are present. To take this into account, one needs to calculate the amount of bound Mg^{2+} at each specific binding site i at every data point of the concentration by using log $K_{A,av}$ and equation (4):

$$\begin{aligned}[Mg^{2+}]_{bound,i} &= [RNA \cdot Mg^{2+}]_i \\ &= \frac{(K_{Ai}[Mg^{2+}]_{tot} + K_{Ai}[RNA]_{tot} + 1)}{2K_{Ai}} \\ &\quad - \frac{\sqrt{(-(K_{Ai}[Mg^{2+}]_{tot} + K_{Ai}[RNA]_{tot} + 1))^2 - 4K_{Ai}^2[Mg^{2+}]_{tot}[RNA]_{tot}}}{2K_{Ai}}\end{aligned} \quad (4)$$

It follows that the amount of Mg^{2+} available for binding at each specific site is given by the total Mg^{2+} concentration minus the amount bound to all other sites (equation (5)):

$$[Mg^{2+}]_{avail,i} = [Mg^{2+}]_{tot} - \Sigma[Mg^{2+}]_{bound,tot} + [Mg^{2+}]_{bound,i} \quad (5)$$

The chemical shift changes are then again plotted versus $[Mg^{2+}]_{avail,i}$ and the curves fitted to a 1:1 binding isotherm to yield a new estimate for binding constants of all protons at this site. Again the obtained affinity constants are averaged to yield a new log $K_{A,av}$ for each binding site (Figure 12b). This procedure is iteratively repeated, until the average affinity constant (log $K_{A,av}$) of each binding site does not change anymore within the error limits [13].

We have automated the iteration procedure using the Matlab Toolbox (Matlab R2006a, Matworks Inc.) in a calculation tool called ISTAR [236], which is available from the authors. As input the chemical shift values from

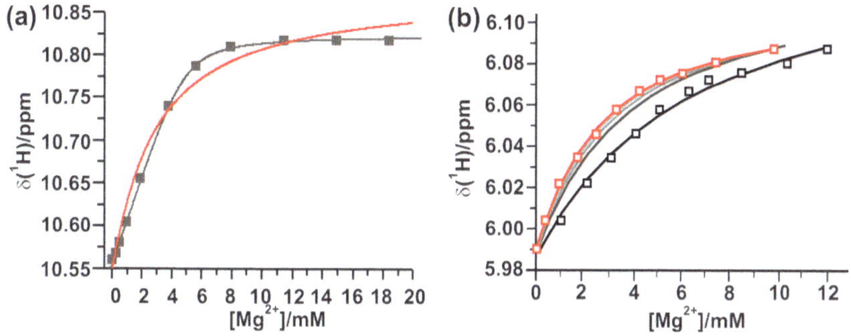

Figure 12. Calculation of affinity constants by using NMR. (**a**) Change in chemical shift of a guanosine N1H proton within D5 of the group II intron *Sc*.ai5γ upon titration with Mg^{2+}. The experimental data can clearly not be fitted to a 1:1 binding model (red line) [237]. (**b**) Correction of available Mg^{2+} for a given binding site and fit of the experimental data with a 1:1 binding model upon iterative calculation with ISTAR. The fit becomes better with every additional iteration round [13].

NMR experiments for as many protons as possible of a titration series with the metal ion of choice are needed.

6.2. Indirect Calculation of K_A Values via Competition

As mentioned in Section 4.3, competition experiments can be used to calculate affinity constants of specific metal ions at their binding sites. Lanthanide(III) cleavage or sulfur rescue with, e.g., Cd^{2+}, will decrease at sites of Mg^{2+} coordination upon addition of increasing amounts of magnesium salts. The procedure shall here be briefly described [237]: First, one needs to calculate the Ln^{3+} affinity at each cleavage site by plotting the change in intensity versus the increasing Ln^{3+} concentration and fitting the data with a 1:1 binding model (unless there is no clear evidence for a cooperative behavior or the like). Note that the single fit to a 1:1 binding model is justified in this case, as the Ln^{3+} usually exceeds the RNA concentration by at least a factor of 1000. This experiment then needs to be repeated at different Mg^{2+} concentrations in order to obtain the affinities of the Ln^{3+} ion under the various conditions. Plotting the change of Ln^{3+} affinities in dependence on the Mg^{2+} concentration and fitting this data to equation (6)

$$K_{\mathrm{app,Ln}} = \frac{K_{\mathrm{Ln}}}{1 + K_{\mathrm{Mg}}[\mathrm{Mg}]_{\mathrm{tot}}} \tag{6}$$

will yield the intrinsic affinities of the Mg^{2+} at the single sites. It should be noted that in order to obtain meaningful values, the ionic strength needs to be equal in all experiments, i.e., the M^+ concentration needs to be reduced while the Mg^{2+} concentration is increased.

A drawback of the method lies in the fact that one can only roughly place the metal ion binding pocket, as in most cases several nucleotides are cleaved. In addition, various parameters need to by carefully adjusted and tested to make sure that the RNA is correctly folded and ribozymes are catalytically active under all applied conditions. The biggest advantage of the method is that in theory there is no size limit and thus even very large RNAs can be investigated.

6.3. Measuring Magnesium(II) Binding Stoichiometries by Equilibrium Dialysis

To measure the total amount of Mg^{2+} bound to RNA, two methods have mainly been exploited: gel filtration and equilibrium dialysis. Gel filtration has the disadvantage that even the smallest columns need a few minutes to run, during which time re-equilibration between free and bound Mg^{2+} will take place. Centricon ultrafiltration devices that are used in "forced equilibrium dialysis" [238] lead to a much faster separation of free and bound metal ions and have therefore given the most consistent results. First, the sample is allowed to equilibrate with a given concentration of solvated $MgCl_2$. The half time for equilibration lies in the order of seconds for a micromolar K_D [238]. The best way to test for sufficient equilibration time is to demonstrate that doubling or halving the incubation time does not alter the final result. Once equilibrium is reached, the sample is added to a pre-washed ultrafiltration device and centrifuged at maximum speed until a small percentage of the total volume has flown through. The metal ion content of the separated samples are analyzed using one of a variety of different techniques. Apart from atomic absorption spectroscopy, fluorescence measurements are most commonly used for that purpose. A number of commercially available fluorophores change their fluorescence intensity upon binding to Mg^{2+} (e.g., from Molecular Probes (Invitrogen) or ISS). Because many of the traditional Mg^{2+} chemosensors also detect Ca^{2+}, finding selective Mg^{2+} sensors is still an active area of research [239,240]. Several Mg^{2+} concentrations encompassing the range used in the final experiment are measured first, in order to establish a standard curve. The Mg^{2+} concentrations in the separated samples is then measured and compared with the standard curve. To avoid interference, the RNA should be denatured by guanidine · HCl before measuring the fluorescence intensity.

This method provides a quick and easy way of measuring the stoichiometry of magnesium ions bound to RNA, but it is not capable of determining, where these binding sites are located [238].

7. CONCLUDING REMARKS AND FUTURE DIRECTIONS

The investigation of metal ion binding to RNAs and the detailed characterization of this interaction is a highly challenging but fascinating task. Many methods as described in this chapter have been developed and optimized in the past, each one tackling the problem from a different angle and thus yielding complementary information. Although we are far from understanding metal ion coordination to RNAs, in the past few years novel approaches have been introduced and an increasingly rapid development has started to take place. It has become obvious that metal ion interactions with nucleic acids cannot easily be categorized: the majority of metal ions surrounding nucleic acids are used for charge screening of the phosphate sugar backbone, but an estimated 10% of divalent metal ions are bound more specifically [5], taking over structural as well as catalytic roles. The differentiation and observation of these site specifically bound metal ions in the presence of a large excess of equal ions (i.e., millimolar concentrations) in solution presents a major difficulty. In addition, the different roles cannot be distinguished clearly from each other because structure and function are tightly linked, and metal ions influence basic properties of nucleotides over a distance of up to 15 Å [44].

Furthermore there seems to be no consensus on how RNAs employ metal ions, especially with respect to folding, but also on how the nucleic acid reacts to changes in metal ion concentration. This is not only true for, e.g., different classes of ribozymes, but often also for two ribozymes of the same kind (but slightly different sequence) stemming from the same gene [241]. Consequently, one has to be very careful when generalizing any new findings as each new RNA may hold unexpected surprises.

In spite of these difficulties, the multitude of methods summarized in this chapter enable us to look at metal ion coordination to RNAs from the most different perspectives in a truly interdisciplinary nature. Their combination, together with a profound reflection on the different physico-chemical properties and vast accumulated knowledge on related interactions with the nucleotide building blocks, provides the tools to investigate the relationship between nucleic acids and metal ions on the atomic level. New methods will certainly arise in the near future, often transferred from metalloproteins or other disciplines. Key directions in this sense are investigations on the dynamics of the metal ion-RNA interaction and experiments on the single molecule level.

ACKNOWLEDGMENTS

Financial support by the 7th European Framework Programme (Marie-Curie postdoctoral fellowship to MCE), the Swiss National Science Foundation (RKOS), the Swiss State Secretariat for Education and Research within the COST Action D39 (RKOS) and the University of Zürich is gratefully acknowledged.

ABBREVIATIONS

AMP^{2-}	adenosine 5′-monophosphate
$AMPS^{2-}$	adenosine 5′-α-thiomonophosphate
Cy3	"cyanine 3″, a fluorescent dye
cw-EPR	continuous-wave electron paramagnetic resonance
DEER	pulsed electron-electron double resonance
DQC-EPR	double quantum coherence electron paramagnetic resonance
EBS	exon binding sites
EDTA	ethylendiammine-N,N,N′,N′-tetracetic acid
ENDOR	electron-nuclear double resonance spectroscopy
EPR	electron paramagnetic resonance
ESEEM	electron-spin echo envelope modulation
EXAFS	extended X-ray absorption fine structure
FRET	Förster resonance energy transfer
FT-IR	Fourier-transform infrared spectroscopy
I	ionic strength
I	nuclear spin
ISTAR	*I*ntrinsic *STA*bilities of *R*NA metal ion complexes
ITC	isothermal calorimetry
GTP	guanosine 5′-triphosphate
HDV	hepatitis delta virus (ribozyme)
HIV-1 TAR	human immunodeficiency virus 1 trans-activation response (element)
HMBC	heteronuclear multiple bond coherence
HPLC	high-performance liquid chromatography
HSQC	heteronuclear single-quantum coherence
K_a	acidity constant

K_A	affinity constant
K_D	dissociation constant
MDB	*M*etalloprotein *D*atabase and *B*rowser
MERNA	*ME*tals in *RNA* database
MINAS	*M*etal *I*ons in *N*ucleic *A*cid*S* database
NAIM	nucleotide analog interference mapping
NDB	Nucleic Acid Database
NL-PB	non-linear Poisson-Boltzman
NMR	nuclear magnetic resonance
NOESY	nuclear Overhauser effect spectroscopy
nt	nucleotide unit
NTP	nucleoside 5′-triphosphate
PAC	perturbed angular correlation (spectroscopy)
PAGE	polyacrylamide gel electrophoresis
PDB	protein databank
PELDOR	pulsed electron-electron double resonance (= DEER)
POM	polyoxometalate
QM/MM	quantum mechanical/molecular mechanical
RNase P	ribonuclease P
S	electron spin
SAXS	small angle X-ray scattering
SDSL	site-directed spin-labeling
SERS	surface-enhanced Raman spectroscopy
snRNA	small nuclear ribonucleic acid
SwS	*S*olvation *W*eb *S*ervice
TRIS	tris(hydroxymethyl)aminomethane
VS	Varkud satellite (ribozyme)
XAS	X-ray absorption spectroscopy
XANES	X-ray absorption near-edge structure

REFERENCES

1. J. L. Boots, M. D. Canny, E. Azimi and A. Pardi, *RNA*, 2008, **14**, 2212–2222.
2. M. Roychowdhury-Saha and D. H. Burke, *RNA*, 2006, **12**, 1846–1852.
3. J. Schnabl and R. K. O. Sigel, *Curr. Opin. Chem. Biol.*, 2010, **14**, 269–275.
4. E. Freisinger and R. K. O. Sigel, *Coord. Chem. Rev.*, 2007, **251**, 1834–1851.
5. R. K. O. Sigel and H. Sigel, *Acc. Chem. Res.*, 2010, **43**, 974–984.
6. M. C. Erat and R. K. O. Sigel, *J. Biol. Inorg. Chem.*, 2008, **13**, 1025–1036.
7. M. Steiner, D. Rueda and R. K. O. Sigel, *Angew. Chem., Int. Ed.*, 2009, **48**, 9739–9742.
8. K. Lang and R. Micura, *Nat. Protoc.*, 2008, **3**, 1457–1466.

9. C. Hoebartner, R. Rieder, C. Kreutz, B. Puffer, K. Lang, A. Polonskaia, A. Serganov and R. Micura, *J. Am. Chem. Soc.*, 2005, **127**, 12035–12045.
10. S. Gallo, M. Furler and R. K. O. Sigel, *CHIMIA*, 2005, **11**, 781–785.
11. B. Francois, A. Lescoute-Phillips, A. Werner and B. Masquida, in *Handbook of RNA Biochemistry*, Ed. R. K. Hartmann, A. Bindereif, A. Schön and E. Westhof, Wiley-VCH, Weinheim, 2008, **Vol. 1**, pp. 438–453.
12. M. C. Erat, in *Advances in Biomedical Spectroscopy*, Ed. S. M. Pascal and A. Dingley, IOS Press, Amsterdam, The Netherlands, Vol. in press.
13. M. C. Erat and R. K. O. Sigel, *Inorg. Chem.*, 2007, **46**, 11224–11234.
14. *NIST Critically Selected Stability Constants of Metal Complexes*, Reference Database 46, Version 8; data collected and selected by R. M. Smith and A. E. Martell, U.S. Department of Commerce, National Institute of Standards and Technology, Gaithersburg, MD, USA, 2001.
15. M. Furler, B. Knobloch and R. K. O. Sigel, *Inorg. Chim. Acta*, 2009, **362**, 771–776.
16. E. M. Bianchi, R. Griesser and H. Sigel, *Helv. Chim. Acta*, 2005, **88**, 406–425.
17. G. G. Liang, N. A. Corfù and H. Sigel, *Z. Naturforsch. B*, 1989, **44**, 538–542.
18. R. Tribolet, R. Malini-Balakrishnan and H. Sigel, *J. Chem. Soc., Dalton Trans., Inorg. Chem.*, 1985, 2291–2303.
19. H. Sigel and R. Griesser, *Chem. Soc. Rev.*, 2005, **34**, 875–900.
20. B. E. Fischer, U. K. Häring, R. Tribolet and H. Sigel, *Eur. J. Biochem.*, 1979, **94**, 523–530.
21. K. L. Buchmueller and K. M. Weeks, *Nucleic Acids Res.*, 2004, **32**, e184.
22. B. L. Golden and C. E. Kundrot, *J. Struct. Biol.*, 2003, **142**, 98–107.
23. V. Tereshko, G. Minasov and M. Egli, *J. Am. Chem. Soc.*, 1999, **121**, 470–471.
24. V. Tereshko, C. J. Wilds, G. Minasov, T. P. Prakash, M. A. Maier, A. Howard, Z. Wawrzak, M. Manoharan and M. Egli, *Nucleic Acids Res.*, 2001, **29**, 1208–1215.
25. D. J. Klein, P. B. Moore and T. A. Steitz, *RNA*, 2004, **10**, 1366–1379.
26. J. K. Frederiksen and J. A. Piccirilli, *Methods*, 2009, **49**, 148–166.
27. H. W. Pley, K. M. Flaherty and D. B. McKay, *Nature*, 1994, **372**, 68–74.
28. J. E. Wedekind and D. B. McKay, *Biochemistry*, 2003, **42**, 9554–9563.
29. T. Pan and O. C. Uhlenbeck, *Biochemistry*, 1992, **31**, 3887–3895.
30. T. Pan and O. C. Uhlenbeck, *Nature*, 1992, **358**, 560–563.
31. T. Pan, B. Dichtl and O. C. Uhlenbeck, *Biochemistry*, 1994, **33**, 9561–9565.
32. N. Sugimoto and T. Ohmichi, *FEBS Lett.*, 1996, **393**, 97–100.
33. W. G. Scott, J. B. Murray, J. R. P. Arnold, B. L. Stoddard and A. Klug, *Science*, 1996, **274**, 2065–2069.
34. W. Yang, J. Y. Lee and M. Nowotny, *Mol. Cell*, 2006, **22**, 5–13.
35. D. E. Draper, D. Grilley and A. M. Soto, *Annu. Rev. Biophys. Biomol. Struct.*, 2005, **34**, 221–243.
36. E. Ennifar, P. Walter and P. Dumas, *Nucleic Acids Res.*, 2003, **31**, 2671–2682.
37. C. G. Hoogstraten, P. Legault and A. Pardi, *J. Mol. Biol.*, 1998, **284**, 337–350.
38. R. K. O. Sigel, D. G. Sashital, D. L. Abramovitz, A. G. Palmer III, S. E. Butcher and A. M. Pyle, *Nat. Struct. Mol. Biol.*, 2004, **11**, 187–192.

39. M. C. Erat, O. Zerbe, T. Fox and R. K. O. Sigel, *ChemBioChem*, 2007, **8**, 306–314.
40. M. C. Erat, *Two Domains of Branching and Catalysis Act as Specific Metal Ion Binding Platforms Within the Group II Itron Ribozyme Core*, Ph.D. Thesis, University of Zürich, Switzerland, 2007, pp. 348.
41. I. Bertini and C. Luchinat, *NMR of Paramagnetic Molecules in Biological Systems*, Benjamin/Cummings, Menlo Park, CA, 1986.
42. V. L. Pecoraro, J. D. Hermes and W. W. Cleland, *Biochemistry*, 1984, **23**, 5262–5271.
43. J. H. Davis, M. Tonelli, L. G. Scott, L. Jaeger, J. R. Williamson and S. E. Butcher, *J. Mol. Biol.*, 2005, **351**, 371–382.
44. R. K. O. Sigel and A. M. Pyle, *Chem. Rev.*, 2007, **107**, 97–113.
45. V. J. DeRose, *Curr. Opin. Struct. Biol.*, 2003, **13**, 317–24.
46. B. Knobloch, D. Suliga, A. Okruszek and R. K. O. Sigel, *Chem. Eur. J.*, 2005, **11**, 4163–4170.
47. M. K. Hansen, J.-P. Simorre, P. Hanson, V. Mokler, L. Bellon, L. Beigelman and A. Pardi, *RNA*, 1999, **5**, 1099–1104.
48. M. D. Tsai, in *Phosphorus-31 NMR: Principles and Applications*, Ed. D. Gorenstein, Academic Press, New York, 1984, pp. 175–196.
49. E. M. Osborne, W. L. Ward, M. Z. Ruehle and V. J. DeRose, *Biochemistry*, 2009, **48**, 10654–10664.
50. M. Maderia, L. M. Hunsicker and V. J. DeRose, *Biochemistry*, 2000, **39**, 12113–12120.
51. M. Maderia, T. E. Horton and V. J. DeRose, *Biochemistry*, 2000, **39**, 8193–8200.
52. R. B. Martin and Y. H. Mariam, *Met. Ions Biol. Syst.*, 1979, **8**, 57–124.
53. T.-D. Son, M. Roux and M. Ellenberger, *Nucleic Acids Res.*, 1975, **2**, 1101–1110.
54. T.-D. Son and J. M. Neumann, *Nucleic Acids Res.*, 1977, **4**, 397–403.
55. G. Slim and M. J. Gait, *Nucleic Acids Res.*, 1991, **19**, 1183–1188.
56. E. Besic, *Investigation of the Structural Changes of the Domain 6 of the Group II Intron Ribozyme ai5γ Due to Metal Ion Binding at the Branch-Point*, Master Thesis, University of Zürich, Switzerland, 2007.
57. K.-I. Suzumura, M. Warashina, K. Yoshinari, Y. Tanaka, T. Kuwabara, M. Orita and K. Taira, *FEBS Lett.*, 2000, **473**, 106–112.
58. Y. Tanaka, C. Kojima, E. H. Morita, Y. Kasai, K. Yamasaki, A. Ono, M. Kainosho and K. Taira, *J. Am. Chem. Soc.*, 2002, **124**, 4595–4601.
59. Y. Tanaka and K. Taira, *Chem. Commun.*, 2005, **16**, 2069–2079.
60. Y. Tanaka and A. Ono, *Dalton Trans.*, 2008, 4965–4974.
61. S. Johannsen, N. Megger, D. Böhme, R. K. O. Sigel and J. Müller, *Nature Chem.*, 2010, **2**, 229–234.
62. E. Kellenbach, B. A. Maler, K. R. Yamamoto, R. Boelens and R. Kaptein, *FEBS Lett.*, 1991, **291**, 367–370.
63. C. Damblon, C. Prosperi, L.-Y. Lian, I. Barsukov, R. P. Soto, M. Galleni, J.-M. Frere and G. C. K. Roberts, *J. Am. Chem. Soc.*, 1999, **121**, 11575–11576.
64. G. Wang, B. L. Gaffney and R. A. Jones, *J. Am. Chem. Soc.*, 2004, **126**, 8908–8909.

65. V. Markowski, G. R. Sullivan and J. D. Roberts, *J. Am. Chem. Soc.*, 1977, **99**, 714–718.
66. C. Wang, H. Gao, B. L. Gaffney and R. A. Jones, *J. Am. Chem. Soc.*, 1991, **113**, 5486–5488.
67. G. A. Bowmaker, R. K. Harris, B. Assadollahzadeh, D. C. Apperley, P. Hodgkinson and P. Amornsakchai, *Mag. Res. Chem.*, 2004, **42**, 819–826.
68. G. C. van Stein, G. van Koten, K. Vrieze, C. Brevard and A. L. Spek, *J. Am. Chem. Soc.*, 1984, **106**, 4486–4492.
69. K. Wüthrich, *NMR of Proteins and Nucleic Acids*, John Wiley & Sons, New York, 1986.
70. J. L. O'Rear, S. Wang, A. L. Feig, L. Beigelman, O. C. Uhlenbeck and D. Herschlag, *RNA*, 2001, **7**, 537–545.
71. E. P. Nikonowicz and J. S. Smith, *Phosphorus, Sulfur and Silicon and the Related Elements*, 1999, **144–146**, 305–308.
72. J. C. Markley, F. Godde and S. T. Sigurdsson, *Biochemistry*, 2001, **40**, 13849–13856.
73. J. H. Davis, T. R. Foster, M. Tonelli and S. E. Butcher, *RNA*, 2007, **13**.
74. F. H.-T. Allain and G. Varani, *Nucleic Acids Res.*, 1995, **23**, 341–350.
75. S. E. Butcher, F. H.-T. Allain and J. Feigon, *Biochemistry*, 2000, **39**, 2174–2182.
76. S. Rüdisser and I. Tinoco Jr., *J. Mol. Biol.*, 2000, **295**, 1211–1223.
77. J. S. Kieft and I. Tinoco Jr., *Structure*, 1997, **5**, 713–721.
78. R. L. Gonzalez Jr and I. Tinoco Jr, *J. Mol. Biol.*, 1999, **289**, 1267–1282.
79. G. Colmenarejo and I. Tinoco Jr, *J. Mol. Biol.*, 1999, **290**, 119–135.
80. A. G. Palmer III, J. Cavanagh, P. E. Wright and M. Rance, *J. Magn. Reson.*, 1991, **93**, 151–170.
81. J. Noeske, C. Richter, M. A. Grundl, H. R. Nasiri, H. Schwalbe and J. Wöhnert, *Proc. Natl. Acad. Sci. USA*, 2005, **102**, 1372–1377.
82. M. C. Erat, H. Kovacs and R. K. O. Sigel, *J. Inorg. Biochem.*, 2010, **104**, 611–613.
83. E. A. Peroza, R. Schmucki, P. Guntert, E. Freisinger and O. Zerbe, *J. Mol. Biol.*, 2009, **387**, 207–218.
84. G. Digilio, C. Bracco, L. Vergani, M. Botta, D. Osella and A. Viarengo, *J. Biol. Inorg. Chem.*, 2009, **14**, 167–178.
85. K. Zangger, G. Oez, I. Armitage and M. J. Otvos, *D. Protein Sci.*, 1999, **8**, 2630–2638.
86. *NMR of Newly Accessible Nuclei: Chemically and Biochemically Important Elements*, Ed. P. Lazlo, Academic Press, New York, 1983.
87. C. V. Grant, V. Frydman, J. S. Harwood and L. Frydman, *J. Am. Chem. Soc.*, 2002, **124**, 4458–4462.
88. D. Rovnyak, M. Baldus, G. Wu, N. V. Hud, J. Feigon and R. G. Griffin, *J. Am. Chem. Soc.*, 2000, **122**, 11423–11429.
89. C. V. Grant, V. Frydman and L. Frydman, *J. Am. Chem. Soc.*, 2000, **122**, 11743–11744.
90. S. S. Reid and J. A. Cowan, *Biochemistry*, 1990, **29**, 6025–6032.
91. V. P. Denisov and B. Halle, *Proc. Natl. Acad. Sci. USA*, 2000, **97**, 629–633.

92. S. Basu, A. A. Szewczak, M. Cocco and S. A. Strobel, *J. Am. Chem. Soc.*, 2000, **122**, 3240–3241.
93. M. L. Gill, S. A. Strobel and J. P. Loria, *J. Am. Chem. Soc.*, 2005, **127**, 16723–16732.
94. X. Zhang, P. Cekan, S. T. Sigurdsson and P. Z. Qin, *Methods Enzymol.*, 2009, **469**, 303–328.
95. A. D. Milov, A. G. Maryasov and Y. D. Tsvetkov, *Appl. Magn. Res.*, 1998, **15**, 107–143.
96. M. Pannier, S. Veit, A. Godt, G. Jeschke and H. W. Spiess, *J. Magn. Res.*, 2000, **142**, 331–340.
97. P. P. Borbat, J. H. Davis, S. E. Butcher and J. H. Freed, *J. Am. Chem. Soc.*, 2004, **126**, 7746–7747.
98. O. Schiemann, *Methods Enzymol.*, 2009, **469**, 329–351.
99. T. E. Edwards and S. T. Sigurdsson, *Biochemistry*, 2005, **44**, 12870–12878.
100. L. Hunsicker-Wang, M. Vogt and V. J. DeRose, *Methods Enzymol.*, 2009, **468**, 335–367.
101. M. G. Santangelo, A. Medina-Molner, A. Schweiger, G. Mitrikas and B. Spingler, *J. Biol. Inorg. Chem.*, 2007, **12**, 767–775.
102. D. Mustafi, A. Bekesi, B. G. Vertessy and M. W. Makinen, *Proc. Natl. Acad. Sci. USA*, 2003, **100**, 5670–5675.
103. A. Danchin and M. Gueron, *Eur. J. Biochem.*, 1970, **16**, 532–536.
104. T. E. Horton, D. R. Clardy and V. J. DeRose, *Biochemistry*, 1998, **37**, 18094–18101.
105. N. Kisseleva, A. Khvorova, E. Westhof and O. Schiemann, *RNA*, 2005, **11**, 1–6.
106. S. R. Morrissey, T. E. Horton and V. J. DeRose, *J. Am. Chem. Soc.*, 2000, **122**, 3473–3481.
107. M. Vogt, S. Lahiri, C. G. Hoogstraten, R. D. Britt and V. J. DeRose, *J. Am. Chem. Soc.*, 2006, **128**, 16764–16770.
108. O. Schiemann, J. Fritscher, N. Kisseleva, S. T. Sigurdsson and T. F. Prisner, *ChemBioChem*, 2003, **4**, 1057–1065.
109. S. R. Morrissey, T. E. Horton, C. V. Grant, C. G. Hoogstraten, R. D. Britt and V. J. DeRose, *J. Am. Chem. Soc.*, 1999, **121**, 9215–9218.
110. C. G. Hoogstraten and R. D. Britt, *RNA*, 2002, **8**, 252–260.
111. W. D. Horrocks Jr, *Methods Enzymol.*, 1993, **226**, 495–538.
112. R. M. Supkowski and W. D. Horrocks Jr, *Met. Ions. Biol. Syst.*, 2003, **40**, 281–322.
113. C. F. Baes Jr and R. E. Mesmer, *The Hydrolysis of Cations*, Krieger Publishing Co., Malabar, Florida, 1976, pp. 1–496.
114. A. L. Feig, M. Panek, W. D. Horrocks Jr and O. C. Uhlenbeck, *Chem. Biol.*, 1999, **6**, 801–810.
115. C. M. Andolina, R. A. Mathews and J. R. Morrow, *Helv. Chim. Acta*, 2009, **92**, 2330–2348.
116. I. Sanchez-Lombardo, C. M. Andolina, J. R. Morrow and A. K. Yatsimirsky, *Dalton Trans.*, 2010, **39**, 864–873.
117. D. Costa, M. L. Ramos, H. D. Burrows, M. J. Tapia and M. d. G. Miguel, *Chem. Phys.*, 2008, **352**, 241–248.

118. N. L. Greenbaum, C. Mundoma and D. R. Peterman, *Biochemistry*, 2001, **40**, 1124–1134.
119. C. Mundoma and N. L. Greenbaum, *J. Am. Chem. Soc.*, 2002, **124**, 3525–3532.
120. C. Mundoma and N. L. Greenbaum, *Biopolymers*, 2003, **69**, 100–109.
121. F. Yuan and N. L. Greenbaum, *Spectrum*, 2007, **20**, 14–17.
122. F. Yuan, L. Griffin, L. Phelps, V. Buschmann, K. Weston and N. L. Greenbaum, *Nucleic Acids Res.*, 2007, **35**, 2833–2845.
123. H. A. Tajmir-Riahi, C. N. N'Soukpoe-Kossi and D. Joly, *Spectroscopy*, 2009, **23**, 81–101.
124. A. Rodriguez-Casado, J. Bartolome, V. Carreno, M. Molina and P. Carmona, *Biophys. Chem.*, 2006, **124**, 73–79.
125. M. Banyay, J. Sandbrink, R. Stroemberg and A. Graeslund, *Biochem. Biophys. Res. Commun.*, 2004, **324**, 634–639.
126. T. Theophanides and H. A. Tajmir-Riahi, *J. Biomol. Struct. Dyn.*, 1985, **2**, 995–1004.
127. S. Nafisi and Z. Norouzi, *DNA Cell Biol.*, 2009, **28**, 469–477.
128. H. A. Tajmir-Riahi and T. Theophanides, *J. Biomol. Struct. Dyn.*, 1985, **3**, 537–542.
129. H. Arakawa, J. F. Neault and H. A. Tajmir-Riahi, *Biophys. J.*, 2001, **81**, 1580–1587.
130. S. Nafisi, A. Sobhanmanesh, K. Alimoghaddam, A. Ghavamzadeh and H.-A. Tajmir-Riahi, *DNA Cell Biol.*, 2005, **24**, 634–640.
131. A. Percot, S. Lecomte, J. Vergne and M.-C. Maurel, *Biopolymers*, 2009, **91**, 384–390.
132. J. G. Duguid, V. A. Bloomfield, J. M. Benevides and G. J. Thomas, *Biophys. J.*, 1993, **64**, A266–A266.
133. J. Duguid, V. A. Bloomfield, J. Benevides and G. J. Thomas, *Biophys. J.*, 1993, **65**, 1916–1928.
134. J. Stangret and R. Savoie, *Phys. Chem. Chem. Phys.*, 2002, **4**, 4770–4773.
135. M. Langlais, H. A. Tajmir-Riahi and R. Savoie, *Biopolymers*, 1990, **30**, 743–752.
136. B. Gong, Y. Chen, E. L. Christian, J. H. Chen, E. Chase, D. M. Chadalavada, R. Yajima, B. L. Golden, P. C. Bevilacqua and P. R. Carey, *J. Am. Chem. Soc.*, 2008, **130**, 9670–9672.
137. E. L. Christian, V. E. Anderson, P. R. Carey and M. E. Harris, *Biochemistry*, 2010, **49**, 2869–2879.
138. C. D. Putnam, M. Hammel, G. L. Hura and J. A. Tainer, *Quart. Rev. Biophys.*, 2007, **40**, 191–285.
139. J. Lipfert and S. Doniach, *Annu. Rev. Biophys. Biomol. Struct.*, 2007, **36**, 307–327.
140. R. P. Rambo and J. A. Tainer, *RNA*, 2010, **16**, 638–646.
141. R. P. Rambo and J. A. Tainer, *Curr. Opin. Struct.Biology*, 2010, **20**, 128–137.
142. J. Lipfert, J. Ouellet, D. G. Norman, S. Doniach and D. M. J. Lilley, *Structure*, 2008, **16**, 1357–1367.
143. J. Lipfert, D. Herschlag and S. Doniach, *Methods Mol. Biol.*, 2009, **540**, 141–159.

144. S. Doniach and J. Lipfert, *Methods Enzymol.*, 2009, **469**, 237–251.
145. L. Pollack and S. Doniach, *Methods Enzymol.*, 2009, **469**, 253–268.
146. S. Moghaddam, G. Caliskan, S. Chauhan, C. Hyeon, R. M. Briber, D. Thirumalai and S. A. Woodson, *J. Mol. Biol.*, 2009, **393**, 753–764.
147. U. Heinz, L. Hemmingsen, M. Kiefer and H.-W. Adolph, *Chem. Eur. J.*, 2009, **15**, 7350–7358.
148. T. Liu, X. Chen, Z. Ma, J. Shokes, L. Hemmingsen, R. A. Scott and D. P. Giedroc, *Biochemistry*, 2008, **47**, 10564–10575.
149. G. Absillis, E. Cartuyvels, R. Van Deun and T. N. Parac-Vogt, *J. Am. Chem. Soc.*, 2008, **130**, 17400–17408.
150. C. Bailly, V. Guerniou, E. Lamour, J.-L. Bernier, F. Villain and H. Vezin, *ChemBioChem*, 2003, **4**, 112–114.
151. K. S. Koutmou, A. Casiano-Negroni, M. M. Getz, S. Pazicni, A. J. Andrews, J. E. Penner-Hahn, H. M. Al-Hashimi and C. A. Fierke, *Proc. Natl. Acad. Sci. USA*, 2010, **107**, 2479–2484.
152. L. Hemmingsen and T. Butz, in *Applications of Physical Methods to Inorganic and Bioinorganic Chemistry*, Ed. R. A. Scott and C. M. Lukehart, John Wiley & Sons Ltd., Chichester, UK, 2007, pp. 423–439.
153. O. Iranzo, T. Jakusch, K.-H. Lee, L. Hemmingsen and V. L. Pecoraro, *Chem. Eur. J.*, 2009, **15**, 3761–3772.
154. O. Iranzo, P. W. Thulstrup, S.-b. Ryu, L. Hemmingsen and V. L. Pecoraro, *Chem. Eur. J.*, 2007, **13**, 9178–9190.
155. A. F. A. Peacock, L. Hemmingsen and V. L. Pecoraro, *Proc. Natl. Acad. Sci. USA*, 2008, **105**, 16566–16571.
156. N. Selevsek, S. Rival, A. Tholey, E. Heinzle, U. Heinz, L. Hemmingsen and H. W. Adolph, *J. Biol. Chem.*, 2009, **284**, 16419–16431.
157. S. Paulus, S. Johannsen, R. K. O. Sigel, P. W. Thulstrup and L. Hemmingsen, *to be submitted.*
158. S. C. Dahm and O. C. Uhlenbeck, *Biochemistry*, 1991, **30**, 9464–9469.
159. M. Forconi, J. Lee, J. K. Lee, J. A. Piccirilli and D. Herschlag, *Biochemistry*, 2008, **47**, 6883–6894.
160. P. M. Gordon, E. J. Sontheimer and J. A. Piccirilli, *RNA*, 2000, **6**, 199–205.
161. R. K. O. Sigel, B. Song and H. Sigel, *J. Am. Chem. Soc.*, 1997, **119**, 744–755.
162. C. P. Da Costa, A. Okruszek and H. Sigel, *ChemBioChem*, 2003, **4**, 593–602.
163. P. M. Gordon and J. A. Piccirilli, *Nat. Struct. Biol.*, 2001, **8**, 893–898.
164. A. D. Griffiths, B. V. Potter and I. C. Eperon, *Nucleic Acids Res.*, 1987, **15**, 4145–4162.
165. P. M. Burgers and F. Eckstein, *Proc. Natl. Acad. Sci. USA*, 1978, **75**, 4798–4800.
166. E. L. Christian, in *Handbook of RNA Biochemistry*, Ed. R. K. Hartmann, A. Bindereif, A. Schön and E. Westhof, Wiley-VCH, Weinheim, 2008, **Vol. 1**, pp. 319–344.
167. K. Lang, M. Erlacher, D. N. Wilson, R. Micura and N. Polacek, *Chem. Biol.*, 2008, **15**, 485–492.
168. J. A. Piccirilli, J. S. Vyle, M. H. Caruthers and T. R. Cech, *Nature*, 1993, **361**, 85–88.

169. L. B. Weinstein, B. C. N. M. Jones, R. Cosstick and T. R. Cech, *Nature*, 1997, **388**, 805–808.
170. E. J. Sontheimer, S. Sun and J. A. Piccirilli, *Nature*, 1997, **388**, 801–805.
171. J. M. Warnecke, J. P. Fuerste, W. D. Hardt, V. A. Erdmann and R. K. Hartmann, *Proc. Natl. Acad. Sci. USA*, 1996, **93**, 8924–8928.
172. K. Kanaori, S. Sakamoto, H. Yoshida, P. Guga, W. Stec, K. Tajima and K. Makino, *Biochemistry*, 2004, **43**, 5672–5679.
173. M. Boudvillain and A. M. Pyle, *EMBO J.*, 1998, **17**, 7091–7104.
174. S. Basu and S. A. Strobel, *RNA*, 1999, **5**, 1399–1407.
175. C. A. Brautigam and T. A. Steitz, *J. Mol. Biol.*, 1998, **277**, 363–377.
176. H. Irving and R. J. P. Williams, *J. Chem. Soc.*, 1953, 3192–3210.
177. H. Sigel and D. B. McCormick, *Acc. Chem. Res.*, 1970, **3**, 201–208.
178. B. Knobloch, B. Nawrot, A. Okruszek and R. K. O. Sigel, *Chem. Eur. J.*, 2008, **14**, 3100–3109.
179. C. Waldsich, *Nature Protocols*, 2008, **3**, 811–823.
180. J. C. Cochrane and S. A. Strobel, *Curr. Protoc. Nucleic Acid Chem.*, 2004, Chapter 6: Unit 6.9.
181. R. Sousa and R. Padilla, *EMBO*, 1995, **14**, 4609–4621.
182. S. Basu, R. P. Rambo, J. Strauss-Soukup, J. H. Cate, A. R. Ferré-D'Amaré, S. A. Strobel and J. A. Doudna, *Nat. Struct. Biol.*, 1998, **5**, 986–992.
183. S. Basu and S. A. Strobel, *Methods*, 2001, **23**, 264–275.
184. C. Waldsich and A. M. Pyle, *Nat. Struct. Mol. Biol.*, 2007, **14**, 37–44.
185. C. Waldsich and A. M. Pyle, *J. Mol. Biol.*, 2008, **375**, 572–580.
186. R. B. Martin and F. S. Richardson, *Quart. Rev. Biophys.*, 1979, **12**, 181–209.
187. *The Lanthanides and Their Interrelation with Biosystems*, Vol. 40 of *Metal Ions in Biological Systems*, Ed. A. Sigel and H. Sigel, Marcel Dekker Inc., New York, Basel, Hong Kong, 2003.
188. R. K. O. Sigel and A. M. Pyle, *Met. Ions Biol. Syst.*, 2003, **40**, 477–512.
189. S. Cotton, *Lanthanides and Actinides*, Macmillan Education Ltd., London, 1991.
190. J. Ciesiolka, T. Marciniec and W. J. Krzyzosiak, *Eur. J. Biochem.*, 1989, **182**, 445–450.
191. N. G. Walter, N. Yang and J. M. Burke, *J. Mol. Biol.*, 2000, **298**, 539–555.
192. S. Dorner and A. Barta, *Biol. Chem.*, 1999, **380**, 243–251.
193. R. K. O. Sigel, A. Vaidya and A. M. Pyle, *Nat. Struct. Biol.*, 2000, **7**, 1111–1116.
194. D. A. Harris and N. G. Walter, in *Handbook of RNA Biochemistry*, Ed. R. K. Hartmann, A. Bindereif, A. Schön and E. Westhof, Wiley-VCH, Weinheim, 2008, **Vol. 1**, pp. 205–213.
195. R. S. Brown, J. C. Dewan and A. Klug, *Biochemistry*, 1985, **24**, 4785–4801.
196. R. S. Brown, B. A. Hingerty, J. C. Dewan and A. Klug, *Nature*, 1983, **303**, 543–546.
197. L. A. Kirsebom and J. Ciesiolka, in *Handbook of RNA Biochemistry*, Ed. R. K. Hartmann, A. Bindereif, A. Schön and E. Westhof, Wiley-VCH, Weinheim, 2008, **Vol. 1**, pp. 214–227.
198. E. Huntzinger, M. Possedko, F. Winter, H. Moine, C. Ehresmann and P. Romby, in *Handbook of RNA Biochemistry*, Ed. R. K. Hartmann, A. Bindereif, A. Schön and E. Westhof, Wiley-VCH, Weinheim, 2008, **Vol. 1**, pp. 151–171.

199. J. Ciesiolka, D. Michalowski, J. Wrzesinski, J. Krajewski and W. J. Krzyzosiak, *J. Mol. Biol.*, 1998, **275**, 211–220.
200. M. Lindell, P. Romby and E. G. H. Wagner, *RNA*, 2002, **8**, 534–541.
201. D. Labuda, K. Nicoghosian and R. Cedergren, *J. Biol. Chem.*, 1985, **260**, 1103–1107.
202. M. Brannvall, N. E. Mikkelsen and L. A. Kirsebom, *Nucleic Acids Res.*, 2001, **29**, 1426–1432.
203. E. E. Regulski and R. R. Breaker, in *Post-Transcriptional Gene Regulation*, Ed. J. Wilusz, Humana Press, Totowa, NJ, 2008, **Vol. 419**, pp. 53–67.
204. G. A. Soukup and R. R. Breaker, *RNA*, 1999, **5**, 1308–1325.
205. D. J. Earnshaw and M. J. Gait, *Nucleic Acids Res.*, 1998, **26**, 5551–5561.
206. M. Hertweck and M. W. Müller, *Eur. J. Biochem.*, 2001, **268**, 4610–4620.
207. H. J. H. Fenton, *Proc. Chem. Soc.*, 1893, **9**, 113.
208. A. Draganescu and T. D. Tullius, *Met. Ions Biol. Syst.*, 1996, **33**, 453–484.
209. A. Flaus, K. Luger, S. Tan and T. J. Richmond, *Proc. Natl. Acad. Sci. USA*, 1996, **93**, 1370–1375.
210. T. D. Tullius, *Free Radic. Res.*, 1991, **12–13**, 521–529.
211. T. D. Tullius, *Nature*, 1988, **332**, 663–664.
212. T. D. Tullius, *Comprehensive Supramolecular Chemistry*, 1996, **5**, 317–342.
213. W. J. Dixon, J. J. Hayes, J. R. Levin, M. F. Weidner, B. A. Dombroski and T. D. Tullius, *Methods Enzymol.*, 1991, **208**, 380–413.
214. T. D. Tullius, *Curr. Opin. Struct. Biol.*, 1991, **1**, 428–434.
215. D. W. Celander and T. R. Cech, *Science*, 1991, **251**, 401–407.
216. J. A. Latham and T. R. Cech, *Science*, 1989, **245**, 276–282.
217. C. Berens, B. Streicher, R. Schroeder and W. Hillen, 1998, **5**, 163–175.
218. I. Hoch, C. Berens, E. Westhof and R. Schroeder, *J. Mol. Biol.*, 1998, **282**, 557–569.
219. G. Bauer and C. Berens, in *Handbook of RNA Biochemistry*, Ed. R. K. Hartmann, A. Bindereif, A. Schön and E. Westhof, Wiley-VCH, Weinheim, 2008, **Vol. 1**, pp. 238–249.
220. Y.-G. Ren, N. Henriksson and A. Virtanen, in *Handbook of RNA Biochemistry*, Ed. R. K. Hartmann, A. Bindereif, A. Schön and E. Westhof, Wiley-VCH, Weinheim, 2008, **Vol. 1**, pp. 345–353.
221. V. K. Misra and D. E. Draper, *J. Mol. Biol.*, 2000, **299**, 813–825.
222. B. Honig and A. Nicholls, *Science*, 1995, **268**, 1144–1149.
223. N. Korolev, A. P. Lyubartsev and L. Nordenskiold, *J. Biomol. Struct. Dyn.*, 2002, **20**, 275–290.
224. I. A. Shkel and M. T. Record, *Biochemistry*, 2004, **43**, 7090–7101.
225. K. Chin, K. A. Sharp, B. Honig and A. M. Pyle, *Nat. Struct. Biol.*, 1999, **6**, 1055–1061.
226. V. B. Chu, Y. Bai, J. Lipfert, D. Herschlag and S. Doniach, *Curr. Opin. Chem. Biol.*, 2008, **12**, 619–625.
227. M. P. Liang, D. R. Banatao, T. E. Klein, D. L. Brutlag and R. B. Altman, *Nucleic Acids Res.*, 2003, **31**, 3324–3327.
228. D. R. Banatao, R. B. Altman and T. E. Klein, *Nucleic Acids Res.*, 2003, **31**, 4450–4460.

229. P. Banas, P. Jurecka, N. G. Walter, J. Sponer and M. Otyepka, *Methods*, 2009, **49**, 202–216.
230. M. A. Ditzler, M. Otyepka, J. Sponer and N. G. Walter, *Acc. Chem. Res.*, 2010, **43**, 40–47.
231. P. Auffinger and Y. Hashem, *Bioinformatics*, 2007, **23**, 1035–1037.
232. J. M. Castagnetto, S. W. Hennessy, V. A. Roberts, E. D. Getzoff, J. A. Tainer and M. E. Pique, *Nucleic Acids Res.*, 2002, **30**, 379–382.
233. L. R. Stefan, R. Zhang, A. G. Levitan, D. K. Hendrix, S. E. Brenner and S. R. Holbrook, *Nucleic Acids Res.*, 2006, **34**, D131–D134.
234. A. M. Pyle, O. Fedorova and C. Waldsich, *Trends Biochem. Sci.*, 2007, **32**, 138–145.
235. R. L. Gonzalez Jr and I. Tinoco Jr, *Methods Enzymol.*, 2001, **338**, 421–443.
236. M. C. Erat, J. Coles, C. Finazzo, B. Knobloch and R. K. O. Sigel, *manuscript in preparation.*
237. R. K. O. Sigel, *details to be published.*
238. A. J. Andrews and C. Fierke, in *Handbook of RNA Biochemistry*, Ed. R. K. Hartmann, A. Bindereif, A. Schön, and E. Westhof, Wiley-VCH, Weinheim,2008, **Vol. 1**, pp. 251–258.
239. Y. Liu, M. Han, H.-Y. Zhang, L.-X. Yang and W. Jiang, *Org. Lett.*, 2008, **10**, 2873–2876.
240. D. Ray and P. K. Bharadwaj, *Inorg. Chem.*, 2008, **47**, 2252–2254.
241. R. K. O. Sigel, *Eur. J. Inorg. Chem.*, 2005, **12**, 2281–2292.
242. R. D. Shannon, *Acta Crystallogr.*, 1976, **A32**, 751–767.
243. W. E. Morf and W. Simon, *Helv. Chim. Acta*, 1971, **54**, 794–810.
244. L. Helm and A. E. Merbach, *Coord. Chem. Rev.*, 1999, **187**, 151–181.
245. S. F. Lincoln, *Helv. Chim. Acta*, 2005, **88**, 523–545.
246. Y. Inada, A. M. Mohammed, H. H. Loeffler and S. Funahashi, *Helv. Chim. Acta*, 2005, **88**, 461–469.
247. H. Sigel, C. P. Da Costa and R. B. Martin, *Coord. Chem. Rev.*, 2001, **219–221**, 435–461.
248. C. P. Da Costa, D. Krajewska, A. Okruszek, W. J. Stec and H. Sigel, *J. Biol. Inorg. Chem.*, 2002, **7**, 405–415.
249. L. Helm and A. E. Merbach, *Chem. Rev.*, 2005, **105**, 1923–1959.
250. A. F. Holleman and E. Wiberg, *Lehrbuch der Anorganischen Chemie*, Walter de Gruyter, Berlin, New York, 1985, pp. 1265–1279.
251. D. Parker, R. S. Dickins, H. Puschmann, C. Crossland and J. A. K. Howard, *Chem. Rev.*, 2002, **102**, 1977–2010.
252. R. J. P. Williams and J. J. R. Fraústo da Silva, *The Natural Selection of the Chemical Elements*, Clarendon Press, Oxford, 1996, pp. 1–646.
253. M. C. Erat, E. Besic, M. Oberhuber and R. K. O. Sigel, *results to be published.*

Met. Ions Life Sci. **2011**, *9*, 101–124

3

Importance of Diffuse Metal Ion Binding to RNA

***Zhi-Jie Tan*[1] *and Shi-Jie Chen*[2]**

[1]Department of Physics and Key Laboratory of Artificial Micro- and Nano-Structures of the Ministry of Education, School of Physics and Technology, Wuhan University, Wuhan 430 072, China

[2]Department of Physics & Astronomy and Department of Biochemistry, University of Missouri, Columbia, MO 65211, USA

< chenshi@missouri.edu >

Metal Ions in Life Sciences, Volume 9 Edited by Astrid Sigel, Helmut Sigel, and Roland K. O. Sigel

Published by the Royal Society of Chemistry, www.rsc.org DOI: 10.1039/9781849732512 00101

ABSTRACT: RNAs are highly charged polyanionic molecules. RNA structure and function are strongly correlated with the ionic condition of the solution. The primary focus of this article is on the role of diffusive ions in RNA folding. Due to the long-range nature of electrostatic interactions, the diffuse ions can contribute significantly to RNA structural stability and folding kinetics. We present an overview of the experimental findings as well as the theoretical developments on the diffuse ion effects in RNA folding. This review places heavy emphasis on the effect of magnesium ions. Magnesium ions play a highly efficient role in stabilizing RNA tertiary structures and promoting tertiary structural folding. The highly efficient role goes beyond the mean-field effect such as the ionic strength. In addition to the effects of specific ion binding and ion dehydration, ion-ion correlation for the diffuse ions can contribute to the efficient role of the multivalent ions such as the magnesium ions in RNA folding.

KEYWORDS: folding thermodynamics · ion binding · kinetics · magnesium ions · metal ions · RNA folding

1. INTRODUCTION

RNA structure and stability are intrinsic to RNA function and RNA-based therapeutic strategies. Recent breathtaking new discoveries on the functions of non-coding RNAs have opened a new door for many potential therapeutic applications of RNAs. This further presses the demand for quantitative understanding and prediction of RNA structure and stability. RNA structural formation is driven by the intramolecular forces, such as base pairing/stacking, ion-mediated electrostatic interactions, and the effects of conformational entropies [1,2]. The primary concern of this chapter is on the role of ion-mediated electrostatic interactions in RNAs.

Metal ions, especially Mg^{2+}, play a critical role in RNA folding. Because each RNA nucleotide carries a unit negative charge on the phosphate, RNA folding causes massive build-up of negative charges. There are two electrostatic effects involved in RNA folding. First, the charge build-up would lead to a strong intrachain Coulombic repulsion to oppose RNA folding. For example, folding of a 400-nt RNA could cause an increase in the electrostatic energy of about 1000 kcal/mol [3]. Second, the charge build-up would attract the counterions in the solution and induce significant ion binding to RNA. Specifically, metal ions in the solution can cluster around the polyanionic RNA to effectively reduce the electrostatic energy and to promote folding and stabilize a folded RNA structure. Therefore, RNA structure and stability are strongly coupled to ion electrostatic interactions [1–8].

The main focus of this chapter is on the effect of diffuse (bound) ions in RNA folding. What is a diffuse (bound) ion? To answer this question, we need first to clarify the use of the phrases "bound ions" and "diffuse ions". Firstly, ion-RNA interaction is predominantly determined by ions in the close vicinity of the RNA surface. The phrase "bound ions" has been used in widely different senses in the literature. Because electrostatic interactions are long-range, all counter ions are "bound" with all nucleic acids in a solution. Therefore, the bound ions for an RNA have been defined as the surrounding ions in excess of the bulk ion concentration. In other literature, bound ions are reserved for interactions that involve specific ion-RNA interactions. In this chapter, we use "bound ion" in the former sense to refer to the diffuse ions accumulated around RNA in excess of the bulk ion concentration away from the RNA.

Secondly, a bound ion can remain hydrated and interact non-specifically with the RNA. These ions are often called "diffuse ions". Other ions can coordinate to specific groups of the RNA. Due to the close interaction with specific RNA groups, a specifically bound ion could be energetically important (in addition to its catalytic role). However, due to the long-range nature of the Coulomb interaction, the (large number of) diffuse ions can result in a significant electrostatic force on the charged (phosphate) groups of RNA and have a critical impact on the structural formation of RNA [4–8].

2. DIFFUSE IONS PROVIDE A SIGNIFICANT STABILIZING FORCE FOR RNA STRUCTURE

Quantitative study of the effects of diffuse ions is difficult due to several reasons: (a) the long-range nature of the electrostatic interaction, (b) the intriguing interplay between ion-ion and ion-RNA interactions, (c) the dependence on the concentration, charge and size of the ions, and (d) the effect of the complex three-dimensional shape of the RNA. These factors directly impact the answers to several key questions about the important roles of the (diffuse) ions. Where do bound ions distribute on the RNA surface? How many ions are bound to the RNA? How large is the ion-dependent stabilizing free energy? How does the free energy vary for different RNA structures? How is the ion electrostatic free energy compared to the other stabilizing and destabilizing free energies in RNA folding? Why do some ions show "unusually" high efficiency in stabilizing RNA structure than other ions? These questions have provided a strong motivation for the measurements of ion distribution and ion electrostatic folding free energies for the different RNA structures.

2.1. Distribution of the Bound Ions

The first key issue in the ion electrostatics is how diffuse ions are distributed around the RNA molecule [9–19]. There have been several useful experimental methods developed for the quantification of the ion atmosphere around nucleic acids. The methods include small angle X-ray scattering (SAXS) [9–12], the ion-counting method [13], and the thermodynamic method [14–19]. These methods have been successfully used to quantify the ion distribution, the competition between the different species of the ions, and the folding thermodynamics [16,17]; see Table 1.

The experimental measurements have led to several important findings. For example, for oligomeric DNA and RNA duplexes in a mixed monovalent/divalent ion solution, SAXS profiles showed that in the titration of the ions, although the number of the bound ions for each species varies due to the competition of the different ions, the shape of the ion distribution (for each species) is invariant [10]. In addition, smaller ionic size causes a stronger ion-RNA interaction and thus, the ion size needs to be explicitly

Table 1. Experimental measurements for RNA (and DNA) ion atmosphere.

RNAs (or DNAs)	Refs.	Ionic conditions	Thermodynamic quantities
25-bp DNA duplex	[9]	divalent ions	SAXS profiles
	[11]	mixed Rb^+/Sr^{2+}	SAXS profiles
	[11]	mixed $Rb^+/(Co(NH_3)_6)^{3+}$	SAXS profiles number of bound ions
25-bp RNA duplex	[12]	Rb^+, Mg^{2+}	SAXS profiles
25-bp DNA duplex	[12]	Rb^+, Mg^{2+}	SAXS profiles
24-bp DNA duplex	[13]	mixed Na^+/Mg^{2+}	number of bound ions
	[13]	mixed monovalent ions	number of bound ions
	[13]	mixed divalent ions	number of bound ions
24-bp DNA triplex	[13]	mixed Na^+/Mg^{2+}	number of bound ions
poly(A·U)	[14]	mixed Na^+/Mg^{2+}	number of bound Mg^{2+}
calf thymus DNA	[15]	mixed Na^+/Mg^{2+}	number of bound Mg^{2+}
BWYV pseudoknot	[16]	Na^+, mixed Na^+/Mg^{2+}	number of bound Mg^{2+}
58-nt rRNA fragment	[17]	mixed K^+/Mg^{2+}	number of bound Mg^{2+}
yeast tRNA	[18]	mixed Na^+/Mg^{2+}	number of bound Mg^{2+}
	[19]	mixed Na^+/Mg^{2+}	number of bound Mg^{2+}

BWYV: beet western yellow virus.

taken into account in the theoretical analysis [10,12,13]. Moreover, experimental data on ion binding for RNA and DNA duplexes showed that metal ions can give more efficient charge neutralization for RNA than for DNA, and the difference likely comes from the higher backbone phosphate charge density for an A-form helix than for a B-form helix [12].

A more direct quantification of bound ions comes from the ion-counting method and thermodynamic method, which have been applied to RNAs and DNAs, including yeast tRNA [18,19], poly(A • U) [14], polymeric calf thymus DNA [15], 58-nt ribosomal RNA fragment [17], BWYV (beet western yellow virus) pseudoknot fragment [16], oligomeric DNA and RNA duplexes [13], and DNA triplex [13]. The experimental data from these experiments have revealed several important features of ion binding:

1. The detailed distribution of the bound ions is highly sensitive to the atomic structure of the RNA [12].
2. Ion-binding of the different (monovalent and divalent) ions shows anti-cooperativity [13].
3. Multivalent ions (e.g., Mg^{2+}) are much more efficient in charge neutralization than monovalent ions. The unusually high efficiency of the multivalent ions over the monovalent ions is beyond the mean-field effect such as the ionic strength. For example, millimolar Mg^{2+} can achieve a similar ion neutralization as molar Na^{+}.
4. The efficiency of multivalent ions (e.g., Mg^{2+}) in neutralization is more pronounced for higher charged nucleic acids, i.e., larger molecules with more compact structures. For example, For a 24-bp DNA duplex and the yeast $tRNA^{Phe}$, 0.4 mM Mg^{2+} can approximately achieve the same ion neutralization as 20 mM and 32 mM Na^{+}, respectively [13,19].

2.2. Stabilization of Helices and Hairpins

Helix and loop are the fundamental segments of RNA secondary structure. A hairpin is the simplest secondary structural motif and plays a variety of structural and functional roles in RNA. The findings about the ion effects in hairpin stability mainly come from the thermodynamic experiments on RNA hairpin folding. Most of the experiments were performed in a Na^{+} solution or a mixed Na^{+}/Mg^{2+} solution; see Table 2 for a brief summary [16,20–32]. These thermodynamic measurements revealed several significant features for ion binding and ion-mediated folding stability for helices and hairpins:

1. Metal ions can stabilize DNA and RNA helices/hairpins in a similar way, with the linear dependence of the folding stability on the

Table 2. Thermodynamic measurements for the ion dependence of RNA and DNA secondary structural stability.

RNAs and DNAs	Refs.	Ionic conditions	Thermodynamic quantities
6-bp DNA duplex	[20]	mixed Na^+/Mg^{2+}	T_m, ΔG
9-bp DNA duplex	[21]	Na^+, Mg^{2+}	T_m, ΔG
12-bp DNA duplex	[22]	Mg^{2+}	T_m
25-bp DNA duplex	[23]	Mg^{2+}	T_m
RNA duplexes (6–14 bp)	[24]	mixed Na^+/Mg^{2+}	ΔG, T_m
DNA duplexes (10–30 bp)	[25]	Na^+	T_m
RNA and DNA hairpins (4–34 nt loop)	[26]	Na^+, Mg^{2+}	T_m
RNA hairpins (49–124 nt)	[27]	K^+, Na^+	ΔG
DNA duplexes	[28]	Na^+, Mg^{2+}	ΔG, T_m
RNA duplexes	[29]	Na^+, Mg^{2+}	ΔG, T_m
DNA and RNA hairpins	[3]	Na^+, Mg^{2+}	ΔG, T_m
BWYV pseudoknot hairpin	[16]	Na^+, Mg^{2+}	ΔH, ΔS
T4 gene pseudoknot hairpin	[31]	mixed K^+/Mg^{2+}	T_m
T2 gene pseudoknot hairpin	[32]	mixed K^+/Mg^{2+}	ΔG

T_m: melting temperature; $\Delta G = \Delta H - T\Delta S$: folding stability.

logarithm of monovalent salt concentration at low salt (e.g., <0.3 M Na^+), and a saturation tendency at a high monovalent concentration (e.g., >0.3 M Na^+).

2. Compared with the monovalent ions, divalent ions (e.g., Mg^{2+}) are more efficient in stabilizing helices/hairpins. The stability in a multivalent ion solution cannot be explained by the mean-field description such as ionic strength. For example, the stabilities for short DNA and RNA oligomers and hairpins in a 10 mM Mg^{2+} solution is approximately equivalent to the stabilities in a 1 M Na^+ solution [20,28–30].
3. Because the binding of Na^+ competes against the (more efficient) binding of the Mg^{2+} ions, a mixed Na^+/Mg^{2+} solution could lead to less stability than a pure Mg^{2+} solution [29].

The thermodynamic parameters have been measured quite extensively at the standard ion condition (1 M Na^+). These parameters form the basis for the predictions of RNA (DNA) secondary structure and folding stabilities [33–37]. For monovalent ion condition other than 1 M NaCl, RNA thermodynamic data for various ionic conditions yields a set of fitted formulas for the enthalpy/entropy parameters for RNA and DNA helices as functions of the different (monovalent) ion concentrations [25,28,29,33]. These empirical extensions have been shown to give good estimates for the thermodynamic stabilities for RNA and DNAs in different Na^+ conditions [25,28,29,38,39].

In contrast to the thermodynamic studies in Na^+ solutions, experimental data on Mg^{2+}-mediated helix/hairpin stability has been quite limited [20–24]. The experiments for the Mg^{2+} effects have motivated the attempt to derive the $[Mg^{2+}]$-dependent thermodynamic parameters [28,29]. Though these empirical formulas lack extensive experimental validations due to the limited experimental data, the theory-experiment agreements for the tests against the available experimental data suggest that the derived formulas may be reliable.

Recently, based on statistical mechanical modeling [29], Tan and Chen [30] derived the hairpin loop stability as functions of $[Na^+]$ and $[Mg^{2+}]$. The results suggest an interplay between the loop entropy and the ion-induced stabilization of the loop. The formation of a loop causes charge build-up of the nucleotide backbone and a stronger ion-RNA interaction. A higher ion concentration would cause a stronger reduction in the electrostatic intrachain repulsion upon loop formation. Therefore, ions help stabilizing the loop. Such ion-induced stability competes against the loss in the conformational entropy upon the loop formation [30].

2.3. Stabilization of Nucleic Acid Helix Assembly

Helix-helix packing is fundamental to tertiary structural folding. Osmotic pressure measurements have led to quantitative characterization for the ion-mediated helix-helix assembly [40,41]. These measurements provided several novel insights for the assembly of long (DNA) helices: (a) multivalent ions, such as Co^{3+}, can induce attraction between the helices, (b) monovalent ions (e.g., Na^+) can only modulate the strength of helix-helix repulsion [40], (c) certain types of divalent ions (e.g., Mn^{2+}) can induce a helix-helix attractive force that results in DNA condensation [42], while other divalent ions (e.g., Ca^{2+}) cannot, and (d) Mg^{2+} ions in the presence of methanol could induce a helix-helix attraction [41]. The different roles of divalent ions might be attributed to different ion-binding affinities to the various groups [1].

Helices in natural RNA structures are usually in the range of several to ten base pairs. In contrast to long helices, short helices have great rotational degrees of freedom and strong end effects. As a result, short helices can have different ion effects than long helices. Aiming to uncover the mechanism for ion-induced helix packing in RNA folding, several experiments have focused on the ion-mediated interactions between short helices [3,43–45]. In addition, SAXS experiments for a system of dispersed short DNA helices suggested that Mg^{2+} ions can significantly reduce the Coulomb repulsion between the helices and induce helix-helix attraction (through end-end base stacking) [11,44,45]. SAXS experiments for a system of loop-tethered short helices revealed a possible weak side-side attraction for helices in a Mg^{2+} solution of concentrations up to 0.6 M [3]. Furthermore, the experiments [3] showed that high-concentration ions (including 1+, 2+ and 3+ ions) can induce a random relaxed state [3] and Mg^{2+} is much more efficient than Na^{+} in causing such a state. A mean-field Poisson-Boltzmann (PB) calculation showed that the PB theory underestimates the efficient role of Mg^{2+} by over 10 times [43]. However, despite the extensive experimental studies, several key issues remain: (a) Is the electrostatic relaxation state a random disordered state or a collapsed compact state? (b) How do the different ions cause the different helix-helix packing states? (c) How does a junction/loop assist ion-induced helix packing?

2.4. Stabilization of Tertiary Structures

Because RNA tertiary structural folding involves massive charge build-up, ion-RNA interaction is significant for tertiary structural folding. In the past decades, extensive experiments have been performed to investigate how metal ions promote RNA tertiary structural folding and stabilize tertiary structures; see Table 3 [16,19,32,46–60]. These experiments have revealed several important roles of metal ions, especially Mg^{2+}, in tertiary structural folding as shown below:

1. The high efficiency of the Mg^{2+} ions (compared to the Na^{+} ions) is significantly more pronounced for tertiary structures than for secondary structures. This is due to the much higher backbone charge density and the much stronger electrostatic effect in tertiary structural folding. For example, 10 mM Mg^{2+} and 1 M Na^{+} can achieve a similar stability for short DNA and RNA duplexes [20,28,29], while for the tertiary structural folding of *Tetrahymena* ribozyme, the transition mid-points are at 0.5 mM and 0.5 M for Mg^{2+} and Na^{+} solutions, respectively [55].
2. Even with high monovalent ion concentration, Mg^{2+} can make a significant contribution to RNA tertiary structural stability. For example, for a 58-nt ribosomal RNA fragment in the background of 1.6 M monovalent ions, Mg^{2+} ions at 0.1 M can contribute about –6 kcal/mol to the tertiary structural folding stability [46].
3. Mg^{2+} can induce much more compact tertiary structures than Na^{+}. For *Tetrahymena* ribozyme, the folded structure in Mg^{2+} solution is about 7 Å more compact than the one in Na^{+} solutions [50].

Table 3. Thermodynamic measurements for the ion dependence of RNA and DNA tertiary structural stability.

RNAs or complexes	Refs.	Ionic conditions	Thermodynamic quantities
BWYV pseudoknot	[16]	Na^+, mixed Na^+/Mg^{2+}	ΔH, ΔS
58-nt rRNA fragment	[46]	mixed NH^+/Mg^{2+}	ΔG
yeast tRNA	[19]	mixed Na^+/Mg^{2+}	ΔG
HIV-1 kiss complex 1	[59]	Na^+, mixed Na^+/Mg^{2+}	T_m
HIV-1 kiss complex 2	[60]	Na^+, mixed Na^+/Mg^{2+}	T_m
T2 pseudoknot	[32]	K^+, mixed K^+/Mg^{2+}	ΔG
MMTV pseudoknot	[47]	Na^+, mixed K^+/Mg^{2+}	T_m
T4-35 pseudoknot	[48]	mixed Na^+/Mg^{2+}	ΔG, T_m
T4-32 pseudoknot	[48]	mixed Na^+/Mg^{2+}	ΔG, T_m
T4-28 pseudoknot	[48]	mixed Na^+/Mg^{2+}	ΔG, T_m
Tetrahymena ribozyme	[55]	K^+, Na^+, Mg^{2+}, spermidine	fraction folded
	[50]	Na^+,Mg^{2+}	R_g
	[51]	Na^+, Mg^{2+}	R_g
	[54]	Mg^{2+}	R_g
	[49]	Mg^{2+}, Ca^{2+}, Sr^{2+}, Ba^{2+}	fraction folded
	[52]	Mg^{2+}, Ca^{2+}, Sr^{2+}, Ba^{2+}	R_g
	[53]	Na^+, Mg^{2+}	l_P
hairpin ribozyme	[57]	Na^+, Mg^{2+}	fraction folded
	[58]	Na^+, Mg^{2+}	fraction folded
A-riboswitch	[56]	Li^+, Na^+, K^+, Rb^+, Cs^+	K_{obs}
Azoarcus ribozyme	[52]	K^+, Na^+, Mg^{2+}, Ca^{2+}, Co^{3+}, spermidine^{3+}	R_g, l_P

R_g: radius of gyration; K_{obs}: equilibrium constant; l_P: persistence length; MMTV: mouse mammary tumor virus.

4. Metal ions with higher charge and smaller size are more efficient than those with lower charge and bulkier size. Ions of higher charge density (charge/volume) would have stronger ability to stabilize RNA tertiary folds [49,55].

Furthermore, Brownian dynamics simulations for a model RNA tertiary structural folding system provided more detailed insights into the role of metal ions in RNA tertiary structural folding [49]: (a) The condensed ions near the RNA surface show a liquid-like correlated distribution; (b) for the RNA molecule (*Tetrahymena* ribozyme) studied, non-specific electrostatic interaction alone can account for the collapsed state and the dependence on the ion charge density [6,49]. However, other experiments suggested that, depending on the sequence, specific interactions (binding) of ions with the RNA could contribute significant stability for RNA tertiary structure [7,16,17]. Such distinctive findings on the roles of specific binding ions suggest the necessity for the further more careful and more extensive studies, especially theoretical studies, on the role of the diffuse ions.

Kissing loop complex is a typical tertiary structural motif formed by base pairing between two hairpin loops. Several experiments have pointed to the critical role of diffuse ions in stabilizing the different structures of kissing loop complexes. For HIV-1 DIS-type kissing loop-loop complexes, the melting temperature for the kissing complex shows much more sensitive ion-dependence than for the corresponding duplex of the same sequence at the kissing interface [59,60]. Moreover, the high efficiency of Mg^{2+} over Na^{+} is much more pronounced for the kissing loop complex than for the duplex [59]. The underlining physics for the above findings may stem from the significantly high charge density of RNA backbone at the kissing interface. For the kissing complex, depending on the RNA sequence, Mg^{2+} may also bind to specific locations at the kissing interface. The specific binding of the ions could further contribute to the highly sensitive $[Mg^{2+}]$-dependence of the kissing thermodynamics [60]. For the Tar-tar RNA complex, a bulkier ion would weaken the folding stability, in accordance with the observations that bulkier ions have lower efficiency in stabilizing RNA tertiary folds [56]

3. DIFFUSE ION IS CRITICAL TO RNA FOLDING KINETICS

The Coulombic interaction between RNA backbone charges causes a tremendous kinetic barrier for RNA folding. By lowering the electrostatic barrier for the kinetics, ions can promote the kinetic process of RNA folding and control the folding rate and pathways. There have been tremendous experimental efforts on the role of metal ions in RNA (and DNA) folding kinetics (see Table 4) at both secondary and tertiary structural levels [20,61–68].

Table 4. Measurements of the ion-dependent folding kinetics of RNAs and DNAs.

RNAs or DNAs	Refs.	Ionic conditions	Folding or unfolding rates
6-bp DNA duplex	[20]	Na^+, Mg^{2+}	folding rates
8-bp DNA duplex	[61]	Na^+	folding and unfolding rates
DNA duplex	[62]	Na^+	folding and unfold rates
DNA hairpins	[63]	Na^+	folding and unfolding rates
hairpin ribozyme	[64]	Na^+, Mg^{2+}	folding and unfolding rates
RNA three-way junction	[65]	Na^+, Mg^{2+}	folding and unfolding rates
Tetrahymena ribozyme	[66]	Mg^{2+}	R_g versus time
	[67]	Na^+, Mg^{2+}, Ba^{2+}	folding rate
	[67]	spermidine, $(Co(NH_3)_6)^{3+}$	folding rate
	[68]	Na^+, Mg^{2+}	folding rates to I and to N

I: the partially unfolded intermediate state; N: the native state.

3.1. Secondary Structural Folding Kinetics

Experiments on the kinetics of helix formation showed that addition of Na^+ ions would accelerate the folding process [20,61,62] and cause nearly no change to the unfolding rate [20,62]. The findings may be interpreted by a model for the kinetic barrier: The ion causes a decrease in the folding entropy loss and thus an increase in the folding rate (if the kinetic barrier for folding is caused by the entropic loss). In contrast, the loop enthalpy would not change with the ion and thus, the unfolding rate is unchanged (if the kinetic barrier for unfolding is caused by the enthalpic increase) [33,39]. In addition, the folding kinetic data suggests an approximate equivalence between 1 M Na^+ and 10 mM Mg^{2+} [20] in promoting folding helix formation.

A hairpin is the simplest secondary structure consisting of a helix and a loop, thus, its folding kinetics is associated with the kinetics of the helix as well as the loop formation [29–30,33]. If the hairpin formation is rate-limited by the slow formation of a base stack in the helix, the dependence of folding and unfolding rates would be similar to the one for helices as described above. If the hairpin folding is rate-limited by the loop formation, addition of metal ions would lead to the acceleration of the loop formation due to the enhanced loop stability and the chain flexibility and thus, a lower kinetic barrier. The unfolding rate decreases with the addition of metal ions probably due to the lowered enthalpy in the folded hairpin state [63].

Recently, a laser temperature-jump spectroscopy experiment showed that for higher ion concentration (e.g., >100 mM Na^+), the loop closure time is not sensitive to the salt concentration. For a low salt solution (e.g., <50 mM Na^+), however, a notable increase in loop closure time has been predicted theoretically [26]. The above ion dependence may arise from the saturation of ion binding at high ion concentrations.

3.2. Tertiary Structural Folding Kinetics

Experimental studies on RNA tertiary structural folding kinetics have provided novel insights into the role of metal ions in folding kinetics. Though these studies focused on different RNA molecules, such as hairpin ribozyme, specific three-way junctions and *Tetrahymena* ribozyme, many of the conclusions drawn from the studies are general.

3.2.1. *Hairpin Ribozyme*

The folding and unfolding rate constants of a minimal hairpin ribozyme have been measured using the fluorescence resonance energy transfer (FRET) method. The rate constants were measured with a broad range of ionic conditions, including Na^+, Mg^{2+}, and mixed Na^+/Mg^{2+} solutions [57,64]. When $[Mg^{2+}]$ increases from 1 mM to 500 mM, the folding rate increases by more than 40-fold, while the unfolding rate remains invariant with $[Mg^{2+}]$ in the range [2 mM, 500 mM] and decreases with $[Mg^{2+}]$ between 1 mM and 2 mM [64]. In contrast, adding Na^+ ions to a pure Na^+ solution or adding Mg^{2+} to a mixed Na^+/Mg^{2+} solution causes an increase in the folding rate and a decrease in the unfolding rate [64]. The distinctive ion dependence can be explained by the more significant ion binding and more compact RNA transition state in a Mg^{2+} solution than in a Na^+ solution. The transition state in a Mg^{2+} solution is highly compact and the folding from the transition state involves a small change in the compactness and no further uptake of ions. In contrast, the transition state in a Na^+ solution is less compact and the transition from the transition state to the folded state involves further uptake of ions.

3.2.2. *A Three-Way Junction*

For an RNA three-way junction in 16S rRNA, Mg^{2+} and protein binding are found to induce similar conformational switches from an extended Y-shaped (unfolded) state to a folded y-shaped state [65]. Therefore, understanding the role of Mg^{2+} may provide useful insights into the different forces that lead to a conformational switch. With the addition of Na^+, the folding rate increases and the unfolding rate decreases, thus, adding Na^+ favors the formation of the tertiary structure. Adding Mg^{2+} to a Na^+ solution causes similar qualitative changes in the rate constants as above for a Na^+ solution. The experiment shows that Mg^{2+} is highly efficient compared to Na^+, indicating the efficient role of Mg^{2+} in driving the formation of the folded (y-)state. When Mg^{2+} exceeds 100 mM, the folding/unfolding rates become saturated in the presence of 50 mM Na^+ [65].

3.2.3. *Tetrahymena* Ribozyme

Tetrahymena ribozyme has a large, complex native structure with multiple helices, cross-linked loops, and tertiary contacts (such as the tetraloop-receptor interaction). Extensive experimental findings suggest that the structural formation of the *Tetrahymena* ribozyme involves multi-state kinetics governed by a highly rugged energy landscape [54]. The folding rates and pathways are sensitive to the ionic condition of the solution [6]. For example, SAXS experiments with millisecond resolution indicate that Mg^{2+} ions induce a rapid collapse in 10 ms with the global compactness decreasing from 75 Å to 55 Å. With the formation of the tertiary contacts, further compaction to 45 Å is achieved at ~100 ms [66].

The ion-induced collapsed intermediate state is critical for the folding speed and kinetic partitioning of the folding pathways [4]. Native gel electrophoresis experiments [67] showed that higher $[Mg^{2+}]$ and metal ions of higher charge density (charge/volume) would significantly stabilize the non-specific collapsed intermediates. The stabilization of the non-specific collapsed state slows down the (productive) folding to the final native state which requires the formation and stabilization of specific native contacts [49,67]. In addition, a recent experiment showed that, depending on the ionic condition of the solution, the kinetic partitioning of the folding flux can be dependent on the initial conformational ensemble [68].

4. THEORETICAL PREDICTIONS FOR THE DIFFUSE ION BINDING TO RNAs

Several theoretical methods have been used to predict the ion distribution and ion-RNA interaction:

1. *Counterion condensation theory* [69,70]. The model successfully predicts the logarithmic dependence of the melting temperature for the DNA

helix on the ion concentration [33,69]. However its prediction on the ion-mediated helix-helix attraction [71] is not consistent with the experimental results [40,44] and computer simulations [72,73].

2. *Poisson-Boltzmann (mean-field) theory* [74–79]. The theory gives an accurate prediction for ion binding to RNA in aqueous/monovalent ion solutions [80]. However, the model, which ignores ion size and electrostatic correlation, cannot predict the multivalent ion-mediated attraction between helices [40–41,44] and underestimates the Mg^{2+}-mediated folding stability.
3. *Models beyond mean-field approximation*
   - *Tightly bound ion model* [81]. The model accounts for ion correlation, finite size, and ion fluctuation effects. Extensive tests against experiments show that the model is highly promising for treating the effects of multivalent (such as Mg^{2+}) diffuse ions in RNA folding; see below for more details.
   - *Size-modified Poisson-Boltzmann model* [82]. By considering the ion size effect, the model with the APBS PB solver [78] gives reliable predictions on monovalent ion binding and the ionic saturation effect at high ion concentration caused by the finite size of the ions. For multivalent ion solutions, such as of Mg^{2+}, further improvement of the model is required to account for the long-range inter-ion correlation.
   - *Modified Poisson-Boltzmann model based on the Kirkwood/Bogoliubov-Born-Green-Yvon hierarchy* [74,83,84]. The model considers the mean effect of ion fluctuation and size exclusion. Tests with the Monte Carlo simulations for simple macroion systems indicate that the model gives improved predictions for multivalent ion distributions than the conventional PB theory. However, the application of the model to treat realistic RNA structures remains computationally challenging [83].
   - *Correlation-corrected Poisson-Boltzmann model* [85]. With an *ad hoc* correlation-corrected inter-ion potential, the model successfully predicts an attractive force between the two planes [85]. Comparisons with computer simulations suggest that the model is able to generate improved predictions on multivalent ion distributions compared to the conventional PB theory. For realistic RNA structures, the model is computationally demanding.
   - *Other models beyond the mean-field approximation* [86–88]. Other theories, such as the density function theory [87], integration theory [86], and local molecular field theory [88], can account for the inter-ion correlation effects. However, the application of these methods to treat realistic three-dimensional RNA structures remains a challenge.

5. CORRELATED DISTRIBUTION OF MULTIVALENT DIFFUSE IONS: THEORY VERSUS EXPERIMENT

Multivalent diffuse bound ions may show a correlated distribution. Consider counterions clustering around a (negatively charged) RNA. Thermal agitation tends to cause disordered distributions of the counterions. However, the Coulombic force between charges tends to bring the ions into an ordered low-energy state. In such a low-energy state, an ion is "networked" ("correlated" or "coupled") with many other ions. For multivalent counterions such as Mg^{2+}, the large ionic charge may cause the Coulombic energy $U_{Coulomb}$ for an ion-pair to out-compete the thermal agitation energy ($k_BT = 1.99$ cal/mol; k is the Boltzmann constant and T is the temperature). As a result, ions can be in an ordered strongly correlated state. For a realistic RNA, such correlation could easily happen. For instance, [Mg^{2+}] near the yeast tRNA^{Phe} surface can be higher than 8 M [80], causing $U_{Coulomb} > 10$ k_BT and hence a strong correlation between the ions. Because of the correlation between the ions, the likelihood of finding an ion at a location is sensitive to the discrete locations of other ions. Therefore, the ion correlation effect is intrinsically tied to the ensemble of discrete ion distributions, i.e., "ion fluctuations" (see Figure 1). To replace the discrete (correlated) ion configurations by an average fluid-like continuous distribution is not appropriate.

Several experimental and theoretical studies have pointed to the potential importance of correlation/fluctuation effects for diffuse ions. For example, SAXS experiments for the ion-induced structural relaxation of DNA duplexes [43] showed that the Poisson-Boltzmann equation, which ignores the correlation effect, overestimates the [Mg^{2+}] midpoint for the transition (structural relaxation) by more than 10 times. Correlation is most likely the reason to cause the discrepancy between PB and the experiment. In addition, computer simulations for metal ions surrounding a model three-dimensional RNA structure showed a liquid-like correlated distribution for the Mg^{2+} ions [49].

Inspired by the potential importance of correlation and fluctuation effects for multivalent counterions such as Mg^{2+} ions, Tan and Chen developed the "Tightly Bound Ion" (TBI) model [81]. Through a classification of the diffuse ions according to the correlation strength, the model considers the different modes for the strongly correlated bound ions while treating the weakly correlated ions using the Poisson-Boltzmann theory. Applications of the TBI model to a broad range of nucleic acid systems such as RNA/DNA helices, RNA/DNA hairpins, DNA helix assembly, DNA bending, RNA three-way junction, and HIV-1 DIS kissing loop complexes [28–30,89–92] provide reliable theoretical predictions for the effect of diffuse ions, especially Mg^{2+} ions, in RNA folding.

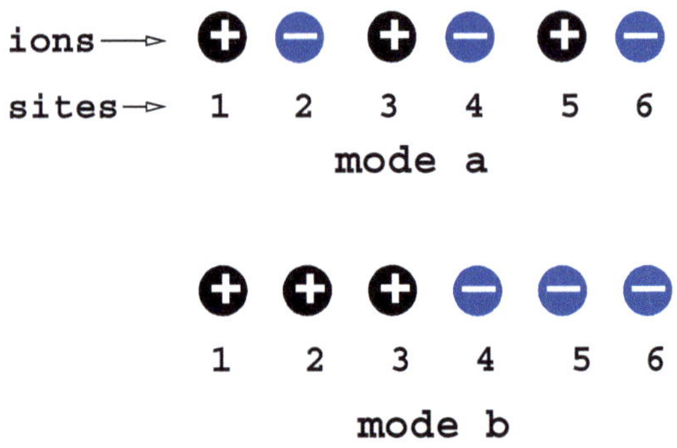

Figure 1. (i) Modes *a* and *b* have an identical net average charge (= 0, i.e., neutral), but mode *a* has a much lower energy than mode *b*, causing the system to have higher tendency to stay in mode *a*. It is important to distinguish the different discrete ion distributions and the use of net average "charge neutralization" may not be appropriate. (ii) The likelihood of finding a positive or negative charge at a site (such as site 2) depends on the discrete distribution (locations) of all the other ions instead of the ensemble average distribution of the ions. (iii) A strongly correlated state (such as mode *a*) can reach much lower energy than a "mean" state. So correlation lowers the energy. Such a correlation-induced low energy effect is more pronounced for a folded RNA structure due to its higher RNA backbone charge density, leading to correlation-enhanced stabilization of the folded structure. Therefore, the unusual efficiency of Mg^{2+} ions, which can be strongly correlated, in RNA stabilization may be attributed to the enhanced stabilization effect due to correlation.

Furthermore, TBI theory-experiment agreements for a wide variety of systems led to a set of highly useful formulas for the ion-dependent thermodynamic parameters such as the enthalpy ΔH, entropy ΔS, free energy ΔG, and the melting temperature of RNA. The following set of formulas give the ion-dependent stability (at 37 °C) for an RNA helix in a mixed Na^+/Mg^{2+} solution [29].

$$\Delta G^{\circ}_{37}[\mathrm{Na}^+/\mathrm{Mg}^{2+}] = \Delta G^{\circ}_{37}(1\ \mathrm{M}) + (N-1)(x_1\Delta g_1 + x_2\Delta g_2) + \Delta g_{12}$$

where x_1 and x_2 denote the fractional contributions of Na^+ and Mg^{2+} to the whole stability, respectively:

$$x_1 = [\mathrm{Na}^+]/([\mathrm{Na}^+] + (8.1 - 32.4/N)(5.2 - \ln[\mathrm{Na}^+])[\mathrm{Mg}^{2+}]);\ x_2 = 1 - x_1.$$

Here Δg_1, Δg_2, and Δg_{12} are given by

$$\begin{aligned}
\Delta g_1 &= a_1 + b_1/N \\
a_1 &= -0.075 \ln[\mathrm{Na}^+] + 0.012 \ln^2[\mathrm{Na}^+] \\
b_1 &= 0.018 \ln^2[\mathrm{Na}^+] \\
\Delta g_2 &= a_2 + b_2/N^2 \\
a_2 &= -0.6/N + 0.025 \ln[\mathrm{Mg}^{2+}] + 0.0068 \ln^2[\mathrm{Mg}^{2+}] \\
b_2 &= \ln[\mathrm{Mg}^{2+}] + 0.38 \ln^2[\mathrm{Mg}^{2+}] \\
\Delta g_{12} &= -0.6 x_1 x_2 \ln[\mathrm{Na}^+] \ln((1/x_1 - 1)[\mathrm{Na}^+])/N.
\end{aligned}$$

Through theory-experiment comparison, the TBI theory also gave a similar set of formulas for the ion-dependent hairpin stability [30].

More recently, a Monte Carlo simulation gives the results for the electrostatic free energy landscape for a system of the two nucleic acid helices [93] (see Figure 2A). In the absence of ions, the electrostatic free energy profile shows a large repulsion between the helices. For example, to push the 10-bp two helices from 40 Å to 22 Å (axis-axis distance) requires a work of $48\,k_BT$. With the addition of ions, such repulsion would be greatly reduced (see Figure 2A). In 0.1 M Na^+ solutions, the work is reduced to about 6.5 k_BT. In divalent ion solution, the ion effect becomes more pronounced. 10 mM of divalent ions can cause the work to be reduced to $0.5\,k_BT$. Moreover, the simulation shows a slight attractive force between the helices in a divalent ion solution. At 26 Å, 10 mM divalent ions can cause a stabilizing electrostatic free energy of $-0.3\,k_BT$.

How can the ions reduce the folding electrostatic repulsion so dramatically? For the helices immersed in ion solutions shown in the bottom panel of Figure 2B, metal ions are diffusively bound around the helices to lower the intramolecular electrostatic energy. When two helices approach each other, such ion binding becomes more pronounced due to the increased charge density for RNA backbone. The ion binding in the region between the two helices causes a significant reduction in the electrostatic repulsion. For multivalent ions, the high affinity of ion binding in the region between two helices causes a high concentration of the ion cloud and, hence, strong correlation between the ions. The correlation between the bound ions tends to bring the system to ordered low-energy states. For multivalent ions, the correlation effect out-competes the Coulomb repulsion. Such a correlation effect is stronger for a closer approach between the helices, resulting in an attractive force between the helices.

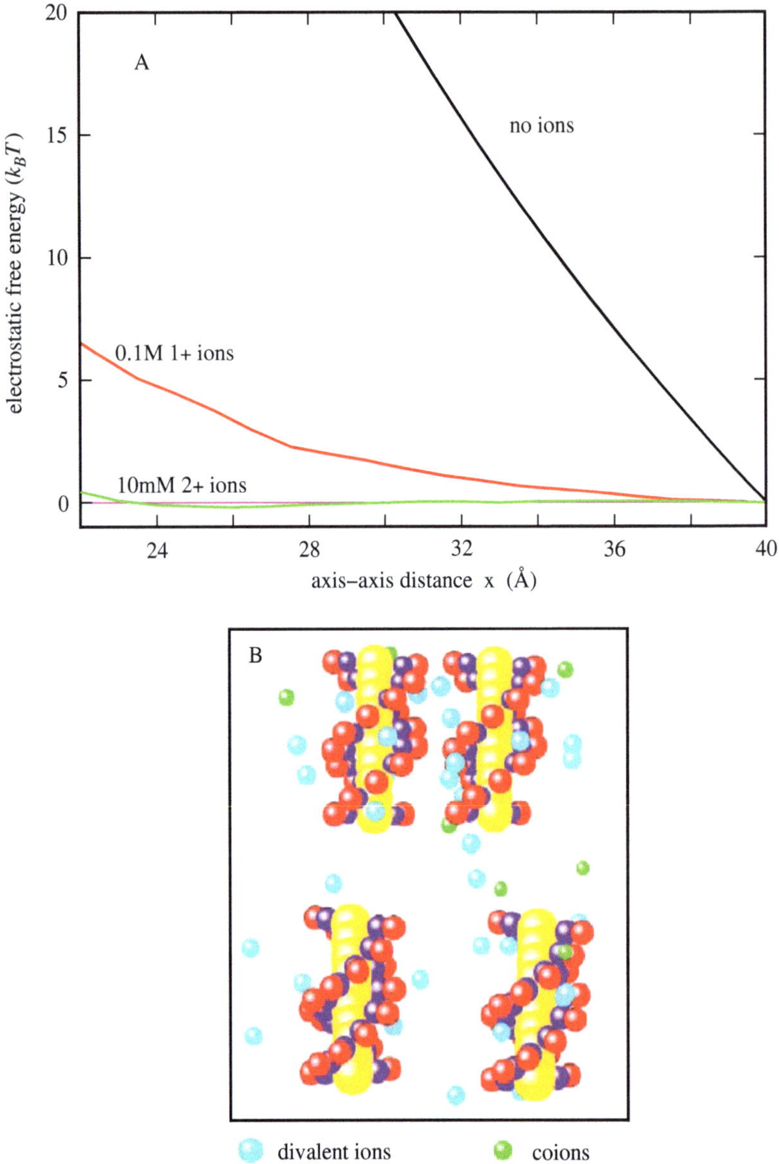

Figure 2. Illustration of the importance of diffuse ion binding in a two-helix system. **A.** The calculated electrostatic free energy $\Delta G(x) = G(x) - G(40\,\text{Å})$ as a function of the axis-axis distance x. **B.** The snapshots for the ion distributions around the helices at $x = 26\,\text{Å}$ (top panel) and $x = 38\,\text{Å}$ (bottom panel), respectively.

6. GENERAL CONCLUSIONS

Extensive theoretical and experimental investigations show that Mg^{2+} plays a role distinctive from that of Na^+, as described below:

1. Mg^{2+} is much more efficient than Na^+ in stabilizing RNA folded structures, including helix, hairpin, helix assembly, and tertiary folds. For the stabilization of secondary structural segments (helices, loops), 10 mM Mg^{2+} is approximately equivalent to 1 M Na^+. Such an efficient role of Mg^{2+} becomes more pronounced for more compact and larger molecules.
2. Mg^{2+} ions are essential for RNA to fold into the native tertiary structure. Insufficient Mg^{2+} would lead to partially folded intermediates or a near-native-like structure. In addition, the high efficiency of Mg^{2+} over Na^+ is much more pronounced for tertiary structural folding than for secondary structural folding. For the folding of the *Tetrahymena* ribozyme, 0.3 mM Mg^{2+} and 0.5 M Na^+ show similar efficiency in stabilizing the tertiary fold [55].
3. For a given RNA, Mg^{2+} ions induce a much more compact (tertiary) structure than Na^+ ions [50].
4. For ion-mediated nucleic acid helices, Na^+ can only modulate the repulsive interaction, while Mg^{2+} could induce a (weak) attractive force.
5. Due to the Mg^{2+}-induced stabilization of misfolded states and the non-specific collapsed state, the most efficient folding occurs likely in the mixture of monovalent Na^+ and Mg^{2+} instead of a pure highly concentrated Mg^{2+} solution [67].

The potential correlation between the bound ions may contribute to the distinctively high efficiency of multivalent ions such as Mg^{2+}. For multivalent ions, the correlation could be strong. The strong ion correlation drives the ions to self-organize (cooperatively) for reaching low-energy states. Such a correlation-induced bias toward the low-energy states enhances the stability of the (bound) ion clouds, which leads to a high affinity for ion binding and a high stability of the RNA structure.

Metal ions can bind to specific groups in RNA either directly or through water as a ligand. Such specific ion binding has been observed in many know structures. For example, the crystal structure of the P4-P6 domain of the group I intron contains 27 specifically bound Mg^{2+} ions [94]. These specifically bound ions can play essential roles in RNA-related catalysis. However, it is not clear how important these specifically bound ions are for RNA stability and folding kinetics. Nevertheless, based on extensive experiments, the following conclusions about specific binding of metal ions such as Mg^{2+}

can be drawn: First, because specific binding of Mg^{2+} has not been observed for many simple RNA systems that show the high efficiency of Mg^{2+}, the distinctively more efficient role of Mg^{2+} compared to Na^{+} is unlikely to stem from specific binding, though specifically bound Mg^{2+} ions can enhance the efficient role of Mg^{2+}. Second, the formation of specific ion binding often involves a high energy barrier arising from ion dehydration. In contrast, the diffuse ions are much more highly populated. Therefore, for most RNAs, the ion stabilizing force may be dominated by the diffuse ions.

In the past two decades, experimental and theoretical studies have provided quantitative insights into the ion (especially the Mg^{2+} ion) effects in RNA folding. However, due to the complexity of the RNA/water/ion system, many questions about the ion effects in RNA folding remain unanswered: Is the efficient role of Mg^{2+} from ion-ion correlation? Can Mg^{2+} induce RNA helix-helix attraction? Can such an attraction cause a compact collapsed state? Do the specifically bound Mg^{2+} ions make important contributions to the folding stability? The answers to these questions will likely come from further synergistic efforts of theory and experiment.

ACKNOWLEDGMENTS

This research was supported by the National Science Foundation (NSF) through grant MCB0920067 and by the National Science Foundation of China (NSFC) through grant No. 10844007 and 10774115, and by the program for "New Century Excellent Talents" in Universities of China (NCET) and the Chutian Scholar Program of Hubei Province.

ABBREVIATIONS

APBS	adaptive Poisson-Boltzmann solver
bp	base pair
BWYV	beet western yellow virus
DIS	dimerization initiation signal
FRET	fluorescence resonance energy transfer
HIV	human immunodeficiency virus
MMTV	mouse mammary tumor virus
nt	nucleotide
PB	Poisson-Boltzmann
Rg	radius of gyration
rRNA	ribosomal ribonucleic acid
SAXS	small angle X-ray scattering

Tar	transactivation response
TBI	tightly bound ion
T_m	melting temperature

REFERENCES

1. D. Bloomfield, M. Crothers and I. Tinoco Jr., *Nucleic Acids: Structure, Properties and Functions*, University Science Books, Sausalito, CA, 2000.
2. *Non-Protein Coding RNAs*, Ed. N. G. Walter, S. A. Woodson and R. T. Batey, Springer-Verlag, Berlin, Heidelberg, 2009.
3. Y. Bai, R. Das, I. S. Millett, D. Herschlag and S. Doniach, *Proc. Natl. Acad. Sci. USA*, 2005, **102**, 1035–1040.
4. S. A. Woodson, *Annu. Rev. Biophys.*, 2010, **39**, 61–77.
5. S. J. Chen, *Annu. Rev. Biophys.*, 2008, **37**, 197–214.
6. V. B. Chu and D. Herschlag, *Curr. Opin. Struct. Biol.*, 2008, **18**, 305–14.
7. D. E. Draper, *Biophys. J.* 2008, **95**, 5489–5495.
8. Z. J. Tan and S. J. Chen, *Methods Enzymol.*, 2009, **469**, 465–487.
9. R. Das, T. T. Mills, L. W. Kwok, G. S. Maskel, I. S. Millett, S. Doniach, K. D. Finkelstein, D. Herschlag and L. Pollack, *Phys. Rev. Lett.*, 2003, **90**, 188103.
10. K. Andresen, R. Das, H. Y. Park, H. Smith, L. W. Kwok, J. S. Lamb, E. J. Kirkland, D. Herschlag, K. D. Finkelstein and L. Pollack, *Phys. Rev. Lett.*, 2004, **93**, 248103.
11. K. Andresen, X. Qiu, S. A. Pabit, J. S. Lamb, H. Y. Park, L. W. Kwok and L. Pollack, *Biophys. J.*, 2008, **95**, 287–295.
12. S. A. Pabit, X. Qiu, J. S. Lamb, L. Li, S. P. Meisburger and L. Pollack, *Nucleic Acids Res.*, 2009, **37**, 3887–3896.
13. Y. Bai, M. Greenfeld, K. J. Travers, V. B. Chu, J. Lipfert, S. Doniach and D. Herschlag, *J. Am. Chem. Soc.*, 2007, **129**, 14981–14988.
14. H. Krakauer, *Biopolymers*, 1971, **10**, 2459–2490.
15. R. M. Clement, J. Sturm and M. P. Daune, *Biopolymers*, 1973, **12**, 405–421.
16. M. Soto, V. Misra and D. E. Draper, *Biochemistry*, 2007, **46**, 2973–2983.
17. D. Grilley, V. Misra, G. Caliskan and D. E. Draper, *Biochemistry*, 2007, **46**, 10266–10278.
18. G. Rialdi, J. Levy and R. Biltonen, *Biochemistry*, 1972, **11**, 2472–2479.
19. R. Romer and R. Hach, *Eur. J. Biochem.*, 1975, **55**, 271–284.
20. P. Williams, C. E. Longfellow, S. M. Freier, R. Kierzek and D. H. Turner, *Biochemistry*, 1989, **28**, 4283–4291.
21. S. Nakano, M. Fujimoto, H. Hara and N. Sugimoto, *Nucleic Acids Res.*, 1999, **27**, 2957–2965.
22. N. Sugimoto, P. Wu, H. Hara and Y. Kawamoto, *Biochemistry*, 2001, **40**, 9396–9405.
23. K. Rippes, N. B. Ramsing and T. M. Jovin, *Biochemistry*, 1989, **28**, 9536–9541.
24. M. J. Serra, J. D. Baird, T. Dale, B. L. Fey, K. Retatagos and E. Westhof, *RNA*, 2002, **8**, 307–323.

25. R. Owczarzy, Y. You, B. G. Moreira, J. A. Manthey, L. Huang, M. A. Behlke and J. A. Walder, *Biochemistry*, 2004, **43**, 3537–3554.
26. S. V. Kuznetsov, C. C. Ren, S. A. Woodson and A. Ansari, *Nucleic Acids Res.*, 2008, **36**, 1098–1112.
27. J. Vieregg, W. Cheng, C. Bustamante and I. Tinoco, Jr., *J. Am. Chem. Soc.*, 2007, **129**, 14966–14973.
28. Z. J. Tan and S. J. Chen, *Biophys. J.*, 2006, **90**, 1175–1190.
29. Z. J. Tan and S. J. Chen, *Biophys. J.*, 2007, **92**, 3615–3632.
30. Z. J. Tan and S. J. Chen, *Biophys. J.*, 2008, **95**, 738–752.
31. P. L. Nixon, C. A. Theimer and D. P. Giedroc, *Biopolymers*, 1999, **50**, 443–458.
32. P. L. Nixon and D. P. Giedroc, *Biochemistry*, 1998, **37**, 16116–16129.
33. J. SantaLucia, Jr., *Proc. Natl. Acad. Sci. USA*, 1998, **95**, 1460–1465.
34. M. Zuker, *Nucleic Acids Res.*, 2003, **31**, 3406–3415.
35. D. H. Mathews and D. H. Turner, *Curr. Opin. Struct. Biol.*, 2006, **16**, 270–278.
36. S. J. Chen and K. A. Dill, *Proc. Natl. Acad. Sci. USA*, 2000, **97**, 646–651.
37. W. B. Zhang and S. J. Chen, *Proc. Natl. Acad. Sci. USA*, 2002, **99**, 1931–1936.
38. J. Santalucia, Jr. and D. Hicks, *Annu Rev Biophys Biomol Struct*, 2004, **33**, 415–440.
39. W. B. Zhang and S. J. Chen, *Biophys J.*, 2006, **90**, 778–787.
40. D. C. Rau and V. A. Parsegian, *Biophys. J.*, 1992, **61**, 246–259.
41. D. C. Rau and V. A. Parsegian, *Biophys. J.*, 1992, **61**, 260–271.
42. C. Ma and V. A. Bloomfield, *Biophys. J.*, 1994, **67**, 1678–1681.
43. Y. Bai, V. B. Chu, J. Lipfert, V. S. Pande, D. Herschlag and S. Doniach, *J. Am. Chem. Soc.*, 2008, **130**, 12334–12341.
44. X. Qiu, K. Andresen, L. W. Kwok, J. S. Lamb, H. Y. Park and L. Pollack, *Phys. Rev. Lett.*, 2007, **99**, 038104.
45. X. Qiu, K. Andresen, J. S. Lamb, L. W. Kwok and L. Pollack, *Phys. Rev. Lett.*, 2008, **101**, 228101.
46. Y. V. Bukhman and D. E. Draper, *J. Mol. Biol.*, 1997, **273**, 1020–1031.
47. A. Theimer and D. P. Giedroc, *RNA*, 2000, **6**, 409–421.
48. H. Qiu, K. Kaluarachchi, Z. Du, D. W. Hoffman and D. P. Giedroc, *Biochemistry*, 1996, **35**, 4176–4186.
49. E. Koculi, C. Hyeon, D. Thirumalai and S. A. Woodson, *J. Am. Chem. Soc.*, 2007, **129**, 2676–2682.
50. K. Takamoto, Q. He, S. Morris, M. R. Chance and M. Brenowitz, *Nature Struct. Biol.*, 2002, **9**, 928–933.
51. K. Takamoto, R. Das, Q. He, S. Doniach, M. Brenowitz, D. Herschlag and M. R. Chance, *J. Mol. Biol.*, 2004, **343**, 1195–1206.
52. S. Moghaddam, G. Caliskan, S. Chauhan, C. Hyeon, R. M. Briber, D. Thirumalai and S. A. Woodson, *J. Mol. Biol.*, 2009, **393**, 753–764.
53. G. Caliskan, C. Hyeon, U. Perez-Salas, R. M. Briber, S. A. Woodson and D. Thirumalai, *Phys. Rev. Lett.*, 2005, **95**, 268303.
54. R. Russell, I. S. Millett, S. Doniach and D. Herschlag, *Nature Struct. Biol.*, 2000, **7**, 367–370.
55. S. L. Heilman-Miller, D. Thirumalai and S. A. Woodson, *J. Mol. Biol.*, 2001, **306**, 1157–1166.

56. D. Lambert, D. Leipply, R. Shiman and D. E. Draper, *J. Mol. Biol.*, 2009, **390**, 791–804.
57. N. G. Walter, J. M. Burke and D. P. Millar, *Nature Struct. Biol.*, 1999, **6**, 544–549.
58. G. Pljevaljcic, D. P. Millar and A. A. Deniz, *Biophys. J.*, 2004, **87**, 457–467.
59. A. Weixlbaumer, A. Werner, C. Flamm, E. Westhof and R. Schroeder, *Nucleic Acids Res.*, 2004, **32**, 5126–5133.
60. C. Lorenz, N. Piganeau and R. Schroeder, *Nucleic Acids Res.*, 2006, **34**, 334–342.
61. W. H. Braunlin and V. A. Bloomfield, *Biochemistry*, 1991, **30**, 754–758.
62. D. Porschke, O. C. Uhlenbeck and F. H. Martin, *Biopolymers*, 1973, **12**, 1313–1335.
63. G. Bonnet, O. Krichevsky and A. Libchaber, *Proc. Natl. Acad. Sci. USA*, 1998, **95**, 8602–8606.
64. G. Bokinsky, D. Rueda, V. K. Misram, M. M. Rhodes, A. Gordus, H. P. Babcock, N. G. Walter and X. Zhuang, *Proc. Natl. Acad. Sci. USA*, 2003, **100**, 9302–9307.
65. D. Kim, G. U. Nienhaus, T. Ha, J. W. Orr, J. R. Williamson and S. Chu, *Proc. Natl. Acad. Sci. USA*, 2002, **99**, 4284–4289.
66. L. W. Kwok, I. Shcherbakova, J. S. Lamb, H. Y. Park, K. Andresen, H. Smith, M. Brenowitz and L. Pollack, *J. Mol. Biol.*, 2006, **355**, 282–293.
67. S. L. Heilman-Miller, J. Pan, D. Thirumalai and S. A. Woodson, *J. Mol. Biol.*, 2001, **309**, 57–68.
68. A. Laederach, I. Shcherbakova, M. A. Jonikas, R. B. Altman and M. Brenowitz, *Proc. Natl. Acad. Sci. USA*, 2007, **104**, 7045–7050.
69. G. S. Manning, *Q. Rev. Biophys.*, 1978, **11**, 179–246.
70. M. J. Schurr, in *Nucleic Acid-Metal Ion Interactions*, ed. N. V. Hud, Royal Society of Chemistry, Cambridge, UK, 2008, pp. 307–344.
71. J. Ray and G. S. Manning, *Macromolecules*, 2000, **33**, 2901–2908.
72. P. Lyubartsev and L. Nordenskiold, *J. Phys. Chem.*, 1995, **99**, 10373–10382.
73. B. Luan and A. Aksimentiev, *J. Am. Chem. Soc.*, 2008, **130**, 15754–15755.
74. S. Carnie and G. M. Torrie, *Adv. Chem. Phys.*, 1984, **56**, 141–253.
75. M. K. Gilson, K. A. Sharp and B. Honig, *J. Comput. Chem.*, 1987, **9**, 327–335.
76. T. J. You and S. C. Harvey, *J. Comput. Chem.*, 1993, **14**, 484–501.
77. H. Boschitsch and M. O. Fenley, *J. Comput. Chem.*, 2007, **28**, 909–921.
78. N. A. Baker, D. Sept, S. Joseph, M. J. Holst and J. A. McCammon, *Proc. Natl. Acad. Sci. USA*, 2000, **98**, 10037–10041.
79. Y. C. Zhou, M. Feig and G. W. Wei, *J. Comput. Chem.*, 2008, **29**, 87–97.
80. V. K. Misra, R. Shiman and D. E. Draper, *Biopolymers*, 2003, **69**, 118–136.
81. Z. J. Tan and S. J. Chen, *J. Chem. Phys.*, 2005, **122**, 044903.
82. V. B. Chu, Y. Bai, J. Lipfert, D. Herschlag and S. Doniach, *Biophys. J.*, 2007, **93**, 3202–3209.
83. S. Gavryushov, *J. Phys. Chem. B*, 2008, **112**, 8955–8965.
84. P. Grochowski and J. Trylska, *Biopolymers*, 2008, **89**, 93–113.
85. J. Forsman, *J. Phys. Chem. B*, 2004, **108**, 9236–9245.
86. V. Vlachy, *Annu. Rev. Phys. Chem.*, 1999, **50**, 145–165.
87. K. Wang, Y. X. Yu and G. H. Gao, *J. Chem. Phys.*, 2008, **128**, 185101.

88. Y. G. Chen and J. D. Weeks, *Proc. Natl. Acad. Sci. USA*, 2006, **103**, 7560–7565.
89. Z. J. Tan and S. J. Chen, *Biophys. J.*, 2006, **91**, 518–536.
90. Z. J. Tan and S. J. Chen, *Nucleic Acids Res.*, 2006, **34**, 6629–6639.
91. Z. J. Tan and S. J. Chen, *Biophys. J.*, 2008, **94**, 3137–3149.
92. G. Chen, Z. J. Tan and S. J. Chen, *Biophys. J.*, 2010, **98**, 111–120.
93. Y. Y. Wu, F. H. Wang and Z. J. Tan, 2010, unpublished.
94. J. H. Cate, A. R. Gooding, E. Podell, K. Zhou, B. L. Golden, C. E. Kundrot, T. R. Cech and J. A. Doudna, *Science*, 1996, **273**, 1678–1685.

Met. Ions Life Sci. **2011**, *9*, 125–139

4

RNA Quadruplexes

Kangkan Halder and Jörg S. Hartig
Department of Chemistry, University of Konstanz, Universitätsstrasse 10, D-78457 Konstanz, Germany
<joerg.hartig@uni-konstanz.de>

ABSTRACT: Opposed to DNA quadruplex sequences, RNA quadruplexes are still less well characterized. On the other hand, RNA quadruplexes are found to be at least as stable as their DNA counterparts. They show the same dependence on metal ions but seem to be much more restricted with respect to the adopted conformations. Other than DNA, which is mostly found to be double-stranded inside cells, RNAs are produced during transcription without its complementary sequence. The absence of a second

Metal Ions in Life Sciences, Volume 9 Edited by Astrid Sigel, Helmut Sigel, and Roland K. O. Sigel

Published by the Royal Society of Chemistry, www.rsc.org DOI: 10.1039/978184973251200125

strand that is able to hybridize and form a duplex makes the folding of RNA quadruplexes a likely event of intramolecular structure formation. Consequently, the formation of RNA quadruplexes in cellular RNAs has recently been suggested and the study of their influence and potential roles in cellular processes has just started. Here we give an overview of the RNA quadruplex field, summarizing issues such as structures, stabilities, and anticipated roles of these interesting four-stranded, guanosine-rich sequences.

KEYWORDS: gene expression · G-quadruplex · G-quartet · guanosine · potassium · transcription · translation · telomere

1. INTRODUCTION TO RNA QUADRUPLEXES

As early as 1962, it was proposed that four guanine monophosphate dianions can assemble coplanarly stabilized by two hydrogen bonds per base [1], later termed as G-tetrad or G-quartet (Figure 1A). These tetramolecular arrangements then prefer to consecutively stack on top of each other due to strong unpolar attractions resulting in a four-stranded helical conformation known as the G-quadruplex motif (Figure 1B). Similar multi-stranded arrangements were proposed for poly-guanylic acid sequences stabilized by Na^+ ions, shortly thereafter in 1963 [2].

Though initially studied for structural information, in the last two decades these higher order nucleic acid conformations have been proposed to be involved in various biological functions such as gene regulation [3,4], nucleosome positioning [5,6], recombination [7], and genomic maintenance [8]. However, studies proving their *in vivo* existence are limited [9,10]. This is

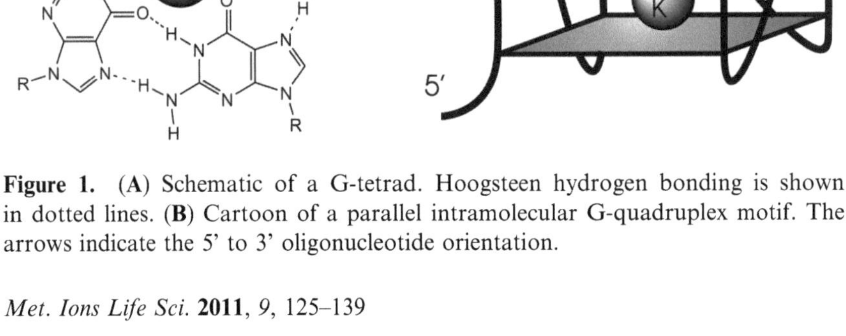

Figure 1. (**A**) Schematic of a G-tetrad. Hoogsteen hydrogen bonding is shown in dotted lines. (**B**) Cartoon of a parallel intramolecular G-quadruplex motif. The arrows indicate the 5' to 3' oligonucleotide orientation.

primarily due to the general acceptance that duplex DNA is the thermodynamically favorable state compared to the G-quadruplex motif in the presence of a complementary strand [11,12]. The single-stranded 3'-overhang of the telomeric repeat, which has been shown to adopt G-quadruplex structures *in vitro* [13], is the most studied G-quadruplex-forming sequence. However, it has been shown that 'capping' of the telomere end involves structures like the 't-loop' [14] as well as protein-ssDNA interactions [15] which are compromised by G-quadruplex formation. On the other hand, single stranded mRNA sequences harboring guanosine-rich stretches could also potentially form four-stranded structures *in vivo*. Proteins capable of inducing or stabilizing DNA and RNA G-quadruplexes have been described, which in turn may have functional relevance [16]. Here we primarily discuss RNA quadruplexes and their biological significance.

2. THERMODYNAMIC STABILITY

2.1. Role of Cations

The G-quartet conformation results in four carbonyl oxygen atoms clustered towards the center (Figure 1A) which necessitates a cation to effectively neutralize the partial negative charge [17]. The size of the inner cavity is an important factor to match with the ionic radii of the cation for suitable coordination interaction. The K^+ ion (ionic radius of 1.33 Å) is the most suitable and studied cation. Interestingly, it is also the monovalent cation with the highest intercellular concentration. However, a wide range of cations has been tested for its role in DNA G-quadruplex formation and stability. The general pattern of stability mediated by monovalent and divalent cations has been found as $K^+ \gg Na^+ \geq Rb^+ > Cs^+ \geq Li^+$ [18,19] and $Sr^{2+} \gg Ba^{2+} > Ca^{2+} > Mg^{2+} > Mn^{2+} > Co^{2+} > Zn^{2+}$ [20–22], respectively. Though a similar thermodynamic hierarchy is expected for RNA quadruplexes, only a few studies are known (Table 1) and further studies are required to substantiate metal ion dependency.

The high affinity of K^+ ions in RNA quadruplexes is exemplified by 10 to 30 °C higher increases in melting temperature ($T_{1/2}$) compared with Na^+ ions under similar conditions [23–25]. The sensitivity towards the cation concentration is well documented for DNA quadruplexes and similar results were found for RNA quadruplexes too, though limited to a few studies [23,25]. A gradual increase of the K^+ ion concentration was found to consistently elevate the melting temperatures ($T_{1/2}$). At physiological concentrations as high as 150 mM K^+ some quadruplex motifs are resistant to denaturing even at 90 °C. It is also worth mentioning that supplementing

Table 1. Melting temperatures of RNA G-quadruplexes.

	RNA			DNA[a]		
G-quadruplex sequence (5′→3′)[a]	$T_{1/2}$ (°C)	Salt (mM)	Ref.	$T_{1/2}$ (°C)[b]	Salt (mM)	Ref.
UAGGGUUAGGGU	63.3; 38.8, 41	K^+ (70); Na^+ (70, 100)	[23], [39]			
U_nGGGU_n (n = 1–3)	50-89	Na^+ (110)	[38], [42]			
UGGAGGU	>90	K^+ (150)	[41]			
$(GGGU)_3GGG$	>90; 73	K^+ (10, 100); Na^+ (100)	[29], [24]		K^+ (100); Na^+ (100)	[24]
$(GGGUU)_3GGG$	73	K^+ (10)	[29]			
$(GGGUUU)_3GGG$	61	K^+ (10)	[29]			
$(GGGUUA)_3GGG$	73; 42	K^+ (100); Na^+ (100)	[24]	65; 47	K^+ (100); Na^+ (100)	[24]
$A(GGGUUA)_3GGGUU$	76	K^+ (100)	[26]	63	K^+ (100)	
$(GGGGUUUU)_3GGG$	>90; 60	K^+ (100); Na^+ (100)	[24]	>90; 65	K^+ (100); Na^+ (100)	[24]
$(GGU)_3GG$	63; 31	K^+ (100); Na^+ (100)	[24]	58; <20	K^+ (100); Na^+ (100)	[24]
GGUUGGUGUGGUUGG	54; <20	K^+ (100); Na^+ (100)	[24]	52; <20	K^+ (100); Na^+ (100)	[24]
GGGUGGGAAGGAGGGUGGGCAUGGG	>90	K^+ (100)	[30]			
GGGUUGGGAAGGAGGGUUGGGAUGGG	78.7	K^+ (100)	[30]			
GGUGGAAGGAGGUGGUUCAUGGG	56.5	K^+ (100)	[30]			
GGUUGGAAGGAGGUUGGUCATGGG	53.7	K^+ (100)	[30]			
GGGUGGGGGGGCGGGGGAGGCCGGGG	79	K^+ (25)	[52]			
GGGAGGGGCGGGUCUGGG			[25]			
UGUGGGGAGGGGCGGGGUCUGGGG	80	K^+ (10)	[25]			
UGUGGGAGGGGCGGGUCUGGG	74	K^+ (10)	[25]			
UGUGGUAGGGUCGGUUCUGGU	45	K^+ (100)	[25]			

[a]DNA oligodeoxynucleotides have T instead of U in the G-quadruplex forming sequence.
[b]Melting temperatures with most similar conditions are reported. The table is not exhaustive.

the buffer with osmolytes (e.g., ethylene glycol), which mimic physiologic conditions better than buffer alone, further increase $T_{1/2}$, showing that supramolecularly crowded environments have stabilizing effects on RNA quadruplex formation [26]. Taken together, formation of an RNA quadruplex structure is very likely under physiological conditions unless other factors that competitively stabilize the unstructured conformation are present.

2.2. Role of Loop Size and Composition

The number and composition of nucleotides in the three loops of intramolecular DNA G-quadruplexes are crucial factors for the preferred topology as well as stability [27,28], however, a systematic study is lacking for RNA quadruplexes. We recently determined the melting temperatures ($T_{1/2}$) of various RNA quadruplexes with different loop sizes and composition [24,29,30]. It shows that the thermodynamic stability is inversely correlated with the loop size, a finding that matches the behavior of similar DNA sequences (Table 1). Moreover, stacking of three G-tetrads is found to be considerably more stable than two G-tetrads, which is expected due to additional stacking interactions in the former conformation [24,30].

2.3. Comparison between DNA and RNA Quadruplexes

It is interesting to compare stability parameters between DNA and RNA quadruplexes especially since various genomic G-quadruplex forming regions are transcribed. In general, slightly increased $T_{1/2}$ values were found for RNA quadruplexes compared to the corresponding DNA counterparts (Table 1). The disparity is believed to arise from better stacking of G-tetrads and the 2'-OH group is able to form additional intramolecular hydrogen bonds in RNA quadruplexes [31]. Moreover, the stability could also result from higher-order scaffolding of RNA quadruplexes [32–34]. Recently, the hydration state has also been implicated in the determination of thermodynamics of DNA quadruplex formation [35]. The formation of the compact structure was found to be accompanied with release of water molecules [36,37] and a similar role is expected in RNA quadruplexes, though to a different extent. Along these lines 6.0 ± 0.5 water molecules were found to be liberated during RNA quadruplex formation of r[A(GGGUUA)$_3$GGGUU] while only 4.0 ± 0.4 water molecules were released during the corresponding DNA quadruplex formation [26].

3. CONFORMATIONAL VARIATIONS

Although there are only few high resolution structures available for RNA quadruplexes (Table 2), they are invariably found to adopt a parallel conformation [23,34,38–42] (Table 3). This is an interesting finding since the DNA counterparts can adopt several different topologies. The reason is believed to be due to the stronger preference of an *anti* conformation of the glycosidic bond in ribonucleotides [43,44]. While in DNA quadruplexes containing antiparallel strand orientations at least some of the glycosidic bonds need to adopt *syn* conformations, the all-parallel-oriented, so called propeller conformation adopted by RNA quadruplexes, only allows *anti* orientations of ribose and guanine nucleobases. For example, the solution structure reported for the human telomeric RNA sequence r[(UAGGG(UUAGGG)$_3$] in the presence of K^+ has been shown to form this all-parallel conformation [23]. This is in contrast to its DNA counterpart, d[AGGG(TTAGGG)$_3$], which has been reported to form parallel as well as antiparallel conformations in both Na^+ and K^+ solutions [45–48]. The 12 nucleotide smaller fragment, r(UAGGGUUAGGGU), also formed a parallel conformation as a dimer in the presence of K^+ [23] and Na^+ [39] ions. Considering other RNA quadruplex sequences (Table 3), the parallel conformation seems universal for RNA quadruplexes. Importantly, this fact could be exploited for differential targeting of DNA and RNA quadruplexes for potential therapeutic purposes.

4. BIOLOGICAL FUNCTION

The biophysical studies documenting thermodynamic stability of RNA quadruplexes near physiological conditions point to the existence of such structures inside cells. Though partially less stable than RNA quadruplexes, as early as 1994 the telomeric DNA quadruplex was found to interfere with

Table 2. Structure of RNA G-quadruplexes.

PDB Id	G-quadruplex sequence ($5' \rightarrow 3'$)	Method	Ref.
2KBP	UAGGGUUAGGGU	NMR	[23]
2GRB	(U^{33})GIGGU	X-Ray	[32]
1MY9	GGAGGUUUUGGAGG	NMR	[34]
1J8G	UGGGGU	X-Ray	[40]
1RAU	UGGGGU	NMR	[68]
2RQJ	GGAGGAGGAGGA	NMR	[69]

Table 3. Conformations adopted by RNA quadruplexes.

	G-quadruplex sequence (5′→3′)	Conformation	Salt (mM)	Ref.
Intramolecular	UAGGGUUAGGGUUAGGGUUAGGG	all-parallel[a]	K^+ (70)	[23]
	UUAGGGUUAGGGUUAGGGUUAGGG	all-parallel	Na^+ (100)	[39]
	GGGUGGGAAGGAGGGUGGGCAUGGG	all-parallel	K^+ (100)	[30]
	GGGUUGGGAAGGAGGGUUGGGAUGGG	all-parallel	K^+ (100)	[30]
	GGUGGAAGGAGGUGGUUCAUGGG	all-parallel	K^+ (100)	[30]
	GGUUGGAAGGAGGUUGGUCATGGG	all-parallel	K^+ (100)	[30]
	$GGGU_nGGGU_nGGGU_nGGG$ (n = 1–3)	all-parallel	K^+ (10)	[29]
	GGGUGGGGGGGCGGGGGAGGCCGGGG	all-parallel	K^+ (25)	[52]
	$(UUAGGG)_{96}$	all-parallel	K^+ (100)	[62]
Intermolecular	UAGGGUUAGGGU	all-parallel	K^+ (70); Na^+ (100)	[23]; [39]
	UGGGGU	all-parallel	Sr^{2+} (40), Ca^{2+} (20), Li^+ (80), Mg^{2+} (20), Na^+ (40)	[40]
	UGGAGGU	all-parallel	K^+ (150)	[41]

[a]As determined by CD spectroscopy (approx. at ~265 nm positive and 240 nm negative peak)

telomerase recognition and activity (telomeric repeat amplification protocol, TRAP assay) [49] and since then the method has been extensively used to evaluate quadruplex formation and stability. A similar scenario could be easily perceived for RNA quadruplexes in mRNAs too; wherein these compact structures would hinder the assembly or scanning of proteins associated with the translation initiation complex. Alternatively, it is also possible that such distinct nucleic acid structures act as recognition sites for specific proteins in regulatory processes.

4.1. RNA Quadruplexes as Translation Inhibitors

The unusual stability of these RNA structures could possibly interfere with the ribosomal assembly and activity, especially when found in 5'-untranslated regions (5'-UTRs, see Figure 2A). Computational searches found a high prevalence of sequences potentially forming RNA quadruplexes in

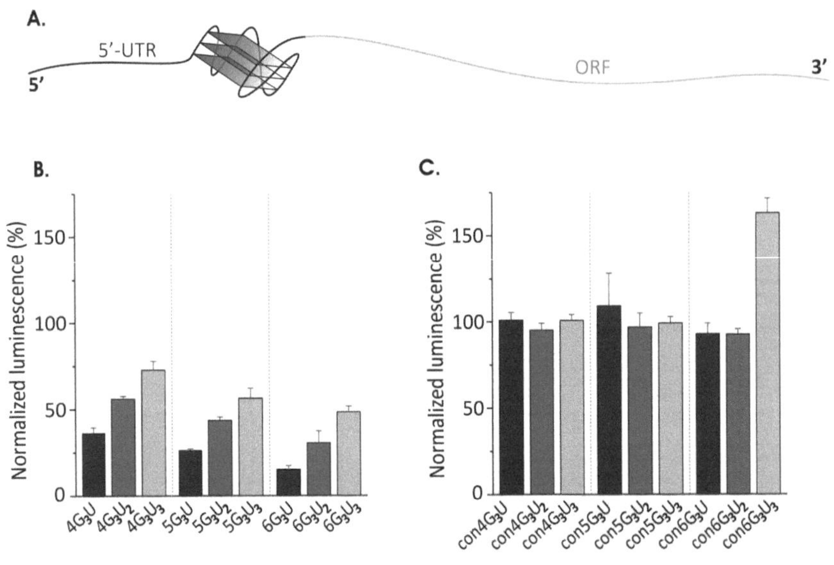

Figure 2. (**A**) Schematic representation of a parallel RNA quadruplex in a 5'-UTR of a mRNA. (**B**) Normalized luciferase luminescence of various 5'-UTRs containing RNA quadruplexes and (**C**) respective control sequences. The normalized luminescence of the wild-type luciferase (not shown) is set to 100% in the above scale. The prefix 4, 5 or 6 in the 'plasmid construct name' represents the total number of GGG repeats in the RNA quadruplex interspersed by 1, 2 or 3 (y) U's. A further prefix of 'con' abbreviates the control sequences, wherein three G positions were mutated to U, such that quadruplex structure formation is compromized [29].

5'-UTRs of human genes [25,29,50,51]. This has prompted us and other groups to investigate the influence of such structures at these positions in detail. Balasubramanian's group first reported that the presence of an RNA quadruplex in the 5'-UTR of the *NRAS* mRNA exhibits an inhibitory effect on translation [50]. The effect was found to be more pronounced when the quadruplex was located near the 5'-end [25], though *in vitro* transcription and *in vitro* translation methods were used to study the influence on gene expression. Shortly thereafter, an *in vivo* functional characterization of an RNA quadruplex in the *ZIC-1* mRNA also concluded a similar inhibitory effect on translation [52]. The above studies were limited to two specific RNA quadruplex motifs. We took a more systematic approach and incorporated an array of G-quadruplexes with varying loop sizes (1 to 3 nucleotides) and GGG-repeats (4 to 6 repeats) along with G to U mutant sequences as structural controls that should disrupt quadruplex formation [29].

The study was initiated in order to investigate whether there is any correlation of the thermodynamic stability of an RNA quadruplex and its effect on gene expression. We found that a gradual increase of loop sizes from 1 to 3 uridines in $4G_3U_y$ RNA quadruplexes is accompanied by thermodynamic destabilization (Table 1). Interestingly, when incorporated into mRNAs the activity of gene expression increased in this series (Figure 2A). In other words, the more the RNA quadruplex is thermodynamically stable, the higher is the level of repression of gene expression. Analogous results were found for the series $5G_3U_y$ and $6G_3U_y$. Additionally, it was found that quadruplexes containing more than the minimal required four G-rich repeats (such as $5G_3U$ and $6G_3U$) are thermodynamically more stable than the corresponding four-repeat sequence ($4G_3U$). Accordingly, $6G_3U$ incorporated into mRNAs displayed less gene expression compared to $5G_3U$ and $4G_3U$. The specificity and correlation of expression with quadruplex formation was confirmed when a minimal change of three G to U mutations for each respective quadruplex fully recovered gene expression (Figure 2B). Apart from these designed, very symmetrical quadruplexes, we also incorporated naturally occurring 5'-UTR RNA quadruplexes with different loop compositions and lengths into our constructs, and the observed expression matched those of similarly composed synthetic constructs. This result indicates that the level of translation inhibition is proportional to structure stability and can be tuned with changes in loop composition and length. In a next step we quantified the levels of reporter gene mRNA for various constructs in order to decipher whether the inhibition of gene expression takes place at the transcriptional or posttranscriptional/translational level. We found that the incorporation of quadruplexes into the 5'-UTR of the reporter genes has no influence on mRNA abundance, ruling out a potential effect on transcription or mRNA stability and indicating that the inhibition of expression takes place at the level of translation initiation. We also found that the insertion position of the G-quadruplex sequence within the 5'-UTR has an effect on the level of suppression [29], a finding that has been reported before but with slightly different position dependencies [25]. However, further studies are necessary to decipher the exact mechanism of how quadruplex formation affects gene expression on the post-transcriptional level.

4.2. RNA Quadruplexes in Post-Transcriptional Regulation

In addition to the occurrence of quadruplexes in 5'-UTRs, *in silico* searches found approximately 3000 potential RNA quadruplexes within the first intron of all human RefSeq genes which show a remarkable degree of conservation [53]. Hence, RNA quadruplexes have been suspected to play a potential role in RNA processing reactions such as splicing. In addition, endonucleolytic cleavage near the 3'-end of the insulin-like growth factor II (IGF-II) mRNA has been found conserved across multiple alternatively spliced variants [54]. Interestingly, in the vicinity of this cleavage site a conserved RNA quadruplex sequence is found [55]. It was suggested that the RNA quadruplex formation may indirectly facilitate the cleavage reaction. Furthermore, dimerization of the human immunodeficiency virus type 1 (HIV-1) RNA genome during the late stage of viral replication was found to be mediated by an RNA quadruplex structure [56].

4.3. Telomeric RNA

Human telomeres consist of tandem 5'-GGGTTA-3' repeats and have been shown to form G-quadruplex motifs under *in vitro* conditions [57,58]. Interestingly, due to the absence of protein-coding genes and the heterochromatin nature of the telomeric and subtelomeric regions, it was believed that telomeres are transcriptionally inactive. In particular, an *eGFP* gene when inserted at the subtelomeric region was found to be silenced, confirming a repressive chromatin architecture [59]. However, in a recent northern blot analysis of whole-cell RNA using a telomere specific probe, a $\sim$100–9,000 bp long telomeric repeat-containing RNA (termed TERRA) was found [60], indicating that telomeres are actively transcribed. Moreover, RNA-FISH (fluorescence *in situ* hybridization) experiments revealed that TERRA is co-localized with the telomere region [60]. Recently, the telomeric RNA repeat has also been shown to form an all-parallel RNA-quadruplex motif *in vitro* [23,61,62]. Although so far the non-coding TERRA has not been assigned any functional role, it is proposed to negatively regulate telomerase activity in a telomere-length-dependent manner [61], potentially by forming an RNA-DNA duplex [39].

4.4. RNA Quadruplex-Interacting Proteins

Several proteins have been identified that specifically bind and stabilize RNA quadruplexes [16]. In line, a recent study determined specific interactions of a relatively small telomeric RNA transcript, consisting of eight consecutive repeats of UUAGGG, with various telomere-associated proteins like telomere repeat factors TRF1 and TRF2, heterochromatin protein 1, histone H3 trimethyl K9, multiple subunits of origin recognition complex (ORC) and DNA-damage-sensing factors [63]. This study also showed that $r(UUAGGG)_8$ can physically bind with ORC1 and TRF2 to form a ternary complex, which in turn is crucial for telomere maintenance and heterochromatin formation. The same group simultaneously showed that Epstein-Barr virus nuclear antigen 1, critical for the replication and maintenance of the Epstein-Barr virus genome, preferentially binds RNA quadruplexes and was inhibited by a G-quadruplex interacting small molecule [64]. Apart from telomere RNA-specific proteins, the coding region of fragile X mental retardation protein (FMRP) has been shown to fold back to form a G-quadruplex motif [65] which was found to be recognized and stabilized by FMRP itself, thereby inhibiting translation in a feedback loop [66].

In addition to the above described studies of RNA quadruplex-stabilizing proteins, an enzyme formerly characterized as an RNA helicase associated with AU-rich sequences (RHAU) was identified to specifically unwind quadruplex structures and hence termed G4 resolvase 1 (G4R1). It binds and resolves tetramolecular RNA quadruplexes with high affinity [67]. Moreover, competitive assays and binding affinities suggested that G4R1/RHAU resolves RNA quadruplexes more efficiently than the DNA counterparts. Especially in light of the influences of RNA quadruplexes on gene expression discussed above, the discovery of an enzymatic activity that specifically resolves such structures is an interesting finding. It is tempting to speculate whether RNA quadruplexes indeed have a distinct role as regulatory motifs or whether their presence is rather accidental with the need to dissolve such disadvantageous structures. The identification and characterization of further proteins specifically interacting with RNA quadruplexes will certainly increase our understanding of the roles and influences of these interesting higher order nucleic acid structures.

5. CONCLUSIONS

RNA quadruplexes are extremely stable conformations implicated in various biological processes. Recent studies illustrated regulatory potential in telomere maintenance and mRNA processing and translation, strongly suggesting their *in vivo* existence and significance. Moreover, small molecules stabilizing RNA quadruplexes and interfering with specific protein interactions further strengthen the importance of RNA quadruplexes. However, understanding the mechanisms that mediate *in vivo* formation and stability of RNA quadruplexes is very important in order to decipher the impact of these structures on cellular processes. In addition, determination of more interacting partners would augment the present repertoire of RNA quadruplex-affected processes. Targeting of specific RNA quadruplex structures and their protein interactions by small molecules may allow for controlling specific events with potential therapeutic uses. Taken together, RNA quadruplexes seem to interfere with or are involved in several RNA-associated functions. Future research along these lines will certainly gain more insight into these unusual four-stranded nucleic acid conformations.

ACKNOWLEDGMENT

KH thanks the Alexander von Humboldt-Foundation for a Fellowship, JSH acknowledges the VolkswagenStiftung and the Fonds der Chemischen Industrie for continuing support.

ABBREVIATIONS

A	adenine
bp	base pair
C	cytidine
CD	circular dichroism
EGFP	enhanced green fluorescent protein gene
FISH	fluorescence *in situ* hybridization
FMRP	fragile X mental retardation protein
G	guanosine
HIV	human immunodeficiency virus type 1
IGF-II	insulin-like growth factor II
mRNA	messenger RNA
NMR	nuclear magnetic resonance
NRAS	neuroblastoma rat sarcoma viral oncogene
ORC	origin recognition complex
ORF	open reading frame
RefSeq	reference sequence
ssDNA	single stranded DNA
T	thymidine
$T_{1/2}$	melting temperature

TRAP	telomeric repeat amplification protocol
TRF	telomere repeat factor
U	uridine
UTR	untranslated region
ZIC-1	zinc finger protein of the cerebellum 1

REFERENCES

1. M. Gellert, M. N. Lipsett and D. R. Davies, *Proc. Natl. Acad. Sci. USA*, 1962, **48**, 2013–2018.
2. J. R. Fresco and J. Massoulie, *J. Am. Chem. Soc.*, 1963, **85**, 1352–1353.
3. Z. Du, Y. Zhao and N. Li, *Genome Res.*, 2008, **18**, 233–241.
4. P. Rawal, V. B. R. Kummarasetti, J. Ravindran, N. Kumar, K. Halder, R. Sharma, M. Mukerji, S. K. Das and S. Chowdhury, *Genome Res.*, 2006, **16**, 644–655.
5. K. Halder, R. Halder and S. Chowdhury, *Mol. BioSyst.*, 2009, **5**, 1703–1712.
6. H. M. Wong and J. L. Huppert, *Mol. BioSyst.*, 2009, **5**, 1713–1719.
7. P. Mani, V. K. Yadav, S. K. Das and S. Chowdhury, *PLoS One*, 2009, **4**, e4399.
8. N. Maizels, *Nat. Struct. Mol. Biol.*, 2006, **13**, 1055–1059.
9. A. Siddiqui-Jain, C. L. Grand, D. J. Bearss and L. H. Hurley, *Proc. Nat. Acad. Sci.*, 2002, **99**, 11593–11598.
10. K. Paeschke, T. Simonsson, J. Postberg, D. Rhodes and H. J. Lipps, *Nat. Struct. Mol. Biol.*, 2005, **12**, 847–854.
11. N. Kumar and S. Maiti, *Nucleic Acids Res.*, 2008, **36**, 5610–5622.
12. A. Risitano and K. R. Fox, *Biochemistry*, 2003, **42**, 6507–6513.
13. D. Sen and W. Gilbert, *Nature*, 1988, **334**, 364–366.
14. J. D. Griffith, L. Comeau, S. Rosenfield, R. M. Stansel, A. Bianchi, H. Moss and T. De Lange, *Cell*, 1999, **97**, 503–514.
15. M. P. Horvath, V. L. Schweiker, J. M. Bevilacqua, J. A. Ruggles and S. C. Schultz, *Cell*, 1998, **95**, 963–974.
16. M. Fry, *Front. Biosci.*, 2007, **12**, 4336–4351.
17. K. Phillips, Z. Dauter, A. I. Murchie, D. M. Lilley and B. Luisi, *J. Mol. Biol.*, 1997, **273**, 171–182.
18. A. Wlodarczyk, P. Grzybowski, A. Patkowski and A. Dobek, *J. Phys. Chem. B*, 2005, **109**, 3594–3605.
19. A. Wong and G. Wu, *J. Am. Chem. Soc.*, 2003, **125**, 13895–13905.
20. D. Miyoshi, A. Nakao, T. Toda and N. Sugimoto, *FEBS Lett.*, 2001, **496**, 128–133.
21. F. M. Chen, *Biochemistry*, 1992, **31**, 3769–3776.
22. E. A. Venczel and D. Sen, *Biochemistry*, 1993, **32**, 6220–6228.
23. H. Martadinata and A. T. Phan, *J. Am. Chem. Soc.*, 2009, **131**, 2570–2578.
24. A. Joachimi, A. Benz and J. S. Hartig, *Bioorg. Med. Chem.*, 2009, **17**, 6811–6815.
25. S. Kumari, A. Bugaut and S. Balasubramanian, *Biochemistry*, 2008, **47**, 12664–12669.
26. A. Arora and S. Maiti, *J. Phys. Chem. B*, 2009, **113**, 10515–10520.

27. A. Bugaut and S. Balasubramanian, *Biochemistry*, 2008, **47**, 689–697.
28. P. Hazel, J. Huppert, S. Balasubramanian and S. Neidle, *J. Am. Chem. Soc.*, 2004, **126**, 16405–16415.
29. K. Halder, M. Wieland and J. S. Hartig, *Nucleic Acids Res.*, 2009, **37**, 6811–6817.
30. M. Wieland and J. S. Hartig, *Chem. Biol.*, 2007, **14**, 757–763.
31. B. Pagano, C. A. Mattia, L. Cavallo, S. Uesugi, C. Giancola and F. Fraternali, *J. Phys. Chem. B*, 2008, **112**, 12115–12123.
32. B. Pan, K. Shi and M. Sundaralingam, *J. Mol. Biol.*, 2006, **363**, 451–459.
33. B. Pan, Y. Xiong, K. Shi and M. Sundaralingam, *Structure*, 2003, **11**, 1423–1430.
34. H. Liu, A. Matsugami, M. Katahira and S. Uesugi, *J. Mol. Biol.*, 2002, **322**, 955–970.
35. D. Miyoshi, H. Karimata and N. Sugimoto, *J. Am. Chem. Soc.*, 2006, **128**, 7957–7963.
36. C. M. Olsen and L. A. Marky, *J. Phys. Chem. B*, 2009, **113**, 9–11.
37. C. M. Olsen, H. T. Lee and L. A. Marky, *J. Phys. Chem. B*, 2009, **113**, 2587–2595.
38. J. L. Mergny, A. De Cian, A. Ghelab, B. Sacca and L. Lacroix, *Nucleic Acids Res.*, 2005, **33**, 81–94.
39. Y. Xu, K. Kaminaga and M. Komiyama, *J. Am. Chem. Soc.*, 2008, **130**, 11179–11184.
40. J. Deng, Y. Xiong and M. Sundaralingam, *Proc. Natl. Acad. Sci. USA*, 2001, **98**, 13665–13670.
41. J. M. Lipay and M. R. Mihailescu, *Mol. Biosyst.*, 2009, **5**, 1347–1355.
42. J. Kim, C. Cheong and P. B. Moore, *Nature*, 1991, **351**, 331–332.
43. E. Westhof and V. Fritsch, *Structure*, 2000, **8**, R55–65.
44. C. F. Tang and R. H. Shafer, *J. Am. Chem. Soc.*, 2006, **128**, 5966–5973.
45. G. N. Parkinson, M. P. Lee and S. Neidle, *Nature*, 2002, **417**, 876–880.
46. J. Li, J. J. Correia, L. Wang, J. O. Trent and J. B. Chaires, *Nucleic Acids Res.*, 2005, **33**, 4649–4659.
47. J. Qi and R. H. Shafer, *Nucleic Acids Res.*, 2005, **33**, 3185–3192.
48. Y. He, R. D. Neumann and I. G. Panyutin, *Nucleic Acids Res.*, 2004, **32**, 5359–5367.
49. N. W. Kim, M. A. Piatyszek, K. R. Prowse, C. B. Harley, M. D. West, P. L. Ho, G. M. Coviello, W. E. Wright, S. L. Weinrich and J. W. Shay, *Science*, 1994, **266**, 2011–2015.
50. S. Kumari, A. Bugaut, J. L. Huppert and S. Balasubramanian, *Nat. Chem. Biol.*, 2007, **25**, 25.
51. J. L. Huppert, A. Bugaut, S. Kumari and S. Balasubramanian, *Nucleic Acids Res.*, 2008, **36**, 6260–6268.
52. A. Arora, M. Dutkiewicz, V. Scaria, M. Hariharan, S. Maiti and J. Kurreck, *RNA*, 2008, **14**, 1290–1296.
53. J. Eddy and N. Maizels, *Nucleic Acids Res, m*, 2008, **36**, 1321–1333.
54. D. Meinsma, P. E. Holthuizen, J. L. Van Den Brande and J. S. Sussenbach, *Biochem. Biophys. Res. Commun.*, 1991, **179**, 1509–1516.
55. J. Christiansen, M. Kofod and F. C. Nielsen, *Nucleic Acids Res.*, 1994, **22**, 5709–5716.

56. W. I. Sundquist and S. Heaphy, *Proc. Natl. Acad. Sci. USA*, 1993, **90**, 3393–3397.
57. G. Parkinson, M. Lee and S. Neidle, *Nature*, 2002, **417**, 876–880.
58. Y. Xu, Y. Noguchi and H. Sugiyama, *Bioorg. Med. Chem.*, 2006, **14**, 5584–5591.
59. C. Koering, A. Pollice, M. Zibella, S. Bauwens, A. Puisieux, M. Brunori, C. Brun, L. Martins, L. Sabatier, J. Pulitzer and E. Gilson, *EMBO Rep.*, 2002, **3**, 1055–1061.
60. C. M. Azzalin, P. Reichenbach, L. Khoriauli, E. Giulotto and J. Lingner, *Science*, 2007, **318**, 798–801.
61. S. Schoeftner and M. A. Blasco, *Nat. Cell Biol.*, 2008, **10**, 228–236.
62. A. Randall and J. D. Griffith, *J. Biol. Chem.*, 2009, **284**, 13980–13986.
63. Z. Deng, J. Norseen, A. Wiedmer, H. Riethman and P. M. Lieberman, *Mol. Cell*, 2009, **35**, 403–413.
64. J. Norseen, F. B. Johnson and P. M. Lieberman, *J. Virol.*, 2009, **83**, 10336–10346.
65. J. C. Darnell, K. B. Jensen, P. Jin, V. Brown, S. T. Warren and R. B. Darnell, *Cell*, 2001, **107**, 489–499.
66. C. Schaeffer, B. Bardoni, J. L. Mandel, B. Ehresmann, C. Ehresmann and H. Moine, *EMBO J.*, 2001, **20**, 4803–4813.
67. S. D. Creacy, E. D. Routh, F. Iwamoto, Y. Nagamine, S. A. Akman and J. P. Vaughn, *J. Biol. Chem.*, 2008, **283**, 34626–34634.
68. C. Cheong and P. B. Moore, *Biochemistry*, 1992, **31**, 8406–8414.
69. T. Mashima, A. Matsugami, F. Nishikawa, S. Nishikawa and M. Katahira, *Nucleic Acids Res.*, 2009, **37**, 6249–6258.

Met. Ions Life Sci. **2011**, *9*, 141–173

5

The Roles of Metal Ions in Regulation by Riboswitches

Adrian R. Ferré-D'Amaré[1] ***and Wade C. Winkler***[2]

[1]Howard Hughes Medical Institute and Division of Basic Sciences, Fred Hutchinson Cancer Research Center, Seattle, WA 98109-1024, USA

[2]Department of Biochemistry, The University of Texas Southwestern Medical Center, Dallas, TX 75390-9038, USA

<aferre@fhcrc.org>

<wadewinkler@utsouthwestern.edu>

Metal Ions in Life Sciences, Volume 9 Edited by Astrid Sigel, Helmut Sigel, and Roland K. O. Sigel

Published by the Royal Society of Chemistry, www.rsc.org DOI: 10.1039/9781849732512000141

ABSTRACT: Metal ions are required by all organisms in order to execute an array of essential molecular functions. They play a critical role in many catalytic mechanisms and structural properties. Proper homeostasis of ions is critical; levels that are aberrantly low or high are deleterious to cellular physiology. To maintain stable intracellular pools, metal ion-sensing regulatory (metalloregulatory) proteins couple metal ion concentration fluctuations with expression of genes encoding for cation transport or sequestration. However, these transcriptional-based regulatory strategies are not the only mechanisms by which organisms coordinate metal ions with gene expression. Intriguingly, a few classes of signal-responsive RNA elements have also been discovered to function as metalloregulatory agents. This suggests that RNA-based regulatory strategies can be precisely tuned to intracellular metal ion pools, functionally akin to metalloregulatory proteins. In addition to these metal-sensing regulatory RNAs, there is a yet broader role for metal ions in directly assisting the structural integrity of other signal-responsive regulatory RNA elements. In this chapter, we discuss how the intimate physicochemical relationship between metal ions and nucleic acids is important for the structure and function of metal ion- and metabolite-sensing regulatory RNAs.

KEYWORDS: *Bacillus subtilis* · magnesium · M-box · riboswitch · RNA folding · RNA structure · *Salmonella*

1. INTRODUCTION

Metal ions serve many important roles in biology. They are required as enzymatic cofactors and help stabilize protein structures and macromolecular interactions. In particular, metal ions exhibit a crucial influence on the structure and function of cellular nucleic acids [1–3]. Indeed, a metal ion 'atmosphere' surrounds nucleic acids, mostly through weak nonspecific interactions with the polymer backbone. Such charge-screening ions function to neutralize the polyanionic backbone, thereby reducing the electrostatic penalty incurred when oligonucleotide strands are brought in close proximity. As a result, nucleic acid secondary structure features are highly dependent on the presence of metal ions. Mg^{2+} and K^{+} are the metal ions with the highest intracellular abundance; correspondingly, they are the predominant non-specific ions closely associated with nucleic acids. In addition to these non-specific interactions, individual metal ions also associate specifically to nucleic acids, sometimes with profound structural consequences accompanying these interactions. Unlike DNA, which exists almost exclusively in B-form double helices, biological RNAs often adopt complex three-dimensional folds. One outcome of these higher-order structural features is the dramatic variation in the local surface electrostatic potential of the folded molecule. This can result in formation of highly electronegative pockets, which can act as metal ion binding sites. Depending on the features of these individual sites, they may localize metal ions nonspecifically but with moderate affinity, or, alternatively, exhibit high selectivity toward a cognate metal ion. For the former, electrostatic interactions alone may govern metal binding. In other instances, metal ions can be bound via hydrogen bonding interactions of RNA functional groups with metal-associated waters. For a subset of RNA-associated metal ions, RNA groups chelate directly to the metal ion. These innersphere coordination interactions result in a high degree of metal ion selectivity. A variety of RNA functional groups have been found to bind metal ions via these innersphere coordinations. The primary atoms that mediate these interactions are nonbridging phosphate oxygens, the 2′ hydroxyl, and a subset of nucleobase nitrogens and oxygens [3].

These different types of metal ion binding sites appear to be generally important for structured RNAs. For example, an analysis of metal ions associated with the large ribosomal subunit revealed the presence of 116 and 88 Mg^{2+} and monovalent ions, respectively [4]. The majority of these metals were found to include at least one innersphere coordination to an RNA functional group. In many such instances, these metal ion-RNA interactions appear to specifically position the backbone and nucleoside atoms in order to facilitate noncanonical interactions or long-range tertiary contacts. These different types of metal ion sites are likely to be generally important for other structured RNAs, in addition to the ribosome. Indeed, three-dimensional structural analyses of an increasingly diverse array of noncoding RNA elements have confirmed this.

One of the fastest growing categories of noncoding RNAs is that of bacterial riboswitches. Riboswitches are *cis*-acting regulatory RNAs, mostly residing within the untranslated regions (UTRs) of mRNAs, which bind specific cellular metabolites and, in response, control gene expression [5,6]. Over 20 riboswitch classes have been identified, which respond to a wide variety of metabolite ligands, including enzyme cofactors, nucleobases, nucleosides, amino acids, a second messenger, and an amino sugar [6,7]. Riboswitches, which have been subjected to extensive structural analyses in recent years, are comprised largely of two portions, a metabolite-sensing domain (aptamer) and downstream regulatory sequences that couple genetic control to the aptamer domain. The riboswitch aptamer domains associate with their target ligands both with high affinity and high selectivity. This is accomplished using an assortment of structural features that together create the riboswitch metabolite-binding pockets. As we discuss in this chapter, metal ions play an indispensable role in promoting aptamer tertiary structures. Moreover, in many instances, RNA-bound metals directly assist formation of the ligand-binding pockets. Study of these structures has expanded the functional repertoire of RNA-associated metal ions and has begun to provide new insights into the mechanisms that guide RNA folding pathways.

Metal ions are required for many cellular functions, in addition to their roles in stabilizing nucleic acid structures, and their absence is deleterious to the cell. However, inappropriately high intracellular concentrations of metal ions are also associated with cellular toxicity [8]. Therefore, proper homeostasis mechanisms are paramount to cell survival, a task that can be daunting given the considerable fluctuations of extracellular metal ion concentrations to which cells are exposed. The ability to accrue or export metals is largely accomplished through membrane-bound transporters [9,10]. Some of these transporter complexes regulate metal ion import through metal-responsive gating mechanisms involving cytoplasmic protein domains. However, as an additional layer of genetic control, microorganisms rely on many individual regulatory factors that control the acquisition, storage, and detoxification of metal ions [11]. These factors are mostly comprised of metal ion sensor proteins that regulate gene expression (metalloregulatory proteins) in response to metal ion fluctuations. The selectivity of these proteins for their target metal ions is achieved largely through recognition of the coordination chemistry of the metal ion [12]. Binding of the appropriate metal ion to the metalloregulatory protein triggers changes in protein structure, which in turn alter the ability of the protein to activate or inhibit gene expression. Many examples of such metalloregulatory proteins have been discovered and characterized [9], mostly for control of transition metal ions. Less is known regarding regulation of other metals, such as magnesium [13].

In bacteria, metalloregulatory proteins are also often coupled to post-transcriptional regulatory circuits. For example, upon conditions of excessive iron the *E. coli* protein Fur (ferric uptake regulator) inhibits expression of iron uptake genes and inhibits the synthesis of certain small regulatory RNAs [14]. The decreased abundance of the latter then promotes de-repression of target mRNAs. However, for some organisms, control of metal ion homeostasis might not require metalloregulatory proteins at all. Instead, in these organisms RNA-based mechanisms may directly sense metal ion concentrations in order to regulate gene expression. As we discuss in this chapter, two categories of riboswitch-like RNA elements have been discovered that regulate magnesium ion transport genes for a variety of bacteria [15,16]. What is unique about these particular riboswitch RNA classes is the fact that they do not require a metabolite ligand for genetic control. Instead, metal ion-induced structural changes are directly coupled to control of downstream gene expression. In general, the discovery and characterization of these RNAs argues that bacteria are likely to have evolved both protein- and RNA-based mechanisms for metal ion homeostasis. In this chapter, we discuss both the roles of metal ions in metabolite-sensing riboswitches and the mechanisms of metal ion-responsive regulation by magnesium-sensing riboswitches.

2. METAL IONS THAT ASSIST RECOGNITION OF RIBOSWITCH LIGANDS

Some of the most compelling evidence for the direct participation of metal ions in the recognition of their small molecule ligands by riboswitches comes from X-ray crystallography. In evaluating the significance of these findings, it is necessary to appreciate their experimental basis. X-ray crystallography reveals the average location and degree of order of the electrons of the molecules that comprise the crystal (the electron density map). The crystallographer interprets features of the electron density map to correspond to, for instance, RNA, small molecule ligands, ordered water molecules, or metal ions. Interpretation is aided by several factors, but is more reliable when high-resolution data (better than ~1.8 Å) are available. First, the chemical composition of the crystal will limit the possibilities. Second, the electron density will be proportional to the number of electrons. Thus, a Mn^{2+} ion will be more electron dense than a water molecule. However, the degree of order (*B*-factor and occupancy) of the ion or molecule producing the electron density has to be taken into account. Third, if the resolution is sufficient to resolve all the ligands of an ion, its coordination geometry (bond lengths and angles) may be diagnostic. Fourth, the anomalous scattering properties (the wavelength-dependent absorption of X-rays) of different ions can be diagnostic. Thus, at typical X-ray wavelengths, Ca^{2+} and K^{+} may have anomalous signals, but not Mg^{2+} or Na^{+}. Fifth, substitution experiments may be informative. For example, Mn^{2+} may be able to replace Mg^{2+} in crystals grown in the presence of the former ion. As with all experiments of this nature, however, the ability of Mn^{2+} to bind to a site in the RNA does not prove it to be a Mg^{2+} binding site, because the two ions have different chemical properties.

2.1. Participation of Magnesium Ion in Binding Pockets of Phosphate-Containing Ligands

Crystal structures indicate that several riboswitches recognize small molecules containing a phosphate moiety through Mg^{2+} ions. In addition to helping neutralize the negative charge of the ligand to facilitate its binding to the anionic riboswitch, the specific chemical properties of the Mg^{2+} ion may play a role in small molecule recognition. At the time of writing, RNAs that appear to use this molecular recognition strategy are the thiamine-pyrophosphate (TPP) binding *thi*-box riboswitch, the *glmS* riboswitch-ribozyme, the flavin mononucleotide (FMN) specific or RFN riboswitch, and the cyclic diguanylate (c-di-GMP) riboswitch.

The 2.05 Å resolution structure of the *E. coli thiM* riboswitch aptamer domain cocrystallized with TPP in the presence of Mg^{2+} revealed that this RNA employs two partially hydrated Mg^{2+} ions to mediate its interactions with the pyrophosphate moiety of its ligand, which is bound in an extended conformation (Figure 1A). The RNA makes only two direct contacts to non-bridging oxygens of the pyrophosphate; all other RNA-pyrophosphate contacts are mediated by the bound Mg^{2+} ions or their waters of hydration [17]. Cocrystallization of the riboswitch in the presence of Mn^{2+}, Ca^{2+}, or Ba^{2+} revealed that these ions can all take the place of the two Mg^{2+} ions that coordinate the pyrophosphate of TPP. The identity of the ions was established by their anomalous diffraction signal [18]. These experiments indicate that the TPP riboswitch structure adapts to accommodate the different ionic radii of TPP-chelating ions, ranging from 0.81 Å (hexacoordinate Mn^{2+}) to 1.49 Å (hexacoordinate Ba^{2+}). Even more dramatic evidence of plasticity emerged from the crystal structure of the riboswitch bound to thiamine-monophosphate (TMP). The single phosphate of this ligand was observed to coordinate to two partially hydrated Mg^{2+} ions. The two metal ions allow the riboswitch to recognize the phosphate and pyrophosphate of TMP and TPP, respectively, using essentially equivalent interactions [18]. Nonetheless, this accommodation results in considerably weaker binding of TMP ($K_d \sim 260$ nM) than TPP ($K_d \sim 8.4$ nM, both measured at 10 mM Mg^{2+} concentration) [19]. The structure of the TPP riboswitch is conserved between bacteria and eukarya, and a 2.25 Å resolution structure confirmed that the eukaryotic riboswitch binds TPP in an identical manner, employing two Mg^{2+} ions to coordinate and recognize the pyrophosphate [20].

In contrast to the plastic TPP riboswitch, crystallographic studies of the *glmS* riboswitch-ribozyme in multiple functional states revealed a rigid, pre-organized RNA [21]. This RNA from Gram-positive bacteria controls gene expression by binding to glucosamine-6-phosphate (GlcN6P). Binding of GlcN6P to the RNA uncovers a latent self-cleavage activity of the riboswitch [22]. The riboswitch is part of the 5′-UTR of the mRNA encoding the

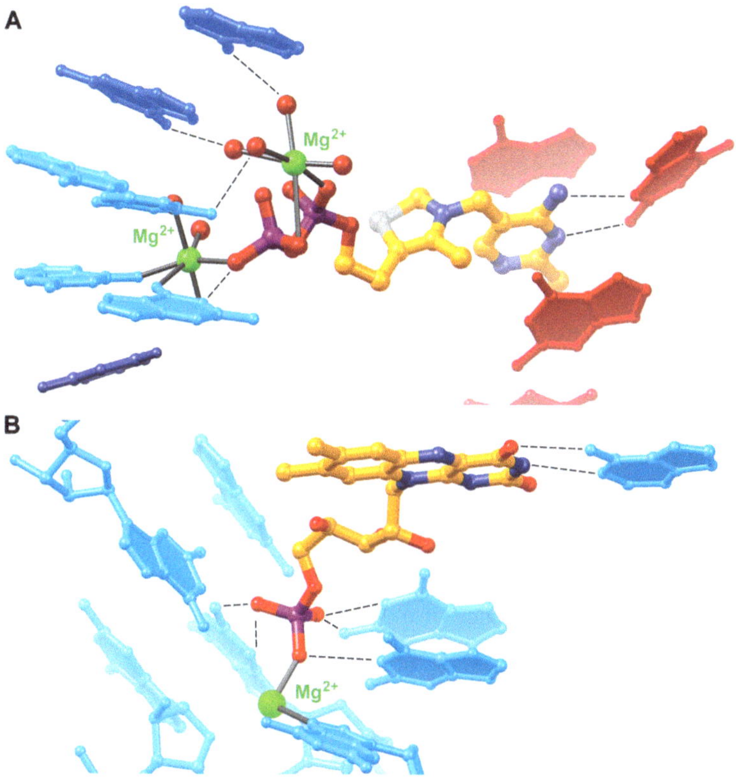

Figure 1. Mg^{2+} ions mediate recognition of phosphate-containing riboswitch ligands. **A**. TPP is bound to its cognate riboswitch chelated to two Mg^{2+} ions [1]. **B**. The phosphate group of FMN binds to a single Mg^{2+} ion in the binding site of the cognate riboswitch [12].

enzyme GlcN6P synthase, and self-cleavage of the riboswitch ultimately results in destruction of the mRNA, thereby completing a negative feedback regulatory loop [23]. GlcN6P binds to the ribozyme in an open, solvent-accessible pocket with its amine group in contact with the scissile phosphate [21]. Biochemical and crystallographic experiments indicate that the amine group, in concert with the G40 nucleobase of the RNA catalyzes the cleavage reaction [21,24,25]. The phosphate group in GlcN6P appears to serve mainly to increase the affinity of the activator to the riboswitch. The phosphate hydrogen bonds to the G1 nucleobase of the riboswitch, and also coordinates fully hydrated Mg^{2+} ions that in turn hydrogen bond to the RNA. Comparison of the GlcN6P complex structures of *glmS* riboswitches from *B. anthracis* [26] and *T. tengcongensis* [27] at 2.5 Å and 1.7 Å resolution, respectively, indicates that the direct interactions of the activator with the RNA are highly conserved. In contrast, the position and number of the Mg^{2+} ions near the phosphate to GlcN6P varies from structure to structure. This, and the fact that all the Mg^{2+} interactions with phosphate of GlcN6P on one hand and the riboswitch on the other are mediated by waters (i.e., are outersphere) suggests that, in this case, the metal ions function in Coulombic stabilization, but not directly in molecular recognition.

The 2.95 Å resolution structure of a FMN riboswitch bound to its cognate ligand revealed a complex RNA fold with internal pseudo-2-fold symmetry. The cofactor is bound asymmetrically, and its phosphate group is coordinated by a Mg^{2+} ion that makes direct (i.e., innersphere) coordination with a non-bridging phosphate oxygen and the N7 nitrogen of a guanine base (Figure 1B). Due to the low resolution, the other ligands of the Mg^{2+} ion, presumed to be water, were not built into the crystallographic model [28]. The phosphate of FMN plays an important role in recognition by the riboswitch. Solution binding studies indicate that riboflavin, the unphosphorylated form of FMN, binds to the RNA ~1000 times more weakly in the presence of physiological monovalent and divalent ion concentrations (100 mM K^+ and 2 mM Mg^{2+}). A crystal structure of riboflavin bound to the riboswitch shows absence of a Mg^{2+} near the location occupied by the ion in the cognate complex [28]. Crystallization experiments show that Ba^{2+}, Ca^{2+}, or Mn^{2+} can replace the Mg^{2+} ion that binds to the phosphate of FMN. However, the monovalent ion Cs^+ (binding of which can be monitored by its anomalous scattering, even at modest resolution) does not support high-affinity binding in solution and does not occupy the position of the Mg^{2+} ion that coordinates the phosphate of the ligand [28]. These experiments argue for the importance of the divalent cation that coordinates the phosphate group of FMN in recognition, and reveal a degree of flexibility of the binding pocket reminiscent of that of the TPP riboswitch. Solution and crystallization experiments show that the FMN riboswitch can even employ cobalt(III) hexammine instead of Mg^{2+} to bind FMN (the respective K_ds differ by less than 2-fold, in the presence of 100 mM K^+) [28]. Cobalt(III) hexammine is isosteric with hexahydrated Mg^{2+}, but the amine ligands are very tightly coordinated to Co^{3+}, and therefore the complex ion is unable to make innersphere coordinations [29]. The ability of the riboswitch to recognize FMN employing cobalt(III) hexammine indicates that the innersphere coordinations seen with Mg^{2+} are not essential for recognition. However, there are steric limits to the flexibility of the binding site: iridium(III) hexammine, which is bulkier than cobalt(III) hexammine, does not support binding and does not bind to the riboswitch-FMN complex near the phosphate of the ligand [28]. In addition, it is important to note that cobalt(III) hexammine is not present in biological systems, so that the riboswitch would not be expected to have evolved selectivity against it.

The c-di-GMP riboswitch is the first known gene-regulatory RNA to bind to a second messenger [30]. Solution studies show that the riboswitch has a strong requirement for Mg^{2+} to adopt the ligand-bound conformation in the presence of saturating c-di-GMP [31]. At present, two cocrystal structures have been reported, at 3.2 and 2.7 Å resolution. Both show an RNA assembled around a three-helix junction, with the symmetric second messenger bound asymmetrically at the junction. The two guanine bases of c-di-GMP are recognized by the riboswitch through canonical and non-canonical base pairing and stacking interactions. The electron density maps of the lower resolution structure were deemed too ambiguous to indicate the presence or position of metal ions bound to the ligand [31]. The authors of the 2.7 Å resolution structure located an iridium(III) hexammine ion coordinating to one of the phosphates of c-di-GMP, and electron density suggestive of a hydrated Mg^{2+} ion near the other phosphate [32]. Definitive identification of Mg^{2+} ion binding sites in this complex awaits higher resolution structures.

2.2. Other Metal Ions that Participate in Ligand Recognition and Selectivity

Crystallographic studies have implicated metal ions in ligand recognition by riboswitches that bind specifically to small molecules that lack a phosphate group. These are the riboswitches responsive to lysine and $preQ_1$. Although it falls beyond the subject matter of this review, an RNA evolved *in vitro* to bind to tetracycline that has been shown to function as an artificial riboswitch *in vivo* has also been found to depend critically on binding of Mg^{2+} to a non-phosphate moiety of its ligand for recognition [33].

The lysine riboswitch has one of the most complex aptamer structures determined to date, and consists of a three-helix bundle and a two-helix bundle radiating from a five-way junction [34]. Solution studies reveal that the overall structure of the riboswitch is insensitive to the presence of metal ions [34–36]. In contrast, binding of lysine is strongly dependent on the presence of K^+ ions. The affinity of the RNA for lysine is decreased by 50–100-fold if Na^+ or Mg^{2+} are present instead [34]. The 1.9 Å resolution structure of the riboswitch reveals that lysine is bound in an extended conformation, with its carboxylate coordinated to a K^+ ion (Figure 2A) [34]. The identity of the ion was confirmed by its anomalous signal. In addition, it could be replaced very efficiently by Tl^+, which is a well-documented mimic of K^+ [37]. Mn^{2+} does not support lysine binding, and this ion was not observed to chelate the carboxylate of bound lysine in crystals. A structure of the lysine-free riboswitch was also determined, and the binding site was

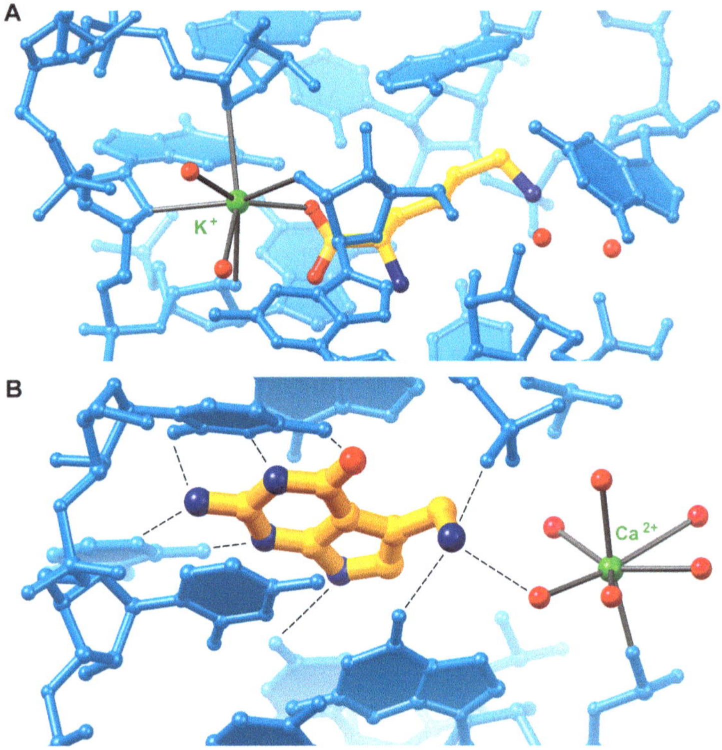

Figure 2. Cation-mediated recognition of riboswitch ligands without phosphate groups. **A.** A K^+ ion coordinates lysine when the amino acid is bound by its cognate riboswitch [18]. **B.** The characteristic exocyclic aminomethyl group of $preQ_1$ engages in hydrogen bonding with a water of hydration of a structural Ca^{2+} ion [26].

found to be devoid both of the lysine and of the K^+ ion. In aggregate, the foregoing experiments provide strong evidence that this riboswitch recognizes its cognate ligand as a K^+ ion chelate.

Several tRNAs of bacteria and eukarya have an essential hypermodified nucleotide called quoeuosine (Q) at the wobble position of the anticodon. The precursor tRNAs are transcribed with G at this position. A specific transglycosylase replaces this guanine with a modified purine base called $preQ_1$. After attachment to tRNA, $preQ_1$ is further elaborated enzymatically to result in Q. Consistent with its being the last small molecule intermediate in the synthesis of Q, $preQ_1$ was found to regulate the expression of enzymes involved in synthesis and transport of Q in bacteria through at least two different classes of riboswitches [38,39]. The ligand binding domain of the class I $preQ_1$ riboswitch is the most compact one known in nature ($\sim$34 nt).

Solution and crystallographic studies revealed it to fold into a canonical pseudoknot [40], with the ligand bound at the interface of two helices [41–43]. The preQ_1 riboswitch discriminates strongly against guanine, from which preQ_1 differs primarily in having an exocyclic aminomethyl group pendant from position 7. The 2.2 Å resolution cocrystal structure of a class I preQ_1 riboswitch suggests that this selectivity depends on a bound metal ion. The structure shows that the G-like portion of the ligand is recognized by Watson-Crick and Hoogsteen base pairing. These interactions would not discriminate against G. The discrimination arises from three hydrogen bonds made by the aminomethyl group that is characteristic of preQ_1, which bonds to the carbonyl group of a G of the RNA, to a non-bridging phosphate oxygen, and to a water of hydration of a tightly-bound Ca^{2+} ion (Figure 2B). This cation appears to be an intrinsic structural component of the RNA, as it is likely to be required to stabilize the pseudoknot fold of the riboswitch [42]. The cation selectivity of the preQ_1 riboswitch remains to be analyzed experimentally.

3. METAL IONS AND RIBOSWITCH FOLDING

Like all structured RNAs, the three-dimensional structures of riboswitches undergo stabilization in the presence of cations. To a first approximation, metal ions can be thought to function in two modes [44]. First, fully solvated metal ions accumulate near the RNA and thereby neutralize its charge (counterion condensation). These ions remain diffuse, thereby not paying the full entropic cost of specific binding, and they remain hydrated, so thereby not paying the enthalpic cost of desolvation. This type of counterion is not readily apparent in high-resolution structures. Second, some ions may be specifically chelated by RNA. These are the types of cations discussed above in Section 2. These bound cations will have lost most of their translational degrees of freedom, and may be partially or fully desolvated. In the study of global folding of RNAs in general, and riboswitches in particular, the former, diffuse class of counterions is more important (see Chapter 3 in this volume), although, as indicated above, ligand recognition by many riboswitches relies on specific chelation of counterions.

3.1. Experimental Considerations

A variety of methods have been employed to study the role of metal ions in RNA folding (see Chapters 2 and 6 in this volume). These can be classified into methods that detect some property that is a surrogate for folding, and those that directly study folding (that is, the change in size and shape of RNA as it adopts its native conformation). In the former category are, for instance, spectroscopic methods and enzymatic and chemical probing methods. In the latter category are, primarily, hydrodynamic and solution scattering techniques. The application of analytical ultracentrifugation to the study of riboswitch folding is discussed in more detail in Section 4. The following emphasizes the results of studies employing small-angle X-ray scattering (SAXS).

An important consideration regarding folding studies of riboswitches is the nature of the RNA construct employed. Riboswitches are not free-standing RNA molecules. Rather, they are functional units embedded within a mRNA. For many riboswitches, it is possible to define a minimal RNA sequence that is capable of specific, high-affinity binding to its cognate ligand. This sequence segment also commonly corresponds to the region of RNA with the highest phylogenetic conservation. This operationally defined segment is the aptamer domain. In order to function in regulating gene expression, this domain must interface with the transcription, translation, or post-transcriptional RNA processing machinery. This often employs downstream sequences that collectively are called the expression platform [45]. Thus, study of the full functional riboswitch requires analysis of ligand- and ion-dependent folding of both moieties. However, both because of experimental expediency and because it is the segment with highest sequence conservation, the majority of riboswitch folding studies to date have studied the aptamer domains in isolation.

3.2. Monovalent Ions and Riboswitch Folding

In vivo, K^+ is the most abundant counterion for nucleic acids. In *E. coli*, its intracellular activity ranges from ~100 mM to ~1 M, depending on the osmolarity of the growth medium [46]. However, there are few studies that examine the role of K^+ in riboswitch folding. Often, K^+ is present as part of the "native" buffer condition to which divalent cations and ligand molecules are added to examine their effects on folding, and therefore the role of K^+ is not explicitly addressed. Another reason for this is that, for a given concentration, monovalent ions are less effective in promoting RNA folding than polyvalent ions. This phenomenon can be understood in terms of the entropy loss incurred by the cation in associating, loosely or tightly, with the nucleic acid. The loss of translational entropy is the same for a monovalent or a polyvalent ion. However, whereas the monovalent ion can only neutralize the charge of one phosphate, the polyvalent ion can neutralize several phosphates [44]. Thus, Mg^{2+} promotes RNA folding better than K^+, and the trivalent cobalt(III) hexammine is even more stabilizing than Mg^{2+}. Monovalent ions at very high (molar) concentration may allow RNAs to fold and even be catalytically active (see e.g., [47]), but the physiological significance of such conditions is open to debate.

A recent study examined the relationship between the size of the monovalent cation and its effect on folding of RNAs by through thermal denaturation. Lambert et al. found that for most RNAs examined, including the purine riboswitch aptamer domain, the smaller Li^+ and Na^+ ions are more stabilizing than the larger Cs^+ ion by as much as 3 kcal/mol [48]. This effect may be explained by the increased ability of smaller cations to approach the RNA surface, especially in the case of intricately folded RNA molecules, such as riboswitches. However, the authors also noted a case in which an RNA structural domain has evolved specifically to coordinate a K^+ ion (analogous to the case of the lysine riboswitch discussed above). This RNA had a preference of the same order for Na^+ or K^+ over smaller or larger ions. Thus, RNAs have the capacity to evolve to be selective for a particular monovalent cation, and riboswitches, being intracellular molecules, presumably have some selectivity for K^+.

3.3. Magnesium Ion and Compaction of Metabolite-Binding Riboswitches

Compaction of several complex RNA molecules induced by addition of Mg^{2+} has been observed employing SAXS, a technique that is sensitive to aggregation state, size and shape of macromolecules free in solution. For instance, studies of the Group I intron in which folding was initiated by addition of Mg^{2+} and followed by time-resolved SAXS indicated that the ribozyme compacts at least 20-fold faster than folding to the native state (revealed by the onset of catalytic activity), indicating that a rapid collapse is followed by local structural rearrangements [49]. Other SAXS studies of this and other RNAs suggest that the most compact state of an RNA is its most stable conformation [49–52]. Because of these precedents, it was expected that riboswitches would be partially stabilized in the presence of physiologic Mg^{2+} concentrations, and that ligand binding would then lead to maximum stability (and maximum compaction, as detected by SAXS, for instance).

SAXS experiments on a variety of riboswitch aptamer domains reveals that this expectation was unfounded (Table 1). Some riboswitches behave in the expected manner.

Thus, the radius of gyration (R_g) of the glycine riboswitch decreases from 55 Å in monovalent ions alone to 45 Å in the presence of Mg^{2+}. Addition of

Table 1. Cation and ligand induced compaction of selected riboswitches.[a]

Riboswitch	R_g (Å)		
	K^+/Na^+ and no/low Mg^{2+}	K^+/Na^+ and Mg^{2+}	K^+, Mg^{2+} and ligand
c-di-GMP [15]	31.6±1	28.5±1	23.9±1
FMN [20]	29.4±1	29.1±1	28.6±1
glycine [37]	55 ±2	45 ±1	39 ±1
lysine [20]	46.8±1	39.1±1	39.2±1
SAM-I [20]	29.9±1	27.5±1	26.4±1
TPP [20]	27.5±1	24.4±1	22.1±1

[a]See references for experimental conditions.

its cognate ligand further compacts the riboswitch to an R_g of 39 Å [53]. The c-di-GMP and TPP riboswitches exhibit similar behavior [31,36]. In contrast the lysine and SAM-I riboswitches compact upon addition of Mg^{2+}, but do not compact further when ligand is added [36]. This suggests that the overall fold of the RNA is insensitive to ligand binding, and only local rearrangements take place upon specific complex formation [34,35]. Even more divergent from expectation is the behavior of the FMN riboswitch. At the level of global conformation analyzed by SAXS, this RNA is insensitive to the presence of Mg^{2+} or ligand. Thus, the overall fold of the RNA is established in monovalent ions alone, and structural accommodation upon ligand binding must be limited to the binding pocket [36]. Enzymatic probing data support this interpretation [28]. The data summarized in Table 1 indicate that the effect of Mg^{2+} on the global folding of riboswitches (strictly, of their aptamer domains) is highly idiosyncratic, as is the effect of their respective cognate ligands. This could indicate that the degree of stabilization of an aptamer domain that results from ligand binding has been fine-tuned to correspond to the regulatory needs of specific genomic loci. Alternatively, the effect of Mg^{2+} and ligand binding on the rates of folding of riboswitch aptamer domains, rather than their stability, may be biologically critical.

4. MAGNESIUM-SENSING RIBOSWITCHES: *SALMONELLA mgtA*

4.1. The Bacterial Metallome

The cellular metallome is defined as the composite of the distribution and equilibrium concentrations of metals within cells [54]. Many of these metals are associated with proteins. Indeed, it has been estimated that approximately one third of intracellular proteins contain a metal cofactor [55]. Most often, these protein-bound metals are transition metals, such as Mn, Cu, Fe, Zn, and Mo, although many cellular proteins associate instead with Mg^{2+}. For example, Fe^{2+} is frequently associated with proteins in the electron transfer chain, bound via iron-sulfur clusters or heme cofactors, while Zn^{2+} associates with DNA and RNA polymerase enzymes, zinc finger proteins, and ribosomes. The other transition metals associate with myriad abundant, but specific, classes of proteins. For example, cobalt is present in adenosylcobalamin-containing proteins and manganese associates with superoxide dismutase enzymes.

Metals in the highest intracellular abundance are K^+, Mg^{2+}, Ca^{2+}, and Na^+. With the exception of sporulating bacteria, Ca^{2+} is generally maintained at submicromolar levels [54]. Mg^{2+} is present at higher levels (10–20 mM free and ~100 mM total) and is required by DNA and RNA polymerases, as well as many ATPases and kinases. K^+ is also retained at high levels, at a total concentration of approximately 100–200 mM, and is important for osmotic balance and pH homeostasis [56,57]. Trace metal ions, which consist mostly of the transition metals plus Zn^{2+}, are maintained at much lower concentrations. The fact that the cytoplasm is a reducing environment alters both the oxidation state of some metals and their intracellular availability.

The predominant alkali (K^+, Na^+) and alkaline earth (Mg^{2+}, Ca^{2+}) metals frequently exist in the free state as they do not bind strongly to ligands. In contrast, the trace metals largely do not occur as free ions inside cells [12,54]. Instead, they mostly remain associated with proteins and certain metabolites. In general, the affinities between proteins and metals correlate with a universal order of preference. For divalents, this order is referred to as the Irving-Williams series (Mg^{2+} and $Ca^{2+} < Mn^{2+} < Fe^{2+} < Co^{2+} < Ni^{2+} < Cu^{2+} > Zn^{2+}$) [58]. Assuming equal accessibility for all of these metals, this order would predict that all proteins would simply bind Cu^{2+}, if given an equal choice. However, that is obviously not the case, and there is an abundance of evidence that collectively argue against equal access to intracellular metals. Therefore, proteins and metabolites that bind metals must employ specialized strategies to exclude metals at the top of the Irving-Williams series. In part, this is accomplished through specific compartmentalization of metals, for example, through sequestration of Cu^{2+} [59]. By corollary, proper localization of proteins can influence which metals can be selected and retained during protein folding. In general, the total numbers of intracellular metals are restricted such that binding of metals to proteins is dictated by their relative affinities to the target proteins [60]. The reduced levels of metals in the cytoplasm are achieved in part by an array of import and export transport proteins, and through sequestration by metallochaperone proteins [9,59]. The quantitative analysis of these respective intracellular metal concentrations, i.e., metallomics, is a rapidly burgeoning research discipline [61]. Some of these emerging data suggest that the metallome is likely to be similar in composition and in overall constancy between microorganisms [61,62]. This fact implies that bacteria are likely to contain elaborate regulatory mechanisms to maintain metal homeostasis. Indeed, many examples of metal-responsive transcription factors have been discovered for control of metal homeostasis [12,59].

4.2. Control of Magnesium Homeostasis and Bacterial Pathogenesis by the *mgtA* RNA

The control of intracellular metal ion pools and other adaptive responses to extracellular metal fluctuations may be particularly important for bacterial pathogens. The human host presents, at minimum, a two-fold challenge to pathogenic bacteria: (1) metal ions are efficiently retained by host cells, such that invading bacteria must use aggressive techniques for metal acquisition, and (2) bacteria oftentimes encounter strongly oxidative and nitrosative stress conditions within host cells, requiring elaborate adaptive techniques for intracellular survival [9,63]. These responses are best characterized in regards to iron acquisition by pathogenic bacteria, which must rely on multiple iron scavenging mechanisms in an attempt to recruit the essential metal, including secretion of iron-gathering proteins and metabolites [64]. Genes encoding for import of other metals, including Cu^{+}, Mn^{2+}, and Zn^{2+}, have also been demonstrated to be required for virulence of certain pathogens [12]. Mg^{2+} too, can be a critical component in the battle between bacterial pathogens and host cells. The importance of metals, including Mg^{2+}, during pathogenesis has been particularly well studied for *Salmonella enterica.*

Salmonella enterica are Gram-negative, facultative anaerobes, which reside in part within the intestinal microbiome, and are a frequent cause of food-born gastroenteritis. *Salmonella* is an intracellular pathogen, and enters both phagocytic and non-phagocytic cells through pathways that require type III secretion systems or through standard phagocytosis [65]. Once internalized, *Salmonella* modify the location and environment of the phagosome to maximize their survival and replication. The modified phagosome, called the *Salmonella*-containing vacuole (SCV), is the product of many bacterial virulence factors, which find passage into the host cell through multiple type I and type III secretion systems [66]. *Salmonella* prevents the fusion of the bacteria-containing vacuole with the lysosome, and ultimately reconfigures the SCV into a modified, microtubule-associated vesicular system called *Salmonella*-induced filaments [67].

In response to bacterial invasion, host cells employ multiple strategies to combat *Salmonella*. This includes utilization of an NADPH oxidase (NOX2) enzyme within the phagosome for production of reactive oxygen species, which is in turn neutralized by a *Salmonella*-produced superoxide dismutase enzyme [68]. In addition to this attempt at creating a "toxic environment" within phagosomes, host cells also try to subject the bacteria to nutrient poor conditions. Such nutrient deprivation strategies are likely to also include access to metals. For example, macrophages produce metal transporters such as Nramp1 and transferrin and, as a consequence, severely reduce the general availability of iron. To counteract the depletion of this and other essential metals, *Salmonella* produces a combination of iron-sequestering siderophores and proteins that transport Mn^{2+}, K^{+}, Zn^{2+}, and Mg^{2+}, which are collectively necessary for optimal virulence [66].

The importance of Mg^{2+} concentrations during *Salmonella* pathogenesis has been particularly well studied, and is influenced in part by a two component regulatory system, PhoP-PhoQ [69]. PhoQ is the cognate sensor kinase for the DNA-binding transcription factor, PhoP. This basic regulatory system is widespread in enteric bacteria, and can affect expression of many operons [70]. Several signals have been reported to activate the system, including low Mg^{2+}, low pH, and antimicrobial peptides [71–73]. Activation of PhoP results in expression of many genes, including those encoding for Mg^{2+} transport, resistance to cationic antimicrobial peptides, modification of the cellular envelope, and other factors that are important for virulence [74,75]. In fact, PhoP regulates ~3% of total genes in *Salmonella enterica* serovar Typhimurium [76]. PhoQ is an integral membrane protein, which projects a sensor domain into the periplasmic space. An acidic patch on the outside face of PhoQ is close to the membrane and has been proposed to mediate the response to metal ions and antimicrobial peptides [77,78]. The interaction of divalent ions with the negatively charged patch has been postulated to alter positioning of the sensor domain relative to the membrane; these conformational changes are believed to control the transmembrane signaling pathway. In *Salmonella enterica*, low Mg^{2+} and low pH activate different sets of genes, in accordance with adaptation to the respective signals [79]. Low-Mg^{2+}-activated PhoP induces expression of many genes including but not limited to *hilA* (transcription factor for invasion genes), *pmrAB* (Fe^{3+}-responsive regulatory system), *prgHIJK* (type III secretion system components), *mgtA* (magnesium transport), *mgtB* (magnesium transport), and *mgtC* (intramacrophage survival) [74].

Although it is a debated subject [80,81], the magnesium level within the phagosome is generally believed to be very low. This is in good agreement with the upregulation of multiple Mg^{2+} transporters upon activation of PhoP, which is required for phagosomal survival. It has been proposed that *Salmonella* ascertains its subcellular location by measuring the surrounding

Mg^{2+} via the PhoP-PhoQ system. Low Mg^{2+} would indicate the intracellular environment within the phagosome; high Mg^{2+} would indicate the extracellular environment. Further demonstrating the importance of Mg^{2+} homeostasis in this microorganism, it was recently discovered that expression of the *mgtA* transcript is controlled by a second Mg^{2+}-responsive pathway, in addition to control of transcription initiation by PhoP [82]. Identification of a second Mg^{2+} regulatory mechanism agreed with earlier data showing that *mgtA* expression was still modulated in response to Mg^{2+} in the context of a constitutively active PhoP mutant; also, the promoter region for *mgtA* was previously found to engender Mg^{2+}-responsive control over a reporter gene even in the absence of the PhoP-PhoQ system [83,84]. Upon further inspection, the second regulatory mechanism was discovered to occur in post-transcription initiation, and to involve a regulatory RNA element. Specifically, the *mgtA* transcript was found to house a long 5′ leader region that regulates expression of the downstream gene [82]. Northern blotting of *mgtA* transcripts revealed that abundance of the leader region decreased as extracellular Mg^{2+} was raised above 0.1 mM, most likely as a reflection of PhoP control over initiation. However, abundance of the downstream *mgtA* coding region decreased at lower Mg^{2+}. When the PhoP-responsive promoter was replaced with a heterologous promoter, high Mg^{2+} still triggered a decrease in abundance for the coding sequence, but not for the leader region [82].

Many different types of *cis*-acting regulatory RNAs have been identified previously, which respond to a diverse range of signals, including temperature, proteins, soluble RNAs, and metabolites [85]. However, a regulatory RNA element that was explicitly evolved for the purposes of sensing metal ions had not yet been observed. Examination of the *mgtA* leader region by a secondary structure prediction program, Mfold, revealed several potential stem-loop structures, which could be involved in genetic control (Figure 3). Importantly, a similar combination of stem-loop structures and interspersed primary sequence elements could be identified upstream of *mgtA* in other enterics. Deletion or mutagenesis of these structural elements resulted in *mgtA* expression that was either constitutively high or constitutively low, depending on which elements were removed. In aggregate, the mutagenesis data suggested that a single stem-loop ("C") was formed during conditions of low Mg^{2+}, conditions that promote downstream expression. In contrast, a mutually exclusive arrangement of two stem-loops ("A", "B") is favored in high Mg^{2+}, conditions that disallow downstream expression (Figure 3). Limited biochemical support for the A, B, and C helices was obtained from interpretation of partial RNase digestion patterns, using RNase T1 for cleavages of unpaired guanosines. These data were interpreted as suggesting that association of Mg^{2+} with the *mgtA* leader RNA *in vitro* directly promoted the A + B conformation.

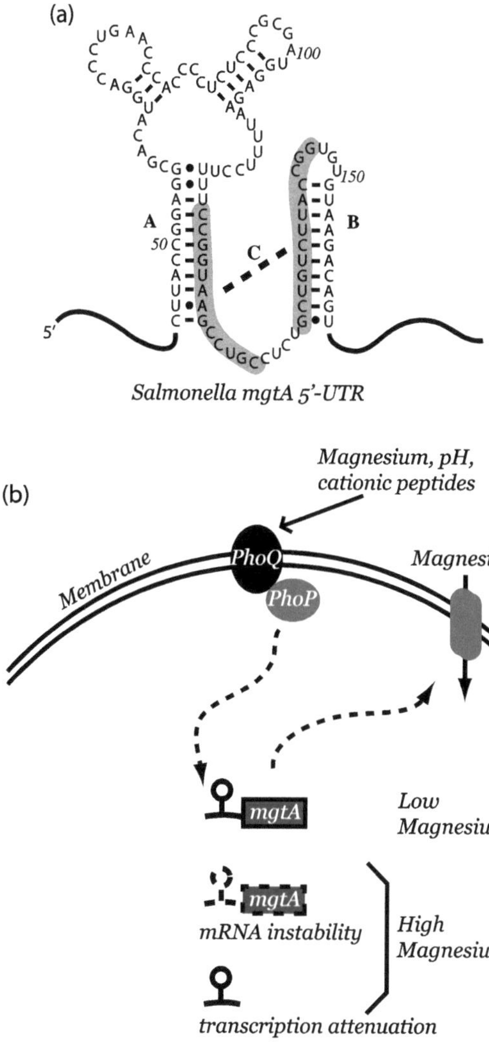

Figure 3. Genetic regulation by the *mgtA* riboswitch. **(a)** Secondary structure of the *Salmonella enterica mgtA* riboswitch. **(b)** Under low Mg^{2+} conditions the PhoP protein activates transcription of the *mgtA* transcript. Under high Mg^{2+} conditions, a riboswitch located within the 5′ leader region activates destabilization of the *mgtA* transcript and reduces the efficiency of transcription elongation of the *mgtA* gene.

4.3. Mechanistic Models for Regulation by the *mgtA* RNA

While it is clear from published data that the *mgtA* riboswitch controls gene expression in a Mg^{2+}-responsive manner, the molecular mechanism through which this is accomplished is unresolved. The riboswitch has been proposed to exert regulatory control over transcription elongation, in a Mg^{2+}-dependent manner [82]. This model is based on *in vitro* transcription reactions using a DNA template encoding the *mgtA* leader region. In these reactions, a truncated transcription intermediate was observed, which appeared to be preferentially synthesized in response to increased Mg^{2+}. However, the exact site of transcription attenuation has not yet been conclusively resolved. Also, the components typically involved in promoting transcription termination (Rho-dependent termination sites and intrinsic terminator hairpins) appear to be absent from the *mgtA* RNA sequence. Therefore, it is still unclear exactly how the *mgtA* riboswitch affects transcription elongation activity *in vitro*. More recently, the *mgtA* riboswitch has been implicated in controlling stability of the overall *mgtA* transcript [86]. Specifically, high Mg^{2+} was shown to stimulate degradation of the *mgtA* transcript *in vivo* through a mechanism that was dependent on both the *mgtA* riboswitch and the global endoribonuclease, RNase E. The structural relationship between the riboswitch and RNase E has not yet been determined. Therefore, it is currently unclear whether the *mgtA* riboswitch controls transcript abundance by regulating transcription elongation, mRNA stability, or both. Nor has the molecular basis of Mg^{2+} recognition been resolved for the riboswitch. It is currently unclear which portions of the *mgtA* leader RNA might comprise a Mg^{2+} aptamer domain, if indeed a discrete aptamer domain exists within the RNA element.

Interestingly, recent data provide some clues as to why *Salmonella* may have evolved two overlapping Mg^{2+}-based regulatory mechanisms. In short, these individual regulatory mechanisms may respond to different ranges of intracellular metals. For example, PhoP-mediated control of *mgtA* transcription initiation occurs over a broad range of extracellular Mg^{2+}, whereas the *mgtA* riboswitch upregulates *mgtA* abundance only when Mg^{2+} is added at $< 50\,\mu M$ [87]. Also, recent data suggest that the *mgtA* RNA element may not even be the only magnesium riboswitch in *Salmonella.* A long 5′ leader region has also been identified upstream of *mgtC*, and was postulated to incorporate another Mg^{2+}-sensing regulatory RNA [16]. MgtC is a protein of unknown function that is somehow involved in adaptation to low-Mg^{2+} environments, and has been particularly implicated in the survival of *Salmonella* inside human cells [80]. Similar to the *mgtA* riboswitch, the mechanistic basis for Mg^{2+}-sensing by the *mgtC* leader region remains to be discovered. Together, the emerging data on regulation of *mgtA* and *mgtC* argue that Mg^{2+}-responsive regulatory mechanisms are essential for survival of *Salmonella* in macrophages, and are likely to occur in other bacteria for control of Mg^{2+} homeostasis. Given that little is known regarding the mechanisms, or even the overall need, for control of Mg^{2+} homeostasis in all three domains of life [13], elucidation of the molecular mechanisms underlying post-transcriptional control of *Salmonella* Mg^{2+} homeostasis remains an important goal.

5. MAGNESIUM-SENSING RIBOSWITCHES: THE M-BOX RNA

5.1. Identification and Distribution of the M-Box RNA

Many different classes of metabolite-sensing riboswitches have been discovered, mostly as an outcome of bioinformatics-based searches for conserved bacterial RNA motifs. Individual riboswitch classes have been discovered that sense nucleobase-containing metabolites, enzyme cofactors, amino acids, a signaling metabolite (cyclic di-guanosine monophosphate), and an amino sugar (glucosamine-6-phosphate) [7]. In all instances, a portion of these regulatory RNAs comprises a discrete aptamer domain, responsible for high affinity association with the target metabolite. Typically, the aptamer domains are followed by sequence and structural elements that coordinate the aptamer with downstream genetic control. The majority of bacterial riboswitches regulate gene expression by either controlling access to a downstream translation initiation site or through formation of a transcription termination site (transcription attenuation). Also, most riboswitches are used for feedback inhibition of downstream genes, although some are used for other regulatory purposes, such as substrate-mediated activation of catabolism genes [7].

In addition to the known classes of riboswitches, there is a rapidly increasingly catalog of 'orphan' riboswitches [7,88–93]. These are evolutionarily conserved, structured RNA elements that resemble riboswitches, but whose metabolite ligands remain unidentified. Thus far, the most successful approach in identifying orphan riboswitch ligands is to examine whether their downstream genes exhibit functional trends consistent with certain signaling cues. For example, TPP-sensing riboswitches are most often utilized for genetic regulation of TPP synthesis and transport; therefore, it is reasonable to assume that a new regulatory RNA upstream of these gene categories is likely to be a TPP riboswitch. Unfortunately, some orphan riboswitches co-occur with complex combinations of gene categories, such that the prediction of metabolite ligand candidates is difficult [94]. For orphan riboswitch classes that are positioned solely upstream of unknown, uncharacterized genes, it is impossible to predict their metabolite ligands. A 2004 search for riboswitches resulted in discovery of several new putative riboswitch classes [88]. Most of the metabolite ligands for these riboswitch classes have been subsequently identified; however, several still remain to be discovered. Interestingly, at least three of the putative riboswitch classes from this study appeared to regulate metal transport genes, suggesting a relationship between metals and their regulatory mechanisms. One of these riboswitches, later renamed the M-box, was found to function as a metal-responsive genetic element [15].

At the time of the preparation of this article, approximately 600 individual M-box sequences from 355 bacterial species had been listed on Rfam, a database of evolutionarily conserved RNA families [92]. Comparative sequence alignments revealed that the M-box RNA is approximately ~160 nucleotides in length. The molecule is characterized by extensive primary sequence conservation and includes five conserved helical portions. It also includes several common secondary structural motifs, including the UAA/GAN motif, and a GNRA tetraloop, which are likely to be essential for stabilization of long-range interactions. The majority of the M-box examples that have been identified are located upstream of putative Mg^{2+} transport genes (Figure 4a).

Mg^{2+} is readily available on Earth, located in mineral deposits, as a salt or a free cation in aqueous solutions. For example, in oceanic environments Mg^{2+} is present at an average concentration of 55 mM [95]. It is also the most abundant divalent cation in living organisms, present at approximately 10–20 mM for the free metals. Due to the abundance of the metal within cells, it was previously presumed that specific regulatory mechanisms would not be required for maintaining intracellular Mg^{2+} concentrations, as they are for trace metals. Recent evidence suggests that this assumption is incorrect; indeed, control of Mg^{2+} homeostasis has now been clearly demonstrated to be essential for growth and bacterial virulence [13]. For example, structural characterization of Mg^{2+} transport proteins has revealed that bacteria employ gating mechanisms to modulate Mg^{2+} transport activity. Furthermore, the discoveries of the PhoP-PhoQ regulatory pathway and the *mgtA* and M-box riboswitches have demonstrated that organisms also utilize dedicated genetic regulatory mechanisms for control of Mg^{2+} pools.

There are three classes of Mg^{2+} transport genes, which are present in all three domains of life [13,96]. The first to be discovered was CorA, an ion channel widely represented in bacterial genomes. It is not subjected to genetic regulatory control in *Salmonella*, leading to the hypothesis that it is usually constitutively expressed in bacteria as a 'housekeeping' Mg^{2+} transporter. MgtE is a Mg^{2+} channel protein that is widespread in bacteria, and has also been clearly identified in eukaryotes (SLC41A1-3). Both MgtE and CorA contain cytoplasmic domains that regulate gating of the ion-conducting pore. However, the two transporters may function at different ion concentrations. CorA transports Mg^{2+}, Co^{2+}, and Ni^{2+} with apparent affinities of 15 μM, 20–40 μM, and 200–400 μM, respectively, and does not

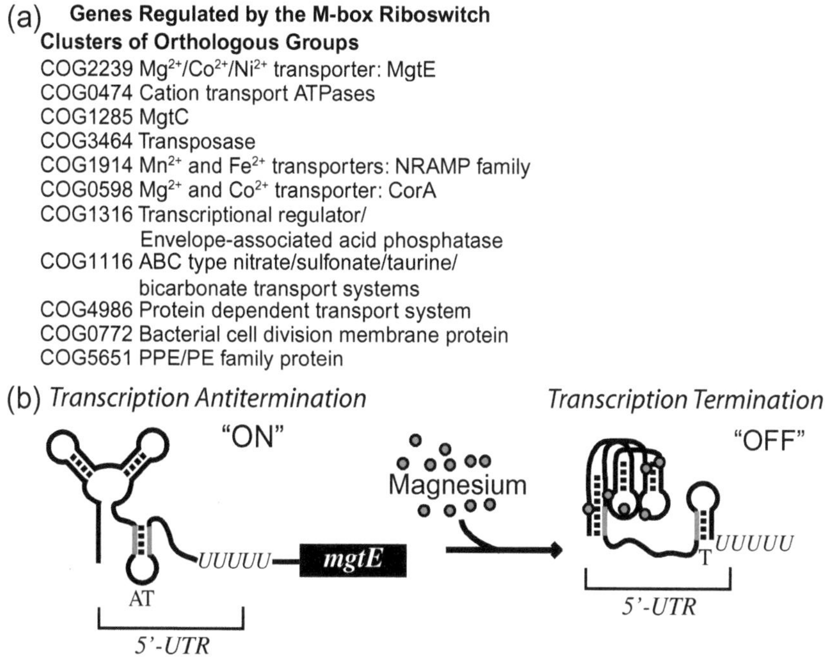

Figure 4. Genetic regulation by the M-box riboswitch. **(a)** Classes of genes that are regulated by the M-box riboswitch. **(b)** Shown in this figure is a schematic depiction of genetic regulation by the M-box riboswitch. Under conditions of high Mg^{2+}, the P1 helix at the base of the aptamer domain is stabilized, which allows for formation of a downstream termination site. Under conditions of low Mg^{2+}, the right half of the P1 helix associates with downstream sequences to prevent terminator formation, and thereby allow for transcription elongation to continue synthesis of downstream genes.

transport Mn^{2+}, Ca^{2+}, Zn^{2+}, or Fe^{2+}. MgtE imports Mg^{2+} and Co^{2+} with an apparent K_m of 300–700 μM for Mg^{2+}, but not Ni^{2+}, which acts as a potent inhibitor of Mg^{2+} transportation. The third class of dedicated Mg^{2+} transporters, called MgtA/B, is a subset of P-type ATPase transporters. In contrast to other P-type ATPase transporters, MgtA/B mediate influx of Mg^{2+} along its electrochemical gradient, rather than against it. Also, they exhibit an apparent affinity of 10 μM for import of the ion, similar in principle to CorA.

Interestingly, all three classes of transporters are subjected to control by M-box riboswitches [97]. However, of the M-box RNAs that have been identified, most appear to be present upstream of homologues of genes for MgtE and MgtA/B [97]. There are only a few examples where the M-box is arranged upstream of CorA homologues. The distribution of these gene categories varies extensively between bacterial species. So, too, does the distribution of M-box RNAs. Most bacteria contain multiple homologues of the different Mg^{2+} transport classes, but only a subset of these genes appear to be subjected to genetic control by an M-box riboswitch. For example, *Listeria monocytogenes* contains three homologues of *corA* and a single copy of *mgtA*; only the *mgtA* gene is regulated by an M-box riboswitch. In contrast, *Bacillus halodurans* contains four *mgtE* homologues, each under control of a separate M-box RNA, along with an *mgtA* that lacks the riboswitch. For some microorganisms, the M-box RNA appears to regulate genes other than Mg^{2+} transporters. For example, in rare instances, the M-box riboswitch is predicted to regulate expression of genes encoding for MgtC, NRAMP family transporters, an unknown transcription factor, uncharacterized ABC type transport genes, cell surface PPE/PE family proteins (in *Mycobacterium* species) and a few unknown, uncharacterized genes [97].

To test whether a regulatory connection is shared between Mg^{2+} pools and expression of M-box-associated genes, a representative *B. subtilis* M-box riboswitch was transcriptionally fused to a *lacZ* reporter gene [15]. Cells expressing this riboswitch-reporter fusion were cultured in media with varying levels of Mg^{2+}. Expression of the *lacZ* reporter was increased upon depletion of Mg^{2+}, but not other divalents, suggesting that the riboswitch indeed specifically responds to Mg^{2+} fluctuations *in vivo*.

5.2. A Mechanistic Model for Metal-Responsive Regulation by the M-Box RNA

Demonstration of Mg^{2+}-responsive regulation by the M-box RNA *in vivo* suggested that one of two possibilities was likely to occur: (1) the M-box aptamer domain constitutes a direct sensor for Mg^{2+} ions, functionally similar to metabolite-sensing aptamers, or (2) the M-box aptamer domain senses a metabolite whose levels fluctuate as a function of intracellular Mg^{2+} pools. All available data suggest the former, that the M-box RNA directly senses intracellular Mg^{2+}, without aid of accessory factors or indirect signaling pathways. This is accomplished using a mechanistic strategy that closely mirrors that of metabolite-sensing riboswitches, and that exploits the propensity of metals to encourage compacted RNA conformations [15,36,98].

To ascertain whether the conserved portion of M-box riboswitches functions as a distinct Mg^{2+} sensory domain, conserved residues were individually altered by site-directed mutagenesis and downstream expression of a *lacZ* reporter was assayed at varying Mg^{2+} concentrations. These mutations disrupted the *in vivo* Mg^{2+} regulatory response, as did disruption of a downstream putative intrinsic terminator site, suggesting that, indeed, the putative aptamer domain coupled a Mg^{2+}-responsive signal with control of transcription attenuation. Indeed, most M-box riboswitches contain putative intrinsic terminator sites downstream of the aptamer domain, suggesting that control of transcription attenuation is the primary mode of genetic regulation for this riboswitch class (Figure 4b). To rule out a role for protein accessory factors in the regulatory mechanism, transcription reactions of the M-box riboswitch were assessed *in vitro*, using only purified components (RNA polymerase, DNA templates, Mg^{2+}, and nucleotides). Mg^{2+}-responsive control of transcription attenuation could be recapitulated in these reactions, demonstrating that Mg^{2+} alone constitutes the signal that is perceived by the M-box aptamer domain. Structural probing of the M-box aptamer domain revealed a concerted, metal-induced conformational change, that occurred with a half-maximal concentration of $\sim$0.5 mM Mg^{2+} [15]. Importantly, cobalt(III) hexammine was unable to elicit this conformational change, suggesting that compaction was dependent on partially dehydrated divalent ions. Furthermore, molar quantities of monovalent ions were unable to trigger the metal-bound conformation alone; divalent ions were required even in the presence of 2.1 M monovalents. A significant decrease in hydrodynamic radius accompanied the metal-induced conformational change, as evidenced by an increase in sedimentation velocity. Indeed, a 'compact' conformation could be physically separated from the population of low-Mg^{2+}, 'extended' conformation(s) by chromatography or nondenaturing polyacrylamide gel electrophoresis.

The 2.6 Å structure of the Mg^{2+}-bound M-box was solved by X-ray crystallography, using crystals grown in the presence of 10 mM Mg^{2+} [15]. This revealed a compact structure, mostly comprising of three closely packed, nearly parallel helices. Inter-helical packing interactions are stabilized by long-range hydrogen bonding interactions between nucleosides, including multiple A-minor motifs (Figures 5 and 6). A total of six well-ordered Mg^{2+} were observed in the structural model, all of which involved at least one innersphere interaction. Indeed, one of the metals included a total of four innersphere contacts to RNA oxygen groups. For each of these metals, the Mg^{2+} ions appeared to assist positioning of neighboring residues in order to promote long-range interactions. Almost all of the RNA-metal contacts in the structural model occurred with nonbridging phosphate oxygens. To functionally probe the importance of these interactions, nonbridging phosphate oxygens were randomly replaced with phosphorothioate groups, by standard nucleotide analog interference methods. The individual phosphorothioate substitutions that were found to restrict the RNA to the extended conformation agreed well with the sites of innersphere interactions

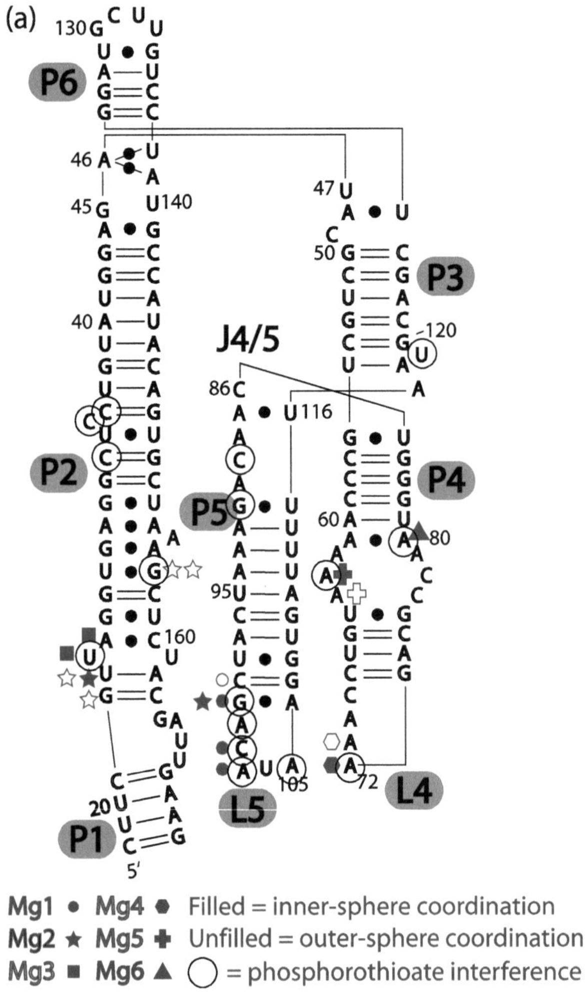

Figure 5. Secondary structure of the *B. subtilis mgtE* M-box aptamer. Symbols denote locations of metal ions that form interactions with the RNA, as observed within the crystallographically determined structural model. Circled residues denote sites of phosphorothioate interferences, which prevent the compacted metal-bound conformation.

that were proposed by the structural model (Figure 5). Therefore, the compact, metal-bound conformation is dependent upon association of a network of Mg^{2+} ions within specific electronegative pockets. Biochemical studies show this transition occurs over a very narrow range of metal concentrations.

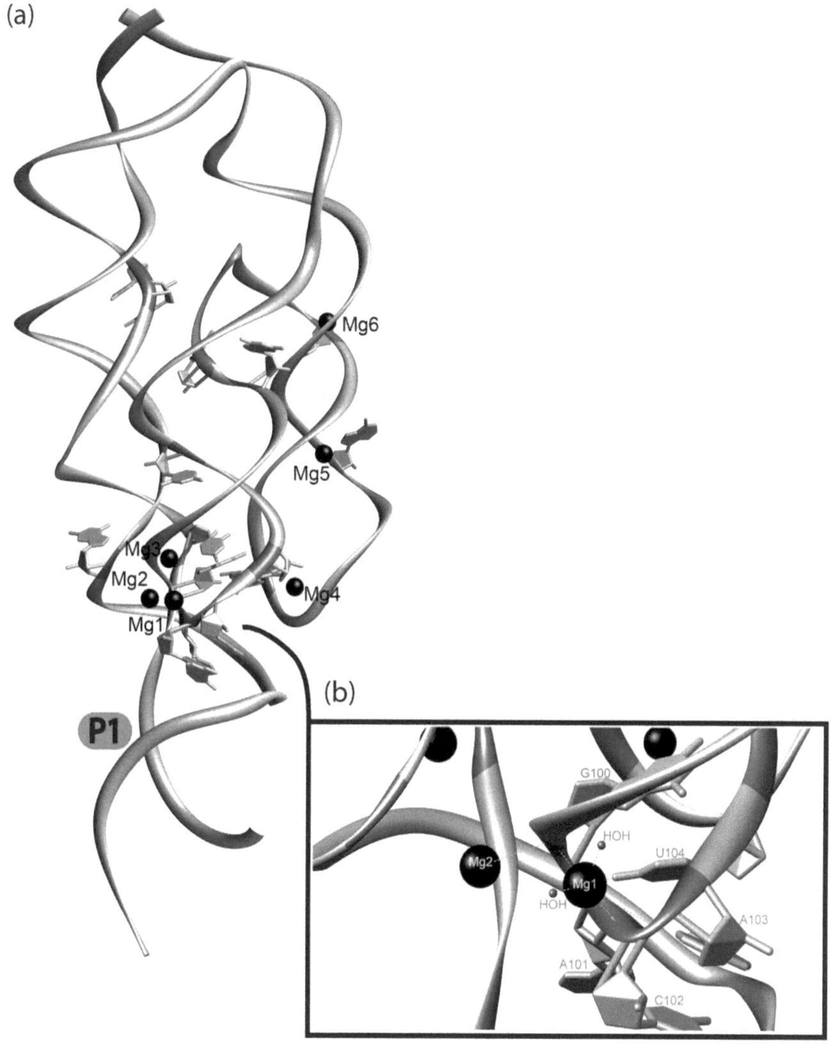

Figure 6. Tertiary structure of the *B. subtilis mgtE* M-box aptamer. Most of the crystallographically observed divalent binding sites occur in a tripartite region wherein two terminal loops (L4, L5) associate with each other and a helix (P2) at the base of the molecule. Long-range base interactions within this region help stabilize the compact tertiary conformation. One of the six RNA-bound Mg^{2+} ions appears to be particular important for the compacted RNA conformation, as evidenced by strong phosphorothioate interferences and four inner-sphere metal coordinations.

The regulatory link between metal-mediated compaction and genetic control is similar in concept to metabolite-sensing riboswitches. In the compact conformation, nucleotides at the 5′ and 3′ termini of the aptamer pair together to form the P1 helix. Specifically, tertiary contacts and base-stacking interactions that further stabilize the P1 helix are formed in the compact conformation. In contrast, the right half of the P1 helix interacts with downstream sequences to form a mutually exclusive helix in low Mg^{2+}. This 'antiterminator' competes with formation of the intrinsic terminator, resulting in control of attenuation. In total, these data demonstrate how an RNA element can function as a metalloregulatory agent, to couple metal concentrations to control of gene expression. Together, the discovery of both the M-box and *Salmonella mgtA* riboswitches suggest that there are likely to be additional mechanisms utilized for control of Mg homeostasis that still await discovery.

6. ARE THERE ADDITIONAL CLASSES OF METAL-SENSING RIBOSWITCHES?

Orphan riboswitches are structured RNA elements encoded in bacterial genomes that share basic features with known riboswitches, but whose metabolite ligands remain unidentified [7,88–93]. Individual orphan riboswitch classes are usually upstream of a common collection of genes, although a few classes are located upstream of genes of diverse functions. Additionally, orphan riboswitches exhibit secondary structure and primary sequence conservation patterns resembling established riboswitches. They are also generally co-transcribed with obvious sequence and structural features that are predicted to regulate translation initiation or transcription termination, again, akin to known riboswitches. Indeed, many of the current, established riboswitches were initially described as orphan riboswitches.

A 2004 search for riboswitches resulted in discovery of several new putative riboswitch classes [88]. Metabolite ligands for at least two of these classes, which appear to regulate expression of metal transport genes, have not yet been identified. For example, the so-called '*yybP*' orphan most often regulates expression of a wide variety of metal transport genes, including Mg^{2+} transporters, an assortment of metal efflux genes, and toxic anion resistance genes. The so-called '*ydaO*' orphan riboswitch most often regulates K^{+} transport genes, uncharacterized transport genes, and cell wall hydrolases. It is not yet clear from existing data whether these orphan riboswitches respond to metal ion-based signals, or to metabolites that fluctuate in concert with certain metal ions. Also, it is possible that a subset of M-box RNAs could function differently from the *B. subtilis* variant that has been examined.

A small subset of M-box RNAs has been predicted to regulate NRAMP genes, which typically transport Fe^{3+} or Mn^{2+} [97]. From this observation, one can speculate that certain M-box RNAs may respond specifically to Mn^{2+}, or, rather that certain microorganisms have evolved a regulatory interrelationship between Mg^{2+} and Mn^{2+} homeostasis.

ACKNOWLEDGMENTS

The authors thank past and current members of the their laboratories for their many contributions. Riboswitch research in the authors' laboratories has been funded by the Bill and Melinda Gates Foundation, the Damon Runyon Cancer Research Foundation, the Howard Hughes Medical Institute (HHMI), the National Institutes of Health (GM63576 and RR15943), and the W. M. Keck Foundation (A.R.F), and the National Institutes of Health (GM081882), the Welch Foundation, and the Searle Scholars Program (W.C.W.). A.R.F. is an Investigator of the HHMI.

ABBREVIATIONS

c-di-GMP	cyclic diguanylate
FMN	flavin mononucleotide
Fur	ferric uptake regulator
GlcN6P	glucosamine-6-phosphate
glmS	glucosamine-6-phosphate synthetase
lacZ	β-galactosidase
mgtA	magnesium transport protein
NADPH	nicotinamide adenine dinucleotide phosphate (reduced)
NOX2	NADPH oxidase
NRAMP	natural resistance-associated macrophage protein
PE	unique *Mycobacterium* genes with Pro-Glu at the N-terminus
PhoP	response regulator phosphorylated by PhoQ
PhoQ	sensor kinase that phosphorylates PhoP
PPE	unique *Mycobacterium* genes with Pro-Pro-Glu at the N-terminus
$preQ_1$	biosynthetic precursor to quoeuosine
Q	quoeuosine
RFN	the flavin mononucleotide riboswitch aptamer domain

R_g	radius of gyration
Rho	transcription termination protein
SAM-I	a class of *S*-adenosylmethionine-sensing riboswitches
SAXS	small-angle X-ray scattering
SCV	*Salmonella*-containing vacuole
TMP	thiamine-monophosphate
TPP	thiamine-pyrophosphate
tRNA	transfer RNA
UTR	untranslated mRNA region

REFERENCES

1. D. E. Draper, D. Grilley and A. M. Soto, *Annu. Rev. Biophys. Biomol. Struct.*, 2005, **34**, 221–243.
2. V. K. Misra, R. Shiman and D. E. Draper, *Biopolymers*, 2003, **69**, 118–136.
3. R. K. O. Sigel and A. M. Pyle, *Chem. Rev.*, 2007, **107**, 97–113.
4. D. J. Klein, P. B. Moore and T. A. Steitz, *RNA*, 2004, **10**, 1366–1379.
5. W. C. Winkler and R. R. Breaker, *Annu. Rev. Microbiol.*, 2005, **59**, 487–517.
6. T. E. Edwards, D. J. Klein and A. R. Ferré-D'Amaré, *Curr. Opin. Struct. Biol.*, 2007, **17**, 273–279.
7. M. D. Dambach and W. C. Winkler, *Curr. Opin. Microbiol.*, 2009, **12**, 161–169.
8. L. A. Finney and T. V. O'Halloran, *Science*, 2003, **300**, 931–936.
9. Z. Ma, F. E. Jacobsen and D. P. Giedroc, *Chem. Rev.*, 2009, **109**, 4644–4681.
10. M. E. Maguire, *Front. Biosci.*, 2006, **11**, 3149–3163.
11. T. V. O'Halloran, *Science*, 1993, **261**, 715–725.
12. D. P. Giedroc and A. I. Arunkumar, *Dalton Trans.*, 2007, **29**, 3107–3120.
13. A. S. Moomaw and M. E. Maguire, *Physiology*, 2008, **23**, 275–285.
14. E. Massé, H. Salvail, G. Desnoyers and M. Arguin, *Curr. Opin. Microbiol.*, 2007, **10**, 140–145.
15. C. E. Dann, C. A. Wakeman, C. L. Sieling, S. C. Baker, I. Irnov and W. C. Winkler, *Cell*, 2007, **130**, 878–892.
16. E. A. Groisman, M. J. Cromie, Y. Shi and T. Latifi, *Cold Spring Harb. Symp. Quant. Biol.*, 2006, **71**, 251–258.
17. A. Serganov, A. Polonskaia, A. T. Phan, R. R. Breaker and D. J. Patel, *Nature*, 2006, **441**, 1167–1171.
18. T. E. Edwards and A. R. Ferré-D'Amaré, *Structure*, 2006, **14**, 1459–1468.
19. N. Kulshina, T. E. Edwards and A. R. Ferré-D'Amaré, *RNA*, 2010, **16**, 186–196.
20. S. Thore, C. Frick and N. Ban, *J. Am. Chem. Soc.*, 2008, **130**, 8116–8117.
21. D. J. Klein and A. R. Ferré-D'Amaré, *Science*, 2006, **313**, 1752–1756.
22. W. C. Winkler, A. Nahvi, A. Roth, J. A. Collins and R. R. Breaker, *Nature*, 2004, **428**, 281–286.
23. J. A. Collins, I. Irnov, S. Baker and W. C. Winkler, *Genes Dev.*, 2007, **21**, 3356–3368.
24. T. J. McCarthy, M. A. Plog, S. A. Floy, J. A. Jansen, J. K. Soukoup and G. A. Soukoup, *Chem. Biol.*, 2005, **12**, 1221–1226.

25. D. J. Klein, M. D. Been and A. R. Ferré-D'Amaré, *J. Am. Chem. Soc.*, 2007, **129**, 14858–14859.
26. J. C. Cochrane, S. V. Lipchock and S. A. Strobel, *Chem. Biol.*, 2007, **14**, 97–105.
27. D. J. Klein, S. R. Wilkinson, M. D. Been and A. R. Ferré-D'Amaré, *J. Mol. Biol.*, 2007, **373**, 178–189.
28. A. Serganov, L. Huang and D. Patel, *Nature*, 2009, **457**, 233–237.
29. J. A. Cowan, *J. Inorg. Biochem.*, 1993, **49**, 171–175.
30. N. Sudarsan, E. R. Lee, Z. Weinberg, R. H. Moy, J. N. Kim, K. H. Link and R. R. Breaker, *Science*, 2008, **321**, 411–413.
31. N. Kulshina, N. J. Baird and A. R. Ferré-D'Amaré, *Nat. Struct. Mol. Biol.*, 2009, **16**, 1212–1217.
32. K. D. Smith, S. V. Lipchock, T. D. Ames, J. Wang, R. R. Breaker and S. A. Strobel, *Nat. Struct. Mol. Biol.*, 2009, **16**, 1218–1223.
33. H. Xiao, T. E. Edwards and A. R. Ferré-D'Amaré, *Chem. Biol.*, 2008, **15**, 1125–1137.
34. A. Serganov, L. Huang and D. Patel, *Nature*, 2008, **455**, 1263–1267.
35. A. D. Garst, A. Héroux, R. P. Rambo and R. T. Batey, *J. Biol. Chem.*, 2008, **283**, 22347–22351.
36. N. J. Baird and A. R. Ferré-D'Amaré, *RNA*, 2010, **16**, 598–609.
37. S. Basu, R. P. Rambo, J. Strauss-Soukoup, J. H. Cate, A. R. Ferré-D'Amaré, S. A. Strobel and J. A. Doudna, *Nature Struct. Biol.*, 1998, **5**, 986–992.
38. A. Roth, W. C. Winkler, E. E. Regulski, B. W. Lee, J. Lim, I. Jona, J. E. Barrick, A. Ritwik, J. N. Kim, R. Welz, D. Iwata-Renyl and R. R. Breaker, *Nat. Struct. Mol. Biol.*, 2007, **14**, 308–317.
39. M. M. Meyer, A. Roth, S. M. Chervin, G. A. Garcia and R. R. Breaker, *RNA*, 2008, **14**, 685–695.
40. U. Rieder, K. Lang, C. Kreutz, N. Polacek and R. Micura, *ChemBioChem*, 2009, **10**, 1141–1144.
41. M. Kang, R. Peterson and J. Feigon, *Mol. Cell*, 2010, **39**, 653–655.
42. D. Klein, T. Edwards and A. R. Ferré-D'Amaré, *Nat. Struct. Mol. Biol.*, 2009, **16**, 343–344.
43. R. Spitale, A. Torelli, J. Krucinska, V. Bandarian and J. Wedekind, *J. Biol. Chem.*, 2009, **284**, 11012–11016.
44. D. E. Draper, *RNA*, 2004, **10**, 335–343.
45. W. C. Winkler, A. Nahvi and R. R. Breaker, *Nature*, 2002, **419**, 952–956.
46. S. Cayley, B. A. Lewis, H. J. Guttman and M. T. Record, *J. Mol. Biol.*, 1991, **222**, 281–300.
47. J. B. Murray, A. A. Seyhan, N. G. Walter, J. M. Burke and W. G. Scott, *Chem. Biol.*, 1998, **5**, 587–595.
48. D. Lambert, D. Leipply, R. Shiman and D. R. Draper, *J. Mol. Biol.*, 2009, **390**, 791–804.
49. R. Russell, I. S. Millett, S. Doniach and D. Herschlag, *Nat. Struct. Biol.*, 2000, **7**, 367–370.
50. X. Fang, K. Littrell, X.-J. Yang, S. J. Henderson, S. Siefert, P. Thiyagarajan, T. Pan and T. R. Sosnick, *Biochemistry*, 2000, **39**, 11107–11113.

51. N. Baird, E. Westhof, H. Qin, T. Pan and T. Sosnick, *J. Mol. Biol.*, 2005, **352**, 712–722.
52. S. Chauhan, G. Caliskan, R. M. Briber, U. Perez-Salas, P. Rangan, D. Thirumalai and S. A. Woodson, *J. Mol. Biol.*, 2005, **353**, 1199–1209.
53. J. Lipfert, R. Das, V. B. Chu, M. Kudaravalli, N. Boyd, D. Herschlag and S. Doniach, *J. Mol. Biol.*, 2007, **365**, 1393–1406.
54. R. J. P. Williams and J. J. R. Frausto da Silva, *Coord. Chem. Rev.*, 2000, **200–202**, 247–348.
55. J. A. Tainer, V. A. Roberts and E. D. Getzoff, *Curr. Opin. Biotechnol.*, 1991, **2**, 582.
56. I. R. Booth, *Microbiol. Rev.*, 1985, **49**, 359–378.
57. J. M. Wood, *Microbiol. Mol. Biol. Rev.*, 1999, **63**, 230–262.
58. H. Irving and R. J. P. Williams, *Nature*, 1948, **162**, 746–747.
59. K. J. Waldron, K. C. Rutherford, D. Ford and N. J. Robinson, *Nature*, 2009, **460**, 823–830.
60. N. J. Robinson, *Nature Chem. Biol.*, 2007, **3**, 692–693.
61. S. Mounicou, J. Szpunar and R. Lobinski, *Chem. Soc. Rev.*, 2009, **38**, 1119–1138.
62. L. L. Barton, F. Goulhen, M. Bruschi, N. A. Woodards, R. M. Plunkett and F. J. Rietmeijer, *Biometals*, 2007, **20**, 291–302.
63. J. R. Forbes and P. Gros, *Trends Microbiol.*, 2001, **9**, 397–403.
64. C. Wandersman and P. Delepelaire, *Annu. Rev. Microbiol.*, 2004, **58**, 611–647.
65. S. P. Salcedo, M. Noursadeghi, J. Cohen and D. W. Holden, *Cell Microbiol.*, 2001, **3**, 587–597.
66. J. A. Ibarra and O. Steele-Mortimer, *Cell Microbiol.*, 2009, **11**, 1579–1586.
67. J. H. Brumell and S. Grinstein, *Curr. Opin. Microbiol.*, 2004, **7**, 78–84.
68. R. Krishnakumar, B. Kim, E. A. Mollo, J. A. Imlay and J. M. Slauch, *J. Bacteriol.*, 2007, **189**, 4343–4352.
69. E. A. Groisman, *J. Bacteriol.*, 2001, **183**, 1835–1842.
70. J. C. Perez, D. Shin, I. Zwir, T. Latifi, T. J. Hadley and E. A. Groisman, *PLoS Genetics*, 2009, **5**, e1000428.
71. E. Garcia Vescovi, F. C. Soncini and E. A. Groisman, *Cell*, 1996, **84**, 165–174.
72. L. R. Prost, M. E. Daley, V. Le Sage, M. W. Bader, H. Le Moual, R. E. Klevit and S. I. Miller, *Mol. Cell*, 2007, **26**, 165–174.
73. M. W. Bader, S. Sanowar, M. E. Daley, A. R. Schneider, U. Cho, W. Xu, R. E. Klevit, H. Le Moual and S. I. Miller, *Cell*, 2005, **122**, 461–472.
74. A. Kato and E. A. Groisman, *Adv. Exp. Med. Biol.*, 2008, **631**, 7–21.
75. L. R. Prost, S. Sanowar and S. I. Miller, *Immunol. Rev.*, 2007, **219**, 55–65.
76. I. Zwir, D. Shin, A. Kato, K. Nishino, T. Latifi, F. Solomon, J. M. Hare, H. Huang and E. A. Groisman, *Proc. Natl. Acad. Sci.*, 2005, **102**, 2862–2867.
77. U. S. Cho, M. W. Bader, M. F. Amaya, M. E. Daley, R. E. Klevit, S. I. Miller and W. Xu, *J. Mol. Biol.*, 2006, **356**, 1193–1206.
78. J. Cheung, C. A. Bingman, M. Reyngold, W. A. Hendrickson and C. D. Waldburger, *J. Biol. Chem.*, 2008, **283**, 13762–13770.
79. E. Choi, E. A. Groisman and D. Shin, *J. Bacteriol.*, 2009, **191**, 7174–7181.
80. E. Alix and A. Blanc-Potard, *Trends Microbiol.*, 2007, **15**, 252–256.
81. E. A. Groisman and C. Mouslim, *Nat. Rev. Microbiol.*, 2006, **4**, 705–709.

82. M. J. Cromie, Y. Shi, T. Latifi and E. A. Groisman, *Cell*, 2006, **125**, 71–84.
83. S. Chamnongpol and E. A. Groisman, *J. Mol. Biol.*, 2000, **300**, 291–305.
84. T. Tao, P. F. Grulich, L. M. Kucharski, R. L. Smith and M. E. Maguire, *Microbiol.*, 1998, **144**, 655–664.
85. Irnov, A. Kertsburg and W. C. Winkler, *Cold Spring Harb. Symp. Quant. Biol.*, 2006, **71**, 239–249.
86. S. V. Spinelli, L. B. Pontel, E. Garcia Vescovi and F. C. Soncini, *FEMS Microbiol. Lett.*, 2008, **280c**, 226–234.
87. M. J. Cromie and E. A. Groisman, *J. Bacteriol.*, 2010, **192**, 604–607.
88. J. E. Barrick, K. A. Corbino, W. C. Winkler, A. Nahvi, M. Mandal, J. Collins, M. Lee, A. Roth, N. Sudarsan, I. Jona, J. K. Wickiser and R. R. Breaker, *Proc. Natl. Acad. Sci. USA*, 2004, **101**, 6421–6426.
89. K. A. Corbino, J. E. Barrick, J. Lim, R. Welz, B. J. Tucker, I. Puskarz, M. Mandal, N. D. Rudnick and R. R. Breaker, *Genome Biol.*, 2005, **6**, R70.
90. Z. Weinberg, J. E. Barrick, Z. Yao, A. Roth, J. N. Kim, J. Gore, J. X. Wang, E. R. Lee, K. F. Block, N. Sudarsan, S. Neph, M. Tompa, W. L. Ruzzo and R. R. Breaker, *Nucleic Acids Res.*, 2007, **35**, 4809–4819.
91. Z. Weinberg, J. X. Wang, J. Bogue, J. Yang, K. Corbino, R. Moy and R. R. Breaker, *Genome Biol.*, 2010, **11**, R31.
92. P. P. Gardner, J. Daub, J. G. Tate, E. P. Nawrocki, D. L. Kolbe, S. Lindgreen, A. C. Wilkinson, R. D. Finn, S. Griffiths-Jones, S. R. Eddy and A. Bateman, *Nucleic Acids Res.*, 2009, **37**, D136–140.
93. X. Xu, Y. Ji and G. D. Stormo, *PLoS Comput. Biol.*, 2009, **5**, e1000338.
94. K. F. Block, M. C. Hammond and R. R. Breaker, *J. Bacteriol.,* Epub ahead of print. Doi:10.1128/JB.00450–10 (2010).
95. M. E. Maguire and J. A. Cowan, *BioMetals*, 2002, **15**, 203–210.
96. D. G. Kehres and M. E. Maguire, *BioMetals*, 2002, **15**, 261–270.
97. A. Ramesh and W. C. Winkler, *RNA Biol.*, 2010, **7**, 77–83.
98. C. A. Wakeman, A. Ramesh and W. C. Winkler, *J. Mol. Biol.*, 2009, **392**, 723–735.

Met. Ions Life Sci. **2011**, *9*, 175–196

6

Metal Ions: Supporting Actors in the Playbook of Small Ribozymes

Alexander E. Johnson-Buck, Sarah E. McDowell, and Nils G. Walter

[a]Department of Chemistry, University of Michigan, 930 N. University, Ann Arbor, MI 48109-1055, USA

<nwalter@umich.edu> <semcdowe@umich.edu> <alebuck@umich.edu>

Metal Ions in Life Sciences, Volume 9 Edited by Astrid Sigel, Helmut Sigel, and Roland K. O. Sigel

Published by the Royal Society of Chemistry, www.rsc.org DOI: 10.1039/9781849732512000175

ABSTRACT: Since the 1980s, several small RNA motifs capable of chemical catalysis have been discovered. These small ribozymes, composed of between approximately 40 and 200 nucleotides, have been found to play vital roles in the replication of subviral and viral pathogens, as well as in gene regulation in prokaryotes, and have recently been discovered in noncoding eukaryotic RNAs. All of the known natural small ribozymes – the hairpin, hammerhead, hepatitis delta virus, Varkud satellite, and *glmS* ribozymes – catalyze the same self-cleavage reaction as RNase A, resulting in two products, one bearing a 2′-3′ cyclic phosphate and the other a 5′-hydroxyl group. Although originally thought to be obligate metalloenzymes like the group I and II self-splicing introns, the small ribozymes are now known to support catalysis in a wide variety of cations that appear to be only indirectly involved in catalysis. Nevertheless, under physiologic conditions, metal ions are essential for the proper folding and function of the small ribozymes, the most effective of these being magnesium. Metal ions contribute to catalysis in the small ribozymes primarily by stabilizing the catalytically active conformation, but in some cases also by activating RNA functional groups for catalysis, directly participating in catalytic acid-base chemistry, and perhaps by neutralizing the developing negative charge of the transition state. Although interactions between the small ribozymes and cations are relatively nonspecific, ribozyme activity is quite sensitive to the types and concentrations of metal ions present in solution, suggesting a close evolutionary relationship between cellular metal ion homeostasis and cation requirements of catalytic RNAs, and perhaps RNA in general.

KEYWORDS: electrostatic screening · general acid-base catalysis · *glmS* ribozyme · hairpin ribozyme · hammerhead ribozyme · hepatitis delta virus ribozyme · Varkud satellite ribozyme

1. INTRODUCTION

Since the discovery that RNA can catalyze chemical reactions [1], RNA enzymes (ribozymes) have been found to perform many essential functions in nature, including protein biosynthesis [2], RNA processing [1,3,4], regulation of gene expression [5], and genomic processing in pathogens [6–10]. While some of these functional RNAs operate within the context of large ribonucleoprotein complexes, many ribozymes can support catalysis without protein cofactors [1,5–10]. The naturally occurring small self-cleaving ribozymes, each comprising fewer than 200 nucleotides, demonstrate the capability of RNA to efficiently and economically catalyze biologically important chemistry. These include the hairpin [7], hammerhead [6,9], hepatitis delta virus (HDV) [8,11,12], Varkud satellite (VS) [10], and *glmS* [5] ribozymes. Although all of these were initially isolated from bacteria [5], viruses [7,11,12], or subviral pathogens [6,9,10], structural and functional homologs of the hammerhead and HDV ribozymes have recently been discovered within the genomes of several eukaryotes, including mammals [13–15], revealing the exciting possibility that small catalytic RNAs may help regulate eukaryotic gene expression. The versatility of small ribozymes as catalysts has been demonstrated by the discovery of many non-natural small ribozymes through *in vitro* selection, including a lead-dependent self-cleaving RNA [16], a ribozyme that catalyzes the synthetically useful Diels-Alder cycloaddition [17], and ribozymes that exploit allosteric binding of particular classes of metal ions [18,19]. The discovery of a tiny 29-nucleotide RNA that catalyzes aminoacyl-RNA synthesis [20] and a recently reported self-replicating RNA enzyme [21] support the notion that RNA could have served as the original catalyst of life [22].

In spite of this versatility, all of the currently known natural small ribozymes catalyze the same internal phosphodiester isomerization reaction as RNase A, resulting in two cleavage products: one bearing a 2′,3′-cyclic phosphate, and the other a 5′-hydroxyl group (see Figure 1). This reaction involves deprotonation of the 2′-OH nucleophile by a general base, attack of the activated 2′-oxyanion on the phosphorous atom of the adjacent phosphate, and

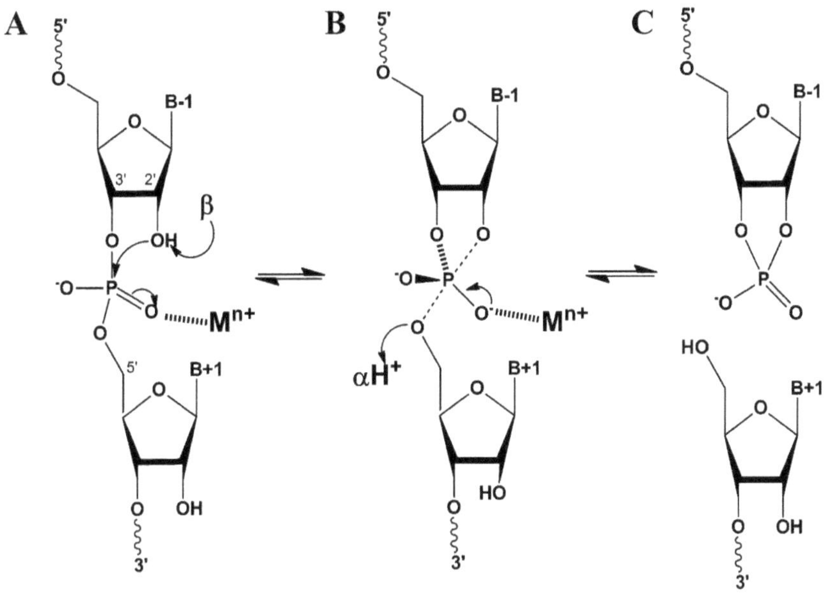

Figure 1. General mechanism of self-cleavage by the natural small ribozymes. **(A)** A Brønsted-Lowry base (β) abstracts a proton to activate the 2′-OH nucleophile, which then attacks the adjacent phosphate, forming a pentacoordinate transition state (**B**) with approximate collinearity between the 2′-oxygen, phosphorus atom, and 5′-oxygen leaving group – the in-line attack geometry. The negative charge of the transition state may be stabilized by one or several metal cations (M^{n+}) that interact through inner- or outersphere contacts with the non-bridging oxygen atoms or by long-distance coulombic stabilization. A Brønsted-Lowry acid (α) donates a proton to the 5′-oxygen leaving group, resulting in a 5′-product bearing a 2′,3′-cyclic phosphate and a 3′-product bearing a 5′-OH (**C**).

protonation of the 5′-oxygen leaving group by a general acid [23–26]. Although the small ribozymes were originally thought to be obligate metalloenzymes like the group I and II self-splicing introns, utilizing site-bound magnesium to directly coordinate and stabilize negatively charged groups in the transition state, this is no longer the prevailing view [27]. However, like all functional RNA, ribozymes require cations to counterbalance the abundant negative charge of their phosphate backbone as they fold into their functional three-dimensional conformations, and may also utilize cations in long-range electrostatic catalysis or general acid-base chemistry. At physiologic ionic strength, multivalent cations are essential for the small ribozymes to adopt their native tertiary structures [27], though a number of other metallic and non-metallic cations support catalysis to varying degrees [19,28–38].

The composition of free metal ions in the cell is well-tuned to the function of small ribozymes. The predominant metal cations in the cytosol are typically K^+, with an activity of ~100 mM [39], and Mg^{2+}, with an activity of ~1 mM [40]. It has recently been revealed that magnesium concentration is regulated by homeostasis in both bacterial [41–44] and eukaryotic [45] cells. In turn, under physiologic conditions, magnesium ions are expected to play the most important role in stabilizing RNA tertiary structure and, by extension, facilitating the function of ribozymes. Perhaps not coincidentally, then, the natural small ribozymes have evolved to be functional at intracellularly available Mg^{2+}concentrations [5,35,46,47], and the likely enhancement of metal binding due to molecular crowding in the cytosol suggests an even tighter correlation [48]. In addition, it can be argued that the total cellular Mg^{2+} concentration of ~20 mM [49] is buffered by the large amounts of nucleic acids and in particular RNAs (typically 1–6% of the cellular mass [50]) that bind the divalent metal ions with typical affinities in the low millimolar range. A picture emerges of an intimate relationship between the metal ion composition of the cell and the metal ion dependence of functional RNAs, analogous to the correlation between intracellular availability of metal ion cofactors and the affinity of protein enzymes for these cofactors [51]. The small ribozymes provide illuminating examples of how RNA structure, function, and dynamics have co-evolved to take advantage of a carefully maintained entourage of metal cations.

2. INTERACTIONS BETWEEN METAL IONS AND SMALL RIBOZYMES

2.1. Modes of Interaction

Near neutral pH, the phosphodiester backbone of RNA carries copious negative charge. In order for functional RNAs to fold into their compact native conformations, this negative charge must be at least partially neutralized. The necessary countercharge is supplied largely by metal ions, whose interactions with RNA span a continuum between two extremes: diffuse interactions, which are transient (with typical residence times thought to be in the millisecond regime [52,53]) and poorly localized, forming a kind of dynamic ionic atmosphere or "cloud" of positive electrostatic potential around the RNA; and specific interactions, which involve relatively tight (and longer-lived) binding to precise sites on the RNA molecule (Figure 2) [54]. Diffuse ion binding to RNA has been described theoretically using Hill-type binding formalisms and continuum treatments such as the nonlinear Poisson-Boltzmann equation [54,55], and accounts for the majority of the electrostatic stabilization in RNA [56]. Consistent with this observation, monovalent ions, which generally bind only weakly to RNA, can induce proper folding and activity in the small ribozymes [35,57,58]. Nevertheless, because the three-dimensional structure of RNA can develop concentrated pockets of negative electrostatic potential (-15 to $-20\,kT/e$ in the major groove, as low as $-100\,kT/e$ at some metal-binding sites) [59], entropy favors stabilization of compact native folds by divalent metal ions at physiologic concentrations [56].

Tightly bound metal ions are observed in crystal structures of all of the small ribozymes (Figure 3), and can associate with RNA through either innersphere interactions involving direct coordination to electronegative RNA functional groups (Figure 2A), or outersphere interactions mediated by water ligands (Figure 2B). Due to the great enthalpic penalty for completely dehydrating the metal ion and RNA, innersphere complexation generally requires a very dense pocket of buried negative charge, such as that provided by close proximity of several negatively charged oxygen atoms at

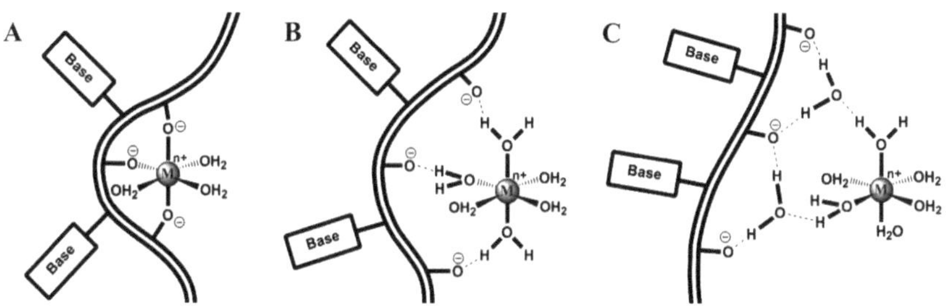

Figure 2. Modes of metal ion binding to RNA. Metal cations (M^{n+}) can associate with RNA via long-lived, specific interactions (**A**, **B**) requiring removal of some water molecules from the first and/or second hydration shell, or transient, diffuse interactions between the solvated RNA and metal ion (**C**). Specific interactions can involve direct chelation of the metal ion by RNA functional groups such as non-bridging phosphate oxygens (**A**), contacts mediated by innersphere water molecules (**B**), or a combination of the two.

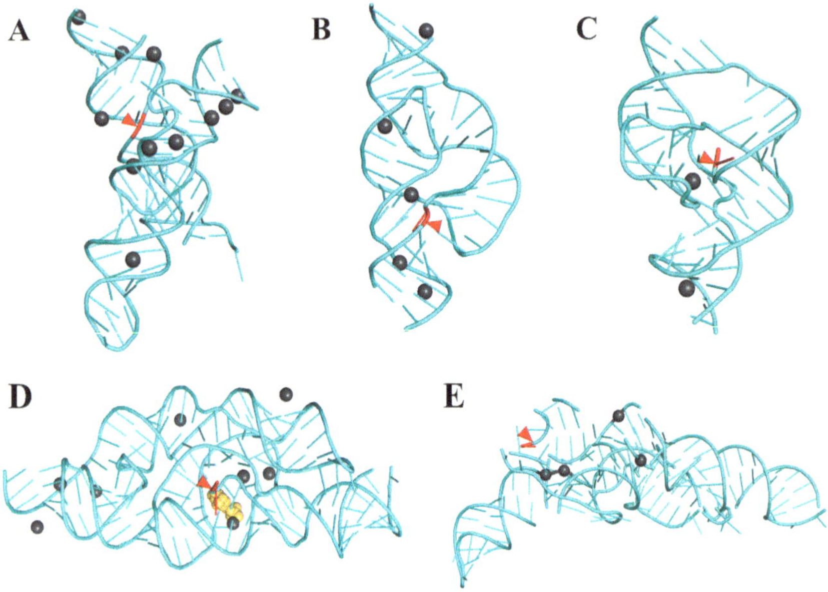

Figure 3. Three-dimensional structures and metal ion binding sites of the natural small ribozymes. The ribozyme structures are shown in cyan, divalent cations or probable binding sites in black, and the cleavage site in each ribozyme is indicated by a red arrow. Crystal structures of (**A**) a hairpin ribozyme in the presence of Ca^{2+} ions [65], (**B**) a hammerhead ribozyme with Mn^{2+} ions [64], (**C**) the HDV ribozyme with Mg^{2+} ions [62], and (**D**) the *glmS* ribozyme in Mg^{2+} ions, with the necessary glucosamine-6-phosphate cofactor shown in yellow [63]. Co-crystallized proteins and protein-binding domains of RNA used for crystallization purposes are not shown in these structures. (**E**) Partial three-dimensional structure of the VS ribozyme derived from two similar low-resolution models [142,143] (courtesy of Richard A. Collins and Ricardo Zamel), with black spheres indicating phosphates having probable direct contacts with divalent metal ions as revealed by phosphorothioate rescue with Mn^{2+} [72].

the interior of an RNA molecule [56]. Direct coordination of divalent cations to phosphoryl oxygens or to the N7 of purine bases is frequently observed, but outersphere interactions are far more common. In fact, all metal ions observed in crystal structures for the hammerhead, HDV, hairpin, and *glmS* ribozymes remain at least partly hydrated, even if they make some innersphere contacts as judged primarily by their distance to potential ligands on the RNA [60–66]. Compared with other divalent ions such as Ca^{2+}, Mn^{2+}, and Zn^{2+}, a magnesium ion has a greater propensity for outersphere interactions, in accordance with its small ionic radius and high charge density that give rise to a large hydration energy and relatively slow rate of water exchange [67,68].

Fully hydrated, tightly bound divalent metal ions have been resolved in X-ray crystal structures of the HDV, hammerhead, *glmS*, and hairpin ribozymes [61–65]. In the latter three of these, the exchange-inert complex cobalt(III) hexammine ($Co(NH_3)_6^{3+}$), used as a rough proxy for fully hydrated divalent ions [69], supports efficient self-cleavage [28,32,36,58], suggesting that outer-sphere coordination of metal ions is sufficient for activity. Usually, though, site-bound divalent ions make at least one innersphere contact with RNA functional groups in crystal structures of the small ribozymes. For instance, all three Mn^{2+} ions that are bound to conserved regions in a crystal structure of the hammerhead ribozyme form innersphere contacts, including an ion at the active site (Figure 3B) [64]. The fact that cobalt(III) hexammine inhibits this ribozyme in the presence of Mn^{2+} [70] suggests that some of these innersphere contacts are functionally important. Although the more biologically available magnesium may be expected not to coordinate to the same functional groups as the softer manganese(II) ion, molecular dynamics (MD) simulations suggest that a Mg^{2+} ion could effectively promote catalysis by occupying nearly the same site as a specific Mn^{2+} ion observed in the active site [71]. In a recently solved X-ray crystal structure of the HDV ribozyme in complex with an inhibitor oligoribonucleotide containing a deoxyribose moiety at the cleavage site, a Mg^{2+} ion was positioned to make innersphere contacts with key atoms at the active site, suggesting a catalytic role as a Lewis acid and/or general base for this ion [144]. While no crystal structure exists for the VS ribozyme, phosphorothioate interference-rescue experiments point to direct metal ion coordination to four phosphate groups in and around the catalytic core (Figure 3E) [72]. This ribozyme cannot efficiently self-cleave in the sole presence of cobalt(III) hexammine, suggesting that innersphere coordination may be important to activity [73]. Interestingly, cobalt(III) hexammine can cooperatively promote VS ribozyme activity in the presence of Mg^{2+} [73], consistent with the presence of at least some orthogonal outersphere and innersphere binding sites.

2.2. Selectivity of Metal Interactions

Most of the small ribozymes bind a variety of cations with different affinities, allowing them to fold and perform efficient self-cleavage with varying maximal rates. The *glmS*, hammerhead, HDV, and hairpin ribozymes can all self-cleave in a variety of divalent cations [5,31,34,74,75].

Consistent with its prevalence in the cell, the magnesium ion is among the most efficient of divalent ions at promoting catalysis in all of the natural small ribozymes, though this preference is only mild in many cases. The hairpin ribozyme cleaves in Mg^{2+} more than twice as efficiently as in Sr^{2+} and ten times as efficiently as in Ca^{2+}, and cannot cleave in Mn^{2+}, Co^{2+}, Ni^{2+}, or Cd^{2+} without facilitation by other cations [34,76]. The hammerhead ribozyme cleaves more rapidly in Mg^{2+} than other group IIA ions, but is actually more strongly activated by certain divalent transition metals (Mn^{2+}, Co^{2+}, Zn^{2+}, and Cd^{2+}) at concentrations of 1 mM divalent ion and 100 mM NaCl [74,75]. One innersphere site in the hammerhead ribozyme appears to accommodate Mg^{2+}, Mn^{2+}, Co^{2+}, and Cd^{2+} ions [64,77,78]. In the HDV ribozyme, a catalytically important site for hydrated Mg^{2+} appears to bind Ca^{2+}, Ba^{2+}, and Sr^{2+} with similar affinity, though catalytic activity is lower in barium and strontium ions [79]. This difference in activity may be tied to a structurally important site of innersphere metal ion coordination, observed biochemically, and shown to have selectivity for a magnesium ion over calcium, barium, and strontium [79], consistent with Raman crystallographic studies showing ~5 direct Mg^{2+}-phosphate contacts per HDV molecule [80]. Interestingly, while the active site of the genomic HDV ribozyme shows a slight preference for binding Mg^{2+} over Ca^{2+}, this preference is reversed in the antigenomic ribozyme, and can be switched by mutation of a single nucleotide [81].

Some binding sites not only accommodate divalent cations other than Mg^{2+}, but also some trivalent ions. While the exchange-inert cobalt(III) hexammine complex effectively binds and supports catalysis in several small ribozymes [28,32,36,58], it is not a perfect substitute for Mg^{2+}, generally giving rise to maximal cleavage rates 10- to 100-fold smaller than in magnesium. The $Co(NH_3)_6^{3+}$ complex can even inhibit some ribozymes, likely by displacing functionally important magnesium ions. For example, cobalt(III) hexammine has been observed to compete for the binding site of an outersphere coordinated magnesium ion at the active site of the HDV ribozyme (Figure 3C) [62], and even displaces some innersphere coordinated metal ions in the HDV and hammerhead ribozymes, consistent with its inhibitory effect on the activity of those ribozymes in the presence of divalent ions [70,82]. In other cases, site-bound ions may inhibit ribozymes by inducing alternate, inactive conformations, as has been suggested in the case of hammerhead, hairpin, and HDV ribozyme inhibition by terbium(III) ions [83–86].

All of the small ribozymes can accept monovalent salts as functional substitutes for divalent cations, as they are almost as active in molar concentrations of NaCl, LiCl, or even the non-metallic NH_4OAc as in millimolar $MgCl_2$ [35,57,58]. Comparison of several modified hammerhead ribozymes suggests that the RNA adopts a similar conformation in monovalent and divalent metal ions, albeit with some subtle differences at the interaction site of a catalytically important divalent ion [87]. Monovalent and divalent cations directly compete for some of the same interactions with the hairpin [36,55] and HDV ribozymes [88], though they act synergistically in promoting self-cleavage of the VS ribozyme [35]. Thus, none of the natural small ribozymes have a strict requirement for Mg^{2+} or other divalent cations, and they exhibit varying degrees of overlap between monovalent and divalent cation binding sites.

Intriguingly, small synthetic ribozymes have been engineered with strong functional selectivity for ions of transition metals or heavy metals over Mg^{2+} [16,19], yet such selectivity is not common in nature. This may be the combined result of the low intracellular activity and frequent toxicity of such metals, which are closely controlled by cellular homeostasis. It has been proposed that the 5S rRNA contains a natural lead-dependent ribozyme that may partly account for the cytotoxicity of lead [89]. The more acidic hydrated cations of transition metals, heavy metals, and lanthanides compared to Mg^{2+} also result in more rapid non-specific degradation of RNA through general base catalysis from their hydroxo complexes [90], making them a poor evolutionary choice for site-specific catalysis in ribozymes. The low free concentration of such ions in the cell and the paucity of natural ribozymes selective for them support the notion of a co-evolution between cellular metal ion composition and cation requirements of functional RNAs.

In summary, metal ions stabilize the structure of small ribozymes by binding diffusely or at specific sites, with generally low structural discrimination between divalent metal ions but stricter ion requirements for efficient catalysis. Water molecules mediate some or all of the contacts between a given metal ion and RNA functional groups because of the very unfavorable enthalpy of dehydration, especially for Mg^{2+}. Direct chelation of metal ions by RNA functional groups is occasionally required for optimal catalysis, and both labile and inert complexes of various metal ions compete for many of the same binding sites.

3. ROLES OF METAL IONS IN SMALL RIBOZYMES

3.1. Structural Roles

3.1.1. Stabilization of the Active Global Conformation

As for all functional RNA, the folding of the small ribozymes into their native conformations may be coarsely viewed as a hierarchical two-step process, where the two steps are distinguished by their temporal and spatial regimes. In the first step, an unfolded RNA rapidly acquires local secondary structure by the formation of hydrogen bonds between nucleobases, resulting in a combination of base-paired helices, junctions, loops, and pseudoknots. In the second, slower step, pre-formed helices and loops establish longer-range interactions in three-dimensional space to form the native tertiary structure, sometimes accompanied by small base-pairing rearrangements [91–94]. Metal ions facilitate both of these processes, but formation of tertiary structure requires much higher ionic strength than that of secondary structure [56,94].

While monovalent and divalent metal cations both facilitate folding by neutralizing the negative charge of the phosphate backbone, they influence the folding pathway somewhat differently. Most obviously, the formation of tertiary structure requires much higher concentrations of monovalent cations. For example, the hairpin ribozyme self-cleaves (and presumably folds) with similar efficiency in 0.5 mM $Co(NH_3)_6^{3+}$, 10 mM Mg^{2+}, or 1 M monovalent salts [35]. More interestingly, in the presence of Mg^{2+}, Na^+ ions actually destabilize secondary structure by preferentially associating with the unfolded random coil, and destabilize tertiary structure by competing with Mg^{2+} [53]. This raises the interesting possibility that monovalent cations may help a ribozyme to find the correct minimum-energy native structure by destabilizing alternative misfolds, as was suggested for the HDV ribozyme [88] as well as for larger group I intron ribozymes [95,96]. Since very dense electronegative pockets can form within the tertiary structure of RNA [59], divalent and trivalent cations, with their high charge density and ability to bridge pairs of negatively charged phosphates, promote the folding of RNA particularly well. Their ability to form stable innersphere contacts with electronegative functional groups, while not conferring much additional stability compared to outersphere electrostatic screening, has been proposed to make a larger range of backbone conformations available to RNA [56]. While the potassium ion has been observed to make stable direct contacts with RNA functional groups at highly electronegative sites [97], even replacing a site-bound Mg^{2+} in the active site of a group I intron [98], such tightly bound monovalent ions have not been routinely noted in the small ribozymes, perhaps in part due to the difficulty of distinguishing fractionally occupied monovalent ion sites from water molecules in X-ray crystal structures [99]. However, one recent crystallographic study of the HDV ribozyme found that two thallium(I) ions bind weakly at a location previously seen to be occupied by a hydrated Mg^{2+}, as predicted by MD simulations [66], and a third Tl^+ binds tightly at a new site with direct coordination to the 2′-OH nucleophile [100]. The similar charge, ionic radius and coordination geometry of the thallium compared to the potassium ion, and the ability of Tl^+ to occupy sites different than Mg^{2+} near the active site, suggest that monovalent ions may play more important structural (and perhaps even catalytic) roles in the natural small ribozymes than is currently appreciated.

Structural stabilization, mostly of an electrostatic nature, may be the most important role for metal ions in the small ribozymes. Most can achieve near-maximal activity (within ~30-fold) in a variety of monovalent, divalent, and trivalent salts [5,35,57]. In addition, the vast majority of well-structured metal ions found in crystal structures of the small ribozymes are located tens of Ångström from the site of cleavage chemistry, including all metal ions observed in the *glmS* and hairpin ribozymes (Figure 3) [61–65], suggesting that these ions contribute to activity either indirectly through structural stabilization or by long-distance electrostatic interactions with the site of cleavage chemistry (see Sections 3.2.1 and 3.2.2, below).

3.1.2. Influence on Conformational Changes

Metal ions have been linked to catalytically important conformational changes in the hairpin, VS, and HDV ribozymes [26]. In the hairpin ribozyme, catalysis requires docking of two internal loops of nucleobases located in separate helical stems [101–104]. At equilibrium, the docked and undocked states are both populated, and the rate constants of their interconversion are sensitive to the concentrations of monovalent and divalent cations. The docking reaction is accompanied by an uptake of sodium and/or magnesium ions, which can compete with each other in promoting this transition [55]. The VS ribozyme exhibits analogous docking behavior that includes the metal cation-dependent formation of a loop-loop "kissing" interaction [105,106] that induces a critical change in the base pairing pattern of the substrate stem-loop in the wild-type ribozyme prior to catalysis [91]. In the HDV ribozyme, the self-cleavage reaction is accompanied by the dissociation of a divalent ion from the active site (Figure 3C) and significant conformational changes that reposition important active site residues [62,85,86,107–109]. Such conformational changes are common, but not universal, features of the folding landscapes of the small ribozymes: for instance, the precatalytic pocket in the *glmS* ribozyme is essentially rigid once it is formed in divalent ion-containing buffer, undergoing little change even upon binding of the glucosamine-6-phosphate cofactor and self-cleavage [63,110]; the addition of Mg^{2+} together with cofactor does, however, induce a catalytically rate-limiting conformational change in this ribozyme [111].

3.1.3. Organization of the Active Site

In some cases, metal ions appear to organize residues or solvent molecules within active sites of the small ribozymes. To achieve self-cleavage the small ribozymes must adopt a so-called in-line attack configuration, with an approximately 180° angle between the 2′-oxyanion nucleophile, the phosphorus atom of the scissile phosphate, and the 5′-OH leaving group (Figure 1) [23,24,26]. In the hammerhead ribozyme, diffusely bound Mg^{2+} ions have been proposed to help properly align the catalytic core from a distance, presumably by twisting its stems I and II that intersect at the core, especially in variants that lack tertiary kissing loop interactions between these stems [53]. In addition, MD simulations suggest that threshold occupancy of a cation-binding pocket near the active site (Figure 3B) is required to sample the correct in-line attack geometry for self-cleavage. The cation facilitates formation of the correct geometry by neutralizing negative charge and possibly by coordinating with particular RNA functional groups [71]. A divalent ion observed crystallographically at this site has also been proposed to organize a network of water molecules that may, in turn, facilitate proton transfer in the cleavage reaction [64]. Such a role for well-ordered water molecules in small ribozymes is an active area of investigation [24]. A cation binding site in the genomic HDV ribozyme could also play a role in organizing the active site, although the geometry around the cleavage site does not appear to depend specifically on the presence of Mg^{2+} [60,62,100].

3.2. Mechanistic Roles

3.2.1. Electrostatic Activation of Catalytic Residues

The small ribozymes are all thought to perform their catalysis by acid-base chemistry in which a general base abstracts a proton from the 2′-OH of the nucleotide 5′ of the cleavage site, activating the nucleophile for attack on the adjacent phosphate, and a general acid donates a proton to the 5′-OH leaving group of the nucleotide 3′ of the cleavage site (Figure 1). For the hammerhead, hairpin, *glmS*, and VS ribozymes, the general acid and base appear to be functional groups of the RNA itself [23,24,112]. However, free nucleobases possess pK_a values far from neutral pH – for example, 3.5–4.2 for adenosine $(N1H)^+$ and cytidine $(N3H)^+$; 9.2–9.5 for guanosine (N1)H and uridine (N3)H, and ~12.5 for the 2′-OH of ribose – at first glance seeming to preclude them as efficient proton donors or acceptors near physiologic conditions [113–116].

One possible way for metal ions to stimulate catalysis in ribozymes is by electrostatic modulation of ground-state active site functional groups so as to shift their effective pK_a values towards neutrality. Electrostatic modulation of catalytic residues is common in protein enzymes: nearby positive charges have been observed to lower the pK_a of serine or cysteine residues in serine and cysteine proteases [117,118], and in ribonuclease H, the binding of a Mg^{2+} cofactor induces a pK_a shift of almost two units in an aspartate residue [119]. These effects can be significant over distances as great as 15 Å [67,120]. The long-distance impact of multiple charges on the acid dissociation constant of an amino acid residue can be partially additive, as well [121]. It is therefore plausible that multiple associated metal cations, or even a diffuse ion atmosphere, could have a significant impact on the reactivity of catalytic residues in some or all of the small ribozymes.

Characterization of electrostatic contributions to catalysis is complicated by the relatively weak binding of most metal ions to RNA, as well as the complex dependence of electrostatic effects on the environment in and around a macromolecule, especially in water with its highly dipolar character [122]. However, metal ion-dependent pK_a values have been observed in some of the small ribozymes. In the VS ribozyme, pH-rate profiles suggest that most or all of the ion-specific rate enhancement may result from differential modulation of nucleobase pK_a by different cations, rather than from effects on the intrinsic bond breaking rate constant [123]. As there is no crystal structure of the VS ribozyme, it is not clear in what manner the metal ions may be modulating the effective pK_a, whether through direct or indirect coordination to RNA functional groups, or long-distance interactions. This phenomenon is not universal, however, as the apparent pK_a of the general base in the hammerhead ribozyme appears to be independent of metal ion identity [30]. Effective pK_a shifts toward neutrality have been observed in a catalytically important adenosine of the hairpin ribozyme [124] and an essential cytosine in the HDV ribozyme [25,57], but there is no evidence of direct metal ion participation in these perturbations. In both of these latter cases, the shifts towards higher pK_a could be mediated by the negative electrostatic environment created by RNA functional groups such as phosphoryl oxygens. Accordingly, in case of the HDV ribozyme Mg^{2+} appears to compete with this pK_a shift [57].

3.2.2. *Direct Participation in Catalysis*

In principle, direct participation by metal ions in the chemical step of self-cleavage in small ribozymes could include (Figure 1): (1) deprotonation of the upstream 2′-OH by a metal hydroxide, (2) electrostatic stabilization of the developing negative charge in the transition state, and/or (3) protonation of the leaving group 5′-oxygen by a hydrated metal ion [23,24,26,125]. In contrast to the group I and II introns [126–129], however, direct participation of metal ions in catalysis by the small ribozymes has not been clearly demonstrated. In all cases, any specific contribution of divalent cations to catalysis is minor, accounting for a modest ~20-30-fold rate enhancement over non-acidic monovalent cations [35,130].

Active site divalent metal ions have been proposed to play non-obligatory, even if important catalytic roles in the HDV and hammerhead ribozymes (Figure 3, B and C). Solution kinetics data are consistent with participation of a single hydrated Mg^{2+} ion as a general base in the HDV ribozyme, and cytosine 75 (C76 in the antigenomic ribozyme) as the general acid [57,131,132], or vice versa [133–135,62]. A Mg^{2+} ion poised for a role as the general base has not been found in the X-ray crystal structures of the cleavage product or non-cleavable mutant forms of the HDV ribozyme [61,62] but was suggested by Raman spectroscopy of a two-stranded HDV ribozyme bearing an inactivating 2′-*O*-methyl modification at the cleavage site

[60]. In a recently solved crystal structure of a pre-cleavage (inhibitor-bound) HDV ribozyme, a Mg^{2+} ion is positioned such that it could directly coordinate to and activate the 2′-OH nucleophile in the active ribozyme, though it cannot be ruled out that the Mg^{2+} instead acts as a general base through a water/hydroxide ligand [144]. In contrast, one of the above X-ray crystal structures shows a hydrated Mg^{2+} ion poised to act as a general acid, although residue 75 is not positioned to act as a general acid or base [62]. Thus, direct participation of a magnesium ion in the chemistry of cleavage by the HDV ribozyme is possible, but not conclusively demonstrated. A crystal structure of the full-length hammerhead ribozyme in Mn^{2+} shows no metal cations in position to participate in acid-base chemistry, but suggest that a divalent ion at the active site may stabilize the transition state by solvent-mediated charge withdrawal or direct coordination to nonbridging oxygens of the scissile phosphate (Figure 3B) [64,67]. While enhancement by charge withdrawal is supported by the crystal structure in Mn^{2+}, MD simulations suggest that Mg^{2+} could facilitate the in-line attack angle by directly coordinating a non-bridging oxygen of the scissile phosphate [71].

An intriguing possibility is that cations may stabilize the negatively charged transition state in small ribozymes through long-distance electrostatic interactions. For instance, although the crystal structure of the hairpin ribozyme showed no divalent cation at the immediate active site [65], it revealed six calcium ions within 16 Å of the scissile bond (Figure 3A), likely close enough to strongly stabilize the transition state [67]. This mode of activation from a distance could also help to explain why aminoglycoside antibiotics and the polyamine spermine support hairpin ribozyme activity approaching that in magnesium ions [33]. A similar function has been proposed for the divalent ion found in the active site of the full-length hammerhead ribozyme [64,67,136], and may also apply to the Mg^{2+} found in the active site of the HDV ribozyme [60,62]. Such long-range stabilization is consistent with the generally small specific rate enhancements conferred by divalent cations, and suggests that transiently bound monovalent cations may even help to stabilize the developing charge of the transition state when present at sufficient concentrations to efficiently populate cation binding sites on the RNA.

3.2.3. Influence of Metal Ions on Reaction Pathways

While it is convenient to conceptualize the self-cleavage of small ribozymes as occurring *via* a unique reaction trajectory, there is evidence that at least some ribozymes may make use of a variety of reaction channels that are differentially populated (and effectively compete with one another) as a function of reaction conditions. An intriguing example is found in the HDV ribozyme, where kinetic studies revealed three possible reaction pathways in the presence of varying concentrations of NaCl and $MgCl_2$ [130]. At very low magnesium ion concentrations ($< 10^{-7}$ M), the rate of self-cleavage is independent of Mg^{2+} concentration, with a pH-rate profile suggesting that solvent and hydroxide ions operate as general and specific bases in the reaction. At intermediate Mg^{2+} concentrations (10^{-7}–10^{-4} M), the observed cleavage rate constant exhibits log-linear dependence on magnesium ion concentration, with pH-rate profiles consistent with the binding of at least one structural divalent cation. Finally, at physiologic Mg^{2+} concentrations and higher, a second metal ion binding site becomes saturated, yielding an inverted pH-rate profile consistent with a role of a metal hydroxide or solvent hydroxide as the base in catalysis. However, a subsequent study found that the cleavage reaction of the HDV ribozyme in 4 M Li^{+} exhibits a similar pH-rate profile in the presence and absence of Mg^{2+}, albeit with a smaller observed rate constant, suggesting that Li^{+} can at least partially substitute for Mg^{2+} in determining pathway preference [29]. Furthermore, due to the modest specific contribution of Mg^{2+} to catalysis [130], the absence of a magnesium hydroxide poised for general base catalysis from the published X-ray crystal structures [62], and the existence of other pH-dependent conformational changes in the absence of divalent ions that affect activity [137], the nature of these apparent reaction channels requires further elucidation.

Scenarios involving multiple metal cation-dependent reaction pathways have also been proposed for other ribozymes. Kinetic characterization of a tertiary-stabilized form of the hammerhead ribozyme suggests that magnesium ions and cobalt(III) hexammine may support separate catalytic pathways with incompatible RNA conformations [138,139]. Furthermore, recent work has demonstrated multiple catalytically active conformations of the hairpin ribozyme [140] as well as the *Tetrahymena* group I intron ribozyme [141] that are all populated near physiological conditions. In the case of the group I intron, interconversion between the different native conformations occurs slowly in the presence Mg^{2+}, but rapidly in its absence. These results raise the interesting possibility that some small ribozymes may operate *via* multiple reaction pathways in a metal ion-dependent fashion. For example, the rate of catalysis by individual subpopulations of ribozymes could be limited by different chemical steps dependent on subtly different conformations or differentially occupied cation binding sites. If this is the case, it will reveal a striking flexibility in the folding and function of ribozymes.

4. CONCLUDING REMARKS AND FUTURE DIRECTIONS

In summary, metal cations are critical to the intramolecular phosphodiester isomerization reaction catalyzed by the small self-cleaving ribozymes known in nature. Much as water is an obligatory solvent for proper folding of many macromolecules, appropriate combinations of metal ions are required for optimal activity in the small ribozymes. They universally facilitate catalysis through structural stabilization, but in certain cases may also help to organize active site functional groups and water molecules through hydrogen bonding, activate the catalytic acid or base, or participate directly in catalysis through acid-base chemistry or transition state stabilization. Proper folding and efficient self-cleavage occur with generally low selectivity in a variety of monovalent and divalent cations, but under physiological conditions Mg^{2+} is the most important of these, and is generally preferred over less naturally abundant divalent cations. While small ribozymes could, in principle, use transition metals and heavy metals for catalysis, their toxicity and generally low free concentrations in the cell preclude these metals from playing an important role as cofactors for the natural small ribozymes.

Future work should further elucidate any direct catalytic roles played by metal ions in the HDV and hammerhead ribozymes, long-distance interactions between the active site and metal ions, potentially overlooked roles of monovalent ions (including the physiologically most relevant K^+), the nature and metal ion-dependence of alternate reaction pathways, and the impact of metal ions on local conformational changes and solvent organization during catalysis. This will require a combination of increasingly sophisticated methods of chemical modification, spectroscopic techniques, and theoretical models. As small ribozymes increasingly appear to be widespread in nature, understanding their manifold interactions with metal ions will yield a more complete understanding of the roles of RNA in life and its origins.

ACKNOWLEDGMENT

The authors gratefully acknowledge support by NIH grant GM062357 to N.G.W. and an NIH Molecular Biophysics Training Fellowship to A.J.B.

ABBREVIATIONS AND DEFINITIONS

divalent cation	an ion with a net electronic charge of +2
enthalpy of dehydration	the enthalpy change accompanying the removal of all water molecules to an infinite distance from a fully hydrated ion
general acid	a functional group or moiety that catalyzes a chemical reaction by donating a proton to a reactive group

general base	a functional group or moiety that catalyzes a chemical reaction by accepting a proton from a reactive group
glmS ribozyme	a ribozyme found in numerous Gram-positive bacteria that self-cleaves in the presence of a glucosamine-6-phosphate cofactor, regulating translation of the *glmS* gene in *E. coli*
HDV	hepatitis delta virus
innersphere interaction	direct interaction between a metal ion and an electronegative ligand such as the oxygen atom of water or an oxygen or nitrogen atom of an RNA molecule
MD	molecular dynamics
monovalent cation	ion with a net electronic charge of +1
outersphere interaction	interaction between a metal ion and another species mediated by water molecules or other ligands of the metal ion
rRNA	ribosomal RNA
secondary structure	the ensemble of hydrogen bonding interactions between nucleobases (base pairs, triples, and occasionally quartets) in an RNA molecule, resulting in the formation of base-paired stems, unpaired loops, and junctions between these
tertiary structure	the three-dimensional structure of an RNA molecule, including all base pairs as well as additional interactions between helical stems and loops
VS	Varkud satellite

REFERENCES

1. K. Kruger, P. J. Grabowski, A. J. Zaug, J. Sands, D. E. Gottschling and T. R. Cech, *Cell*, 1982, **31**, 147–157.
2. T. R. Cech, *Science*, 2000, **289**, 878–879.
3. C. Guerrier-Takada, K. Gardiner, T. Marsh, N. Pace and S. Altman, *Cell*, 1983, **35**, 849–857.
4. N. Toor, K. S. Keating, S. D. Taylor and A. M. Pyle, *Science*, 2008, **320**, 77–82.
5. W. C. Winkler, A. Nahvi, A. Roth, J. A. Collins and R. R. Breaker, *Nature*, 2004, **428**, 281–286.
6. A. C. Forster and R. H. Symons, *Cell*, 1987, **49**, 211–220.

7. A. Hampel and R. Tritz, *Biochemistry*, 1989, **28**, 4929–4933.
8. M. M. C. Lai, *Annu. Rev. Biochem.*, 1995, **64**, 259–286.
9. G. A. Prody, J. T. Bakos, J. M. Buzayan, I. R. Schneider and G. Bruening, *Science*, 1986, **231**, 1577–1580.
10. B. J. Saville and R. A. Collins, *Cell*, 1990, **61**, 685–696.
11. M. Y. P. Kuo, L. Sharmeen, G. Dintergottlieb and J. Taylor, *J. Virol.*, 1988, **62**, 4439–4444.
12. H. N. Wu, Y. J. Lin, F. P. Lin, S. Makino, M. F. Chang and M. M. C. Lai, *Proc. Natl. Acad. Sci. USA*, 1989, **86**, 1831–1835.
13. M. Martick, L. H. Horan, H. F. Noller and W. G. Scott, *Nature*, 2008, **454**, 899–U857.
14. C. H. T. Webb, N. J. Riccitelli, D. J. Ruminski and A. Luptak, *Science*, 2009, **326**, 953–953.
15. K. Salehi-Ashtiani, A. Luptak, A. Litovchick and J. W. Szostak, *Science*, 2006, **313**, 1788–1792.
16. T. Pan and O. C. Uhlenbeck, *Biochemistry*, 1992, **31**, 3887–3895.
17. J. J. Agresti, B. T. Kelly, A. Jaschke and A. D. Griffiths, *Proc. Natl. Acad. Sci. USA*, 2005, **102**, 16170–16175.
18. C. Reymond, J. D. Beaudoin and J. P. Perreault, *Cell. Mol. Life Sci.*, 2009, **66**, 3937–3950.
19. M. Zivarts, Y. Liu and R. R. Breaker, *Nucleic Acids Res.*, 2005, **33**, 622–631.
20. M. Illangasekare and M. Yarus, *RNA*, 1999, **5**, 1482–1489.
21. T. A. Lincoln and G. F. Joyce, *Science*, 2009, **323**, 1229–1232.
22. W. Gilbert, *Nature*, 1986, **319**, 618–618.
23. J. C. Cochrane and S. A. Strobel, *Acc. Chem. Res.*, 2008, **41**, 1027–1035.
24. N. G. Walter, *Mol. Cell*, 2007, **28**, 923–929.
25. A. K. Oyelere, J. R. Kardon and S. A. Strobel, *Biochemistry*, 2002, **41**, 3667–3675.
26. N. G. Walter and S. Perumal, in *Non-Protein Coding RNAs*, Ed. N. G. Walter, S. A. Woodson and R. T. Batey, Springer, Berlin, 2009, pp. 103–127.
27. M. J. Fedor, *Curr. Opin. Struct. Biol.*, 2002, **12**, 289–295.
28. E. A. Curtis and D. P. Bartel, *RNA*, 2001, **7**, 546–552.
29. A. T. Perrotta and M. D. Been, *Biochemistry*, 2006, **45**, 11357–11365.
30. M. Roychowdhury-Saha and D. H. Burke, *RNA*, 2006, **12**, 1846–1852.
31. Y. A. Suh, P. K. R. Kumar, K. Taira and S. Nishikawa, *Nucleic Acids Res.*, 1993, **21**, 3277–3280.
32. K. J. Young, F. Gill and J. A. Grasby, *Nucleic Acids Res.*, 1997, **25**, 3760–3766.
33. D. J. Earnshaw and M. J. Gait, *Nucleic Acids Res.*, 1998, **26**, 5551–5561.
34. B. M. Chowrira, A. Berzal-Herranz and J. M. Burke, *Biochemistry*, 1993, **32**, 1088–1095.
35. J. B. Murray, A. A. Seyhan, N. G. Walter, J. M. Burke and W. G. Scott, *Chem. Biol.*, 1998, **5**, 587–595.
36. A. Hampel and J. A. Cowan, *Chem. Biol.*, 1997, **4**, 513–517.
37. S. Nesbitt, L. A. Hegg and M. J. Fedor, *Chem. Biol.*, 1997, **4**, 619–630.
38. N. G. Walter and J. M. Burke, *Curr. Opin. Chem. Biol.*, 1998, **2**, 303–303.

39. S. Cayley, B. A. Lewis, H. J. Guttman and M. T. Record, *J. Mol. Biol.*, 1991, **222**, 281–300.
40. E. M. Froschauer, M. Kolisek, F. Dieterich, M. Schweigel and R. J. Schweyen, *FEMS Microbiol. Lett.*, 2004, **237**, 49–55.
41. M. J. Cromie, Y. X. Shi, T. Latifi and E. A. Groisman, *Cell*, 2006, **125**, 71–84.
42. C. E. Dann, C. A. Wakeman, C. L. Sieling, S. C. Baker, I. Irnov and W. C. Winkler, *Cell*, 2007, **130**, 878–892.
43. A. Ramesh and W. C. Winkler, *RNA Biol.*, 2010, 7, in press.
44. C. A. Wakeman, A. Ramesh and W. C. Winkler, *J. Mol. Biol.*, 2009, **392**, 723–735.
45. T. Voets, B. Nilius, S. Hoefs, A. W. C. M. van der Kemp, G. Droogmans, R. J. M. Bindels and J. G. J. Hoenderop, *J. Biol. Chem.*, 2004, **279**, 19–25.
46. A. Khvorova, A. Lescoute, E. Westhof and S. D. Jayasena, *Nat. Struct. Biol.*, 2003, **10**, 708–712.
47. M. De la Peña, S. Gago and R. Flores, *EMBO J.*, 2003, **22**, 5561–5570.
48. S. Nakano, H. T. Karimata, Y. Kitagawa and N. Sugimoto, *J. Am. Chem. Soc.*, 2009, **131**, 16881–16888.
49. L. J. Dai and L. J. Quamme, *J. Clin. Invest.*, 1991, **88**, 1255–1264.
50. B. Alberts, D. Bray, J. Lewis, M. Raff, K. Roberts and J. D. Watson, *Molecular Biology of the Cell*, Garland Publishing, New York, 1994.
51. J. A. Cowan, *Chem. Rev.*, 1998, **98**, 1067–1087.
52. V. L. Pecoraro, J. D. Hermes and W. W. Cleland, *Biochemistry*, 1984, **23**, 5262–5271.
53. D. Rueda, K. Wick, S. E. McDowell and N. G. Walter, *Biochemistry*, 2003, **42**, 9924–9936.
54. D. E. Draper, D. Grilley and A. M. Soto, *Annu. Rev. Biophys. Biom.*, 2005, **34**, 221–243.
55. G. Bokinsky, D. Rueda, V. K. Misra, M. M. Rhodes, A. Gordus, H. P. Babcock, N. G. Walter and X. W. Zhuang, *P. Natl. Acad. Sci. USA*, 2003, **100**, 9302–9307.
56. D. E. Draper, *Biophys. J.*, 2008, **95**, 5489–5495.
57. S. Nakano, D. M. Chadalavada and P. C. Bevilacqua, *Science*, 2000, **287**, 1493–1497.
58. A. Roth, A. Nahvi, M. Lee, I. Jona and R. R. Breaker, *RNA*, 2006, **12**, 607–619.
59. K. Chin, K. A. Sharp, B. Honig and A. M. Pyle, *Nat. Struct. Biol.*, 1999, **6**, 1055–1061.
60. J. H. Chen, B. Gong, P. C. Bevilacqua, P. R. Carey and B. L. Golden, *Biochemistry*, 2009, **48**, 1498–1507.
61. A. R. Ferre-D'Amare, K. H. Zhou and J. A. Doudna, *Nature*, 1998, **395**, 567–574.
62. A. L. Ke, K. H. Zhou, F. Ding, J. H. D. Cate and J. A. Doudna, *Nature*, 2004, **429**, 201–205.
63. D. J. Klein and A. R. Ferre-D'Amare, *Science*, 2006, **313**, 1752–1756.
64. M. Martick, T. S. Lee, D. M. York and W. G. Scott, *Chem. Biol.*, 2008, **15**, 332–342.
65. P. B. Rupert and A. R. Ferré-D'Amaré, *Nature*, 2001, **410**, 780–786.

66. M. V. Krasovska, J. Sefcikova, K. Reblova, B. Schneider, N. G. Walter and J. Sponer, *Biophys. J.*, 2006, **91**, 626–638.
67. R. K. O. Sigel and A. M. Pyle, *Chem. Rev.*, 2007, **107**, 97–113.
68. S. F. Lincoln, *Helv. Chim. Acta*, 2005, **88**, 523–545.
69. J. A. Cowan, *J. Inorg. Biochem.*, 1993, **49**, 171–175.
70. T. E. Horton and V. J. DeRose, *Biochemistry*, 2000, **39**, 11408–11416.
71. T. S. Lee, G. M. Giambasu, C. P. Sosa, M. Martick, W. G. Scott and D. M. York, *J. Mol. Biol.*, 2009, **388**, 195–206.
72. V. D. Sood, T. L. Beattie and R. A. Collins, *J. Mol. Biol.*, 1998, **282**, 741–750.
73. J. L. Maguire and R. A. Collins, *J. Mol. Biol.*, 2001, **309**, 45–56.
74. S. C. Dahm and O. C. Uhlenbeck, *Biochemistry*, 1991, **30**, 9464–9469.
75. J. L. Boots, M. D. Canny, E. Azimi and A. Pardi, *RNA*, 2008, **14**, 2212–2222.
76. F. Walter, A. I. H. Murchie, J. B. Thomson and D. M. J. Lilley, *Biochemistry*, 1998, **37**, 14195–14203.
77. H. W. Pley, K. M. Flaherty and D. B. Mckay, *Nature*, 1994, **372**, 68–74.
78. W. G. Scott, J. B. Murray, J. R. P. Arnold, B. L. Stoddard and A. Klug, *Science*, 1996, **274**, 2065–2069.
79. S. Nakano, A. L. Cerrone and P. C. Bevilacqua, *Biochemistry*, 2003, **42**, 2982–2994.
80. B. Gong, Y. Chen, E. L. Christian, J. H. Chen, E. Chase, D. M. Chadalavada, R. Yajima, B. L. Golden, P. C. Bevilacqua and P. R. Carey, *J. Am. Chem. Soc.*, 2008, **130**, 9670–9672.
81. A. T. Perrotta and M. D. Been, *Biochemistry*, 2007, **46**, 5124–5130.
82. B. Gong, J. H. Chen, P. C. Bevilacqua, B. L. Golden and P. R. Carey, *Biochemistry*, 2009, **48**, 11961–11970.
83. N. G. Walter, N. Yang and J. M. Burke, *J. Mol. Biol.*, 2000, **298**, 539–555.
84. A. L. Feig, W. G. Scott and O. C. Uhlenbeck, *Science*, 1998, **279**, 81–84.
85. S. Jeong, J. Sefcikova, R. A. Tinsley, D. Rueda and N. G. Walter, *Biochemistry*, 2003, **42**, 7727–7740.
86. D. A. Harris, R. A. Tinsley and N. G. Walter, *J. Mol. Biol.*, 2004, **341**, 389–403.
87. J. L. O'Rear, S. L. Wang, A. L. Feig, L. Beigelman, O. C. Uhlenbeck and D. Herschlag, *RNA*, 2001, **7**, 537–545.
88. T. S. Brown, D. M. Chadalavada and P. C. Bevilacqua, *J. Mol. Biol.*, 2004, **341**, 695–712.
89. M. Z. Barciszewska, E. Wyszko, R. Bald, V. A. Erdmann and J. Barciszewski, *J. Biochem.*, 2003, **133**, 309–315.
90. J. W. Huff, K. S. Sastry, W. E. C. Wacker and M. P. Gordon, *Biochemistry*, 1964, **3**, 501–506.
91. A. A. Andersen and R. A. Collins, *Mol. Cell*, 2000, **5**, 469–478.
92. P. Brion and E. Westhof, *Annu. Rev. Biophys. Biom.*, 1997, **26**, 113–137.
93. B. Onoa and I. Tinoco, *Curr. Opin. Chem. Biol.*, 2004, **14**, 374–379.
94. I. Tinoco and C. Bustamante, *J. Mol. Biol.*, 1999, **293**, 271–281.
95. Y. F. Jiang, M. Xiao, P. Yin and Y. Zhang, *RNA*, 2006, **12**, 561–566.
96. R. Russell, R. Das, H. Suh, K. J. Traver, A. Laederach, M. A. Engelhardt and D. Herschlag, *J. Mol. Biol.*, 2006, **363**, 531–544.

97. G. L. Conn, A. G. Gittis, E. E. Lattman, V. K. Misra and D. E. Draper, *J. Mol. Biol.*, 2002, **318**, 963–973.
98. P. L. Adams, M. R. Stahley, A. B. Kosek, J. Wang and S. A. Strobel, *Nature*, 2004, **430**, 45–50.
99. V. Tereshko, C. J. Wilds, G. Minasov, T. P. Prakash, M. A. Maier, A. Howard, Z. Wawrzak, M. Manoharan and M. Egli, *Nucleic Acids Res.*, 2001, **29**, 1208–1215.
100. A. L. Ke, F. Ding, J. D. Batchelor and J. A. Doudna, *Structure*, 2007, **15**, 281–287.
101. K. J. Hampel, N. G. Walter and J. M. Burke, *Biochemistry*, 1998, **37**, 14672–14682.
102. N. G. Walter, K. J. Hampel, K. M. Brown and J. M. Burke, *EMBO J.*, 1998, **17**, 2378–2391.
103. N. G. Walter, D. A. Harris, M. J. B. Pereira and D. Rueda, *Biopolymers*, 2001, **61**, 224–242.
104. X. W. Zhuang, H. Kim, M. J. B. Pereira, H. P. Babcock, N. G. Walter and S. Chu, *Science*, 2002, **296**, 1473–1476.
105. M. J. B. Pereira, E. N. Nikolova, S. L. Hiley, D. Jaikaran, R. A. Collins and N. G. Walter, *J. Mol. Biol.*, 2008, **382**, 496–509.
106. T. Rastogi, T. L. Beattie, J. E. Olive and R. A. Collins, *EMBO J.*, 2006, **15**, 2820–2825.
107. R. A. Tinsley, D. A. Harris and N. G. Walter, *Biochemistry*, 2004, **43**, 8935–8945.
108. D. A. Harris, D. Rueda and N. G. Walter, *Biochemistry*, 2002, **41**, 12051–12061.
109. M. J. B. Pereira, D. A. Harris, D. Rueda and N. G. Walter, *Biochemistry*, 2002, **41**, 730–740.
110. R. A. Tinsley, J. R. W. Furchak and N. G. Walter, *RNA*, 2007, **13**, 468–477.
111. K. M. Brooks and K. J. Hampel, *Biochemistry*, 2009, **48**, 5669–5678.
112. M. A. Ditzler, J. Sponer and N. G. Walter, *RNA*, 2009, **15**, 560–575.
113. C. P. Da Costa and H. Sigel, *Inorg. Chem.*, 2003, **42**, 3475–3482.
114. B. Knobloch, W. Linert and H. Sigel, *Proc. Natl. Acad. Sci. USA*, 2005, **102**, 7459–7464.
115. J. J. Christen, J. H. Rytting and R. M. Izatt, *J. Chem. Soc. B*, 1970, 4907–4913.
116. J. J. Christen, J. H. Rytting and R. M. Izatt, *J. Chem. Soc. B*, 1970, 1643–1646.
117. S. D. Lewis, F. A. Johnson and J. A. Shafer, *Biochemistry*, 1981, **20**, 48–51.
118. A. C. Storer and R. Menard, *Methods Enzymol.*, 1994, **244**, 486–500.
119. Y. Oda, T. Yamazaki, K. Nagayama, S. Kanaya, Y. Kuroda and H. Nakamura, *Biochemistry*, 1994, **33**, 5275–5284.
120. S. E. Jackson and A. R. Fersht, *Biochemistry*, 1993, **32**, 13909–13916.
121. A. J. Russell and A. R. Fersht, *Nature*, 1987, **328**, 496–500.
122. A. R. Fersht and M. J. E. Sternberg, *Protein Eng.*, 1989, **2**, 527–530.
123. M. D. Smith, R. Mehdizadeh, J. E. Olive and R. A. Collins, *RNA*, 2008, **14**, 1942–1949.
124. M. Guo, R. C. Spitale, R. Volpini, J. Krucinska, G. Cristalli, P. R. Carey and J. E. Wedekind, *J. Am. Chem. Soc.*, 2009, **131**, 12908–12909.
125. G. M. Emilsson, S. Nakamura, A. Roth and R. R. Breaker, *RNA*, 2003, **9**, 907–918.

126. P. M. Gordon, R. Fong and J. A. Piccirilli, *Chem. Biol.*, 2007, **14**, 607–612.
127. S. O. Shan, A. Yoshida, S. G. Sun, J. A. Piccirilli and D. Herschlag, *Proc. Natl. Acad. Sci. USA*, 1999, **96**, 12299–12304.
128. M. R. Stahley and S. A. Strobel, *Science*, 2005, **309**, 1587–1590.
129. L. Zhang and J. A. Doudna, *Science*, 2002, **295**, 2084–2088.
130. S. Nakano, D. J. Proctor and P. C. Bevilacqua, *Biochemistry*, 2001, **40**, 12022–12038.
131. A. L. Cerrone-Szakal, N. A. Siegfried and P. C. Bevilacqua, *J. Am. Chem. Soc.*, 2008, **130**, 14504–14520.
132. S. R. Das and J. A. Piccirilli, *Nat. Chem. Biol.*, 2005, **1**, 45–52.
133. P. Banas, L. Rulisek, V. Hanosova, D. Svozil, N. G. Walter, J. Sponer and M. Otyepka, *J. Phys. Chem. B*, 2008, **112**, 11177–11187.
134. M. V. Krasovska, J. Sefcikova, N. Spackova, J. Sponer and N. G. Walter, *J. Mol. Biol.*, 2005, **351**, 731–748.
135. A. T. Perrotta, I. H. Shih and M. D. Been, *Science*, 1999, **286**, 123–126.
136. M. Martick and W. G. Scott, *Cell*, 2006, **126**, 309–320.
137. T. S. Wadkins, I. H. Shih, A. T. Perrotta and M. D. Been, *J. Mol. Biol.*, 2001, **305**, 1045–1055.
138. M. Roychowdhury-Saha and D. H. Burke, *RNA*, 2007, **13**, 841–848.
139. Y. Takagi, A. Inoue and K. Taira, *J. Am. Chem. Soc.*, 2004, **126**, 12856–12864.
140. M. A. Ditzler, D. Rueda, J. Mo, K. Håkansson and N. G. Walter, *Nucleic Acids Res.*, 2008, **36**, 7088–7099.
141. S. V. Solomatin, M. Greenfeld, S. Chu and D. Herschlag, *Nature*, 2010, **463**, 681–684.
142. S. L. Hiley and R. A. Collins, *EMBO J.*, 2001, **20**, 5461–5469.
143. D. A. Lafontaine, D. G. Norman and D. M. Lilley, *EMBO J.*, 2002, **21**, 2461–2471.
144. J.-H. Chen, R. Yajima, D. M. Chadalavada, E. Chase, P. C. Bevilacqua and B. L. Golden, *Biochemistry*, 2010, **49**, 6508–6518.

Met. Ions Life Sci. **2011**, *9*, 197–234

7

Multiple Roles of Metal Ions in Large Ribozymes

Daniela Donghi and Joachim Schnabl
Institute of Inorganic Chemistry, University of Zürich, Winterthurerstrasse 190, CH-8057, Zürich, Switzerland
<daniela.donghi@aci.uzh.ch>

Metal Ions in Life Sciences, Volume 9 Edited by Astrid Sigel, Helmut Sigel, and Roland K. O. Sigel

Published by the Royal Society of Chemistry, www.rsc.org DOI: 10.1039/978184973251200197

ABSTRACT: Since the discovery of catalytic RNA molecules (ribozymes), intense research has been devoted to understand their structure and activity. Among RNA molecules, the large ribozymes, namely group I and group II introns and RNase P, are of special importance. The first two ribozymes are known for their ability to perform self-splicing while RNase P is responsible for the 5′-end maturation of tRNA in bacteria, archea, and eukaryotes. All three groups of ribozymes show a significant requirement for metal ions in order to establish the active tertiary structure that enables catalysis. The primary role of both monovalent and divalent metal ions is to screen the negative charge associated with the phosphate sugar backbone, but the metal ions also play an active role in catalysis. Biochemical and biophysical investigations, supported by recent findings from X-ray crystal structures, allow clarifying and rationalizing both the structural and catalytic roles of metal ions in large ribozymes. In particular, the "two-metal-ion mechanism", describing how metal ions in the active center take part in catalysis, has been largely corroborated.

KEYWORDS: catalysis · folding · group I introns · group II introns · metal ions · ribozymes · RNase P

1. INTRODUCTION

The word *ribozyme*, coined more than twenty years ago, refers to catalytic RNAs, i.e., RNA molecules that can perform catalysis without the aid of a protein [1–3]. The reactions catalyzed by ribozymes include mainly cleaving or splicing of RNA/DNA sequences [4], but other catalytic functions have also been described [5,6] (see also further chapters in this volume). Different types of ribozymes have been identified, but in general, ribozymes can be divided into two classes according to their size: the small, cleaving ribozymes (see Chapter 6) and the large, splicing (with the exception of RNase P) ribozymes.

Ribozymes need the help of metal ions for both folding and catalysis [1,7–14]. The polyanionic nature of the RNA backbone is compensated by monovalent and divalent metal ions that primarily act as charge-screening agents. The majority of M^{n+} present is thus diffusely bound to RNA in an unspecific way [15,16]. On the other hand, X-ray structures of large RNAs show that roughly 10% of the negative charge is neutralized by specifically bound M^{n+} [17].

In the process of folding, RNAs create pockets of high negative potential that is neutralized by interaction with divalent [18–22] or monovalent [20,22,23] metal ions. This neutralization allows the RNA molecule to fold into the complex active structure that is crucial for catalysis. Moreover, recent crystal structures show that divalent metal ions are located within the catalytic core, suggesting their direct involvement in catalysis [21,24,25].

The key players in RNA catalysis are limited: sugar, phosphodiester groups, and the four nucleobases; very few, when compared to proteins, where a broader chemical diversity is granted by numerous amino acids. Therefore, RNA needs a complex landscape of folding and activation in order to create specific active sites. Cofactors, such as metal ions, are crucial to create the active site and perform catalysis. The assistance of such "co-actors" allows RNA to perform catalysis despite the restricted variety of functional groups, whose pK_a values are far away from neutrality, that is, the physiological pH range [12,22].

Metal ions can promote catalysis in different ways: they can stabilize the active tertiary structure through electrostatic interactions, they can directly participate in acid–base catalysis by activating the nucleophile or by stabilizing the leaving group or they can modify the pK_a of a functional group [26–28]. For example, their coordination to nucleobases causes large pK_a shifts both in amino and imino protons [26,28–31]. Such acidified protons can therefore participate actively in catalysis: hydrogen bonds can be strengthened and different tautomers stabilized, so changing the aspect of the catalytic core of the ribozyme.

The best way to describe the direct participation of metal ions in catalysis of RNA molecules is the "two-metal-ion mechanism" proposed years ago for polymerases [32–34], kinases [35], the hydrolysis of ATP and other nucleoside 5′-triphosphates [36] (depending on the positioning of the metal ions at the triphosphate residue either a kinase-type [35,36,188] or a polymerase-type [36,188–190] reaction occurs [36,188]), and ribozymes [37].

Due to the strict requirement of metal ions in both folding and activity, ribozymes can be considered as obligate metalloenzymes [38]. This was opposed for a long time, as there are some cases in which Mg^{2+} (the usual cofactor in RNA catalysis) can be substituted by high concentrations of monovalent metal ions [22,39,40]. However, recently it was reviewed that metal ions do not need to be directly coordinated to the atoms participating in catalysis but they can exert their influence also from a long distance [41], strengthening the hypothesis that ribozymes are in fact metalloenzymes.

Along the same line, dependence of the reaction rate on metal ion concentration has been shown for large ribozymes, and it has recently been reviewed how ribozyme activity can be tuned by metal ions [42]. A distinction between the structural and catalytic role of metal ions is not always straightforward as these two processes are directly linked to each other: many spectroscopic and biochemical methods have been developed in order to distinguish the multiple roles of metal ions in small and large ribozymes, each of them providing different information [43–45] (see also Chapter 2 of this volume).

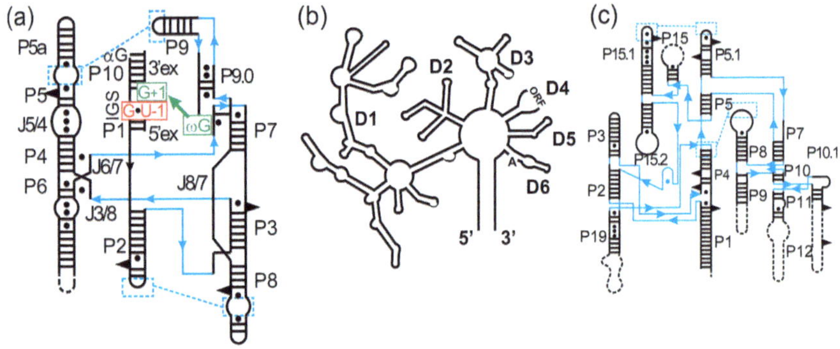

Figure 1. Schematic secondary structures of (**a**) group I and (**b**) group II intron ribozymes, as well as (**c**) RNAse P.

In Chapter 6 of this volume, the interaction of metal ions with small ribozymes is extensively described; Chapter 8 deals with the spliceosome, and Chapter 9 with the ribosome. The present chapter provides a concise overview on the role played by metal ions in large ribozymes, i.e., group I and II introns and RNase P (Figure 1), as it is emerging from recent X-ray crystal structure data. Indeed, the first X-ray structure of a group II intron ribozyme from *Oceanobacillus iheyensis* by Pyle and coworkers in 2008 [46] gave a first picture of this class of large ribozymes. Moreover, due to the similarity of group II intron ribozymes to the eukaryotic spliceosome, the recently solved crystal structure allows to extend theories and hypotheses concerning the spliceosome structure and function [47–51] (see also Chapter 8).

This chapter is divided into four parts, followed by short concluding remarks. In the first part a brief basic description of the role of metal ions in folding and catalysis of RNAs is given. The following three parts are dedicated to the three known large ribozymes, group I and II introns and RNase P, respectively (Figure 1). A strong focus will be directed to studies reported in the last few years, in order to confirm (or not) the already established properties of these fascinating molecules.

2. METAL IONS IN FOLDING AND CATALYSIS: A BRIEF OVERVIEW

2.1. Folding Pathways

In the last few years many papers, reviews, and book chapters have focused on the understanding of the thermodynamic and kinetic aspects of RNA folding [14,52–55].

RNA folding is generally hierarchical, meaning that the tertiary structure formation takes place only after the secondary structure has been established [55,56] (Figure 2). The secondary structure comprises all the hydrogen bonded base pairs within the RNA sequence (the helical/stem part) as well as the unpaired nucleotides, arranged in loops and bulges. On the other hand, the tertiary structure describes all three-dimensional contacts that normally take place between various secondary structure elements [57,58]. One of the most common tertiary contacts in RNA architecture is the so-called tetraloop/tetraloop receptor interaction, which has been found to be extremely dependent on metal ion concentration [59].

The secondary structure is usually stable in a large variety of experimental conditions and normally requires only monovalent ions in concentrations ranging from 50 to 100 mM. Because of the electrostatic repulsions within the phosphodiester backbone, the absence of metal ions hampers the formation of a secondary structure, but a very low ionic strength, less than 1 mM, is mostly enough to allow the formation of the secondary structure. The folding landscape to the active tertiary structure, usually reached only in the presence of divalent metal ions, is quite rugged for many RNA molecules: many intermediates can occur before the molecule adopts the active form, and the rate of unfolding of these off-pathway intermediates (kinetic traps) determines the effectiveness of the folding [60]. Interestingly, it has been shown that several RNA molecules follow a folding pathway without kinetic traps [61,62].

K^+ is the most common among the monovalent metal ions used *in vitro* [63], and only in a few cases it is substituted by Na^+ and Li^+ [64]. Whereas monovalent metal ions are needed for formation of the secondary structure, divalent metal ions are essential to reach the tertiary structure (Figure 2), and only rarely they can be substituted by a molar concentration of monovalent metal ions [22]. Compared to monovalent metal ions, fewer

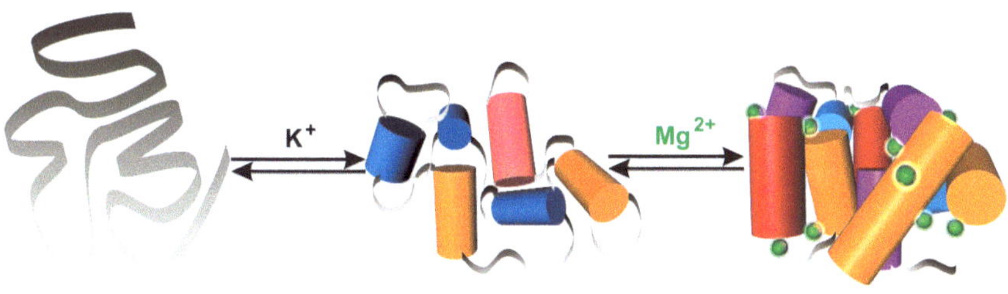

Figure 2. Cartoon representing the folding pathway of large RNAs. Starting from a single strand structure, monovalent metal ions acting as charge screening units allow the establishment of the secondary structure. Tertiary contacts promoted by divalent metal ions lead to the folding into the compact tertiary structure (adapted from [22]).

di/multivalent metal ions are needed to compensate the high negative charge accumulated in specific regions of folded RNA, as originally predicted by the counterion condensation theory developed by Manning [65].

The most common divalent metal ion found as cofactor both in folding and catalysis (see below) is Mg^{2+}. This ion is very abundant in nature and plays the role of a cofactor in many biomacromolecules. In a few cases, it can be replaced by Ca^{2+}, Mn^{2+}, Zn^{2+}, Cd^{2+}, and Pb^{2+}, but it remains unique in its way of interaction with RNA, that makes it the best candidate for RNA folding (and catalysis, see below). Mg^{2+} can interact with RNA in three different ways: (i) by simply screening the negative charge through electrostatic interactions, (ii) by creating a net of hydrogen bonds between its coordinated water molecules and specific RNA motifs in so-called outersphere interactions, (iii) by directly interacting with phosphoryl oxygens, purine N7, guanosine O6, uracil O4, and ribose 2′-OH sites, through so-called innersphere interaction. For a more detailed description of the different metal ion binding modes refer to Chapter 1 of the present volume.

Being a divalent ion, Mg^{2+} better satisfies the requirements for positive charges within folded RNAs compared to monovalent metal ions [66]. Moreover, small cations usually provide better stabilization of the phosphate sugar backbone by packing it more closely [67].

In folded RNA, metal ions are normally found in the major groove, where a significant negative charge is accumulated [68]. In RNA tertiary structure many specific RNA-binding motifs have been identified. Among them are the tandem G–U pair motif [18], the loop E motif (particularly suitable for Mg^{2+} binding) [69], and the A platform motif (specific for K^+ binding) [23] (see also Chapter 1).

The presence of Mg^{2+} in the active site of many ribozymes, as postulated by biochemical studies and confirmed by crystallographic data (see below), explains why the active tertiary structure is often established only in the presence of Mg^{2+}. It also suggests that, very often, structural and catalytic roles of metal ions are hardly separable.

2.2. Metal Ion Involvement in Catalysis

Large ribozymes catalyze trans-esterification reactions (Figure 3) through a S_N2 mechanism that yields the formation of a 5′-phosphate and a 3′-hydroxyl terminus in the first step. The reaction is accompanied by a configuration inversion of the non-bridging phosphate oxygens that are stereochemically distinct. The nucleophile employed in the reaction is different for each type of ribozyme. Group I and II introns perform two subsequent trans-esterification reactions that end up with the ligation of two exons and the excision of the intron. These reactions can be accelerated by

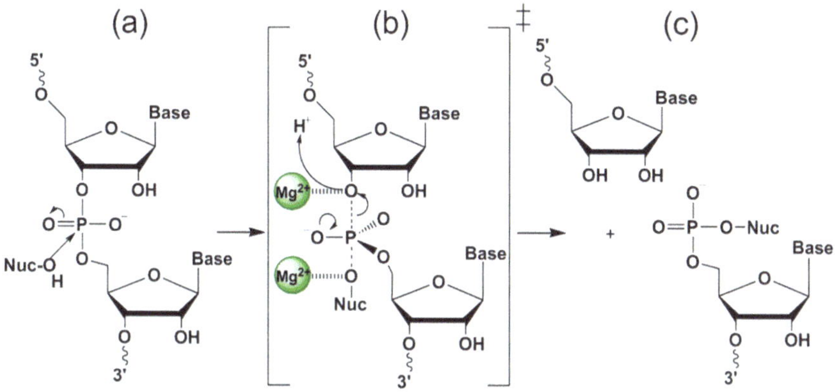

Figure 3. Phosphodiester transfer reaction catalyzed by large ribozymes. (**a**) The nucleophile (Nuc-OH) attacks the scissile phosphate. (**b**) The transition state is stabilized by the presence of two metal ions: one activates the nucleophile while the other stabilizes the leaving group, following a "two-metal-ion mechanism". (**c**) A 5′-phosphate and a 3′-hydroxyl terminus are formed.

the presence of metal ions. But which metal ions show the best performances in catalysis of biological relevance? Mg^{2+}, together with Na^{+}, K^{+}, and Ca^{2+} are the most abundant metal ions in cells. Mn^{2+} has similar physico-chemical properties as Mg^{2+}, but it is not as abundant as the latter and it can also be redox-active. Metal ions are normally needed at millimolar concentration in order to perform a structural and/or a catalytic role, therefore their solubility in water at neutral pH is important, and a large number of metal ions are thus excluded from the use in catalysis under biological conditions. A metal ion involved in catalysis should be able to coordinate, stabilize, and activate different ligands, with kinetic and thermodynamic properties that match with the catalyzed reaction. To act as catalyst, a metal ion must form a thermodynamically favorable interaction with the ligand but at the same time it must be sufficiently dynamic to adapt to the coordination requirements within the catalytic reaction pathway. Depending on the kinetics of ligand exchange, metal ions have been divided into three groups (see for example [45]). The metals in the first and the third group show too high or too low mobility, respectively, to be active in catalysis. The majority of alkali and alkaline earth metals belongs to the first group, whilst metal ions like Al^{3+}, Be^{2+}, Cr^{3+}, and Co^{3+} belong to the third group. The second group contains those metal ions whose exchange rates ($k_{ex}(H_2O)$) fall into an intermediate range, and Mg^{2+} is one of them.

Another important parameter to consider when describing metal ions is their binding affinity to different ligands. In general, metals can be grouped into two classes: the hard and the soft ones. The hard metals, small and difficult to polarize, preferentially interact with hard ligands, such as oxygen sites. On the other hand, the soft ones, being large and easily polarizable, prefer, e.g., sulfur residues. The biologically relevant metal ions Na^{+}, K^{+}, Mg^{2+}, and Ca^{2+} belong to the first class, and are consequently good partners for interactions with the hard phosphate oxygens within the active site of the ribozymes. Nevertheless, the overall physico-chemical properties of Mg^{2+} make it the best adapted cofactor for catalysis performed by large ribozymes.

The role of metal ions in enhancing catalysis has been extensively studied [10,11,45]. For this reason, only the different ways of action of metal ions in ribozyme catalysis are described shortly here. There are two methods by which metal ions enhance catalysis in large ribozymes: a direct and an indirect way. The direct way consists of four different subgroups. Accordingly, metal ions can intervene by (i) stabilization of the leaving group, (ii) activation of the nucleophile, (iii) coordination to the non-bridging phosphoryl oxygen, and (iv) a concurrent interaction with the nucleophile and the non-bridging phosphoryl oxygen. An accurate localization of the metal ions within the catalytic core (i.e., a description of their coordination pattern) allows to understand how these metal ions facilitate splicing. Both, biochemical and structural data indicate the simultaneous presence of two or three catalytic metal ions within the active core of large ribozymes (see below). Up to now, the best mechanistic description of how these metal ions participate in catalysis is the "two-metal-ion mechanism", proposed to be valid also for ribozymes in 1993 by T. A. Steitz and J. A. Steitz [37].

In this mechanism one metal ion is responsible for the activation of the entering nucleophile while the other stabilizes the leaving group, in a general acid–base catalysis (Figure 3). In the case of ribozymes, the coordination of Mg^{2+} to the oxygens at the active site should activate the nucleophile and stabilize the scissile phosphate group, while the other Mg^{2+} should screen and stabilize the negative charge of the leaving group.

Finally, the indirect effect of metal ions in catalysis also deserves mentioning. It was shown that breaking a metal–RNA interaction can lead to a strong decrease in the catalytic activity, even if the metal is far away from the active site [23,70]. In a recent review [41] the authors invite the readers to broaden the definition of metal ion-assisted catalysis. In addition to the direct coordination of divalent metal ions in catalysis, they list at least three other mechanisms by which metal ions can assist and enhance catalysis: (i) they can perform electrostatic and structural stabilization of the transition state via outersphere interactions; (ii) they can stabilize the transition state by long-range electrostatic contacts, and (iii) they can stimulate the shift of the pK_a of some nucleotides, thereby rendering nucleobases or sugar residues able to actively participate in catalysis. In principle, given the latter three ways of interaction, the catalytic role of divalent metal ions could be fulfilled by different cations, including monovalent metal ions or polyamines. Nevertheless, it is probable that ribozymes preferentially employ Mg^{2+} (1–2 mM) and K^{+} (around 100 mM), being the most abundant metal ions within the cell, by using different mechanisms.

Mg^{2+} is considered the natural cofactor of large ribozymes, but in a few cases it can be substituted by other divalent ions. This substitution causes a modification in ribozyme activity, but rarely clear correlations between the physico-chemical properties of the metal ions and the activity of the ribozymes have been found [42]. Such a correlation, as found for the Hammerhead ribozyme [42], allows to better understand the mechanism employed by different metal ions in catalysis.

3. GROUP I INTRON RIBOZYMES

3.1. General Remarks on Group I Introns

Introns are non-coding sequences of a pre-RNA that need to be removed from the primary transcript to yield a functional RNA. Most genes are in fact primarily transcribed into RNA precursors that contain both coding and non-coding sequences. The non-coding sequences must be removed in order to obtain the mature RNAs. Several genes can contain different introns, and the alternative splicing of these introns is one reason for the incredible complexity of the eukaryotic system [48]. Group I introns are self-splicing introns, i.e., large ribozymes discovered more than twenty years ago that are able to splice themselves without the aid of a protein. They occur naturally in bacteria and bacteriophages [71], but also in the mitochondria of some animals [3].

Self-splicing introns are extremely interesting: they can migrate, inserting themselves at different positions of the host genome, thereby becoming mobile genetic elements responsible for genome diversification [72–75]. In this sense, they could potentially be engineered and used for biomedical applications [76]. Moreover, their existence strengthens the hypothesis of the "RNA world": a single molecule able to transport genetic information and to perform chemical reactions [77].

The first group I intron ribozyme (*Tetrahymena thermophila*) was discovered in 1982, by Cech and coworkers, and in 1986 it was shown to be a true metalloenzyme [78,79]. Group I intron ribozymes are abundant in nature: more than 2000 sequences have been discovered so far [80]. In spite of the poor sequence conservation, group I introns have been classified into 13 subgroups [80].

They generally comprise 200–1500 nucleotides and present a secondary structure which consists of ten base-paired elements (P1–P10) capped by loops and connected by junctions (Figure 1a). Two helical domains can be recognized, P4–P6 and P3–P9, while the region P1, where the 5′-splice site is located, docks at the interface between the two domains. Helix P10 is formed after the first step of splicing [80,81].

Group I intron splicing consists of two subsequent trans-esterification reactions between which a conformational change occurs. They use an exogenous guanosine or guanosine triphosphate as nucleophile to cleave the 5′-splice site (Figure 4). In the first splicing step, the 3′-OH of the external guanosine attacks the phosphodiester bond at the 5′-splice site, at the P1 site. A conserved G–U wobble pair located in P1 contributes to the recognition of the 5′-splice site. Helix P10 is formed after the first step of splicing and includes base pairs between the intron and the 3′-exon. After an internal rearrangement, a second trans-esterification reaction leads to the excision of the intron and the re-ligation of the adjacent exons.

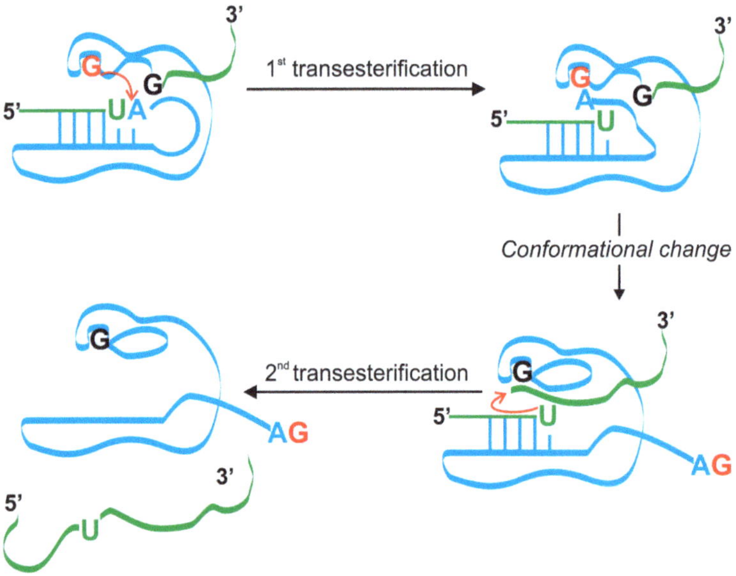

Figure 4. Schematic representation of the self-splicing of group I intron ribozymes. An external guanosine (ωG; shown in red) acts as the nucleophile in the first step of the splicing. After a conformational change, the conserved 3′-terminal nucleotide of the intron (ΩG in the text) replaces ωG in the binding pocket and prepares the 3′-exon for ligation. The second trans-esterification leads to the ligation of the two exons and the excision of the intron (adapted from [2]).

Recent studies have provided many data and information on the mechanism of folding and catalysis [80–82]. As expected, group I introns require divalent metal ions for both correct folding and catalysis [18–20, 83–88]. In the last twenty years, much research was dedicated to the elucidation of crystal structures of group I intron ribozymes to obtain further insights into both structure and activity and to verify the data coming from biochemical studies. Between 2004 and 2005, three different groups solved crystal structures of group I introns, originating from different organisms. The structures of group I introns from the purple bacterium *Azoarcus* [24,89], the ciliate *Tetrahymena thermophila* [90], and the *Staphylococcus aureus* bacteriophage *Twort* [91] were published. Further studies were performed by Lipchock and Strobel, in order to crystallize the intron at different stages of the splicing/ligation process [92]. A close analysis of these structures together with the already available data from the crystal structure of the non-catalytic P4–P6 domain [93] of *Tetrahymena thermophila* allowed to understand the organization of the molecule around its catalytic core, as well as to elucidate the role of metal ions in this complex picture [80,82] (see also Section 3.3).

3.2. Metal Ions in Folding

The role of metal ions in folding and stabilization of the tertiary structure of the group I introns is by now universally acknowledged [14,81]. For example, it was shown that the *Tetrahymena* group I intron folds very slowly in the presence of Mg^{2+} from an unfolded state, at low ionic strength, but, if the ribozyme is pre-folded in the presence of Na^{+}, the Mg^{2+}-mediated refolding is very fast [94]. In the case of the *Azoarcus* group I intron, submillimolar concentrations of Mg^{2+} are sufficient for the assembly of the internal helices, while millimolar concentrations of Mg^{2+} are required for tertiary structure stabilization. This amount of Mg^{2+} almost coincides with activity requirements [95]. It was also shown that, even if many cations could promote folding and assembly, no active tertiary structure was formed in the absence of Mg^{2+} [84].

Recently, the influence of the divalent alkaline earth ions in the folding of the *Tetrahymena* group I intron by non-denaturing gel electrophoresis and Brownian dynamic simulations was described [85]. The authors show that the stabilization of the RNA tertiary structure can be described in terms of charge density of the cation used as cofactor. This parameter indeed accounts for both the ability of the metal ion to approach the negatively charged RNA (electrostatic effects) and the excluded volume between the cations. The higher the charge density is, the better is the stabilization. Consequently, the ribozyme is better stabilized by the small Mg^{2+}. The authors suggest that non-specific polyelectrolyte effects play a crucial role in folding and recommend to the readers to take into account size and shape of the cations in the stabilization of RNA tertiary structures. These findings confirm previous studies in which metal ions were replaced by polyamines [96].

In a successive paper [97], the same authors studied the counterion-dependent collapse of *Azoarcus* and *Tetrahymena* group I introns by means of small-angle X-ray scattering (SAXS). This technique allows to follow changes in size and shape of biomolecules. They evaluated the variation of the radius of gyration of the two ribozymes in the presence of several metal ions at different concentrations confirming the folding pathways of the two ribozymes: the *Tetrahymena* group I intron follows a three-state model (the native structure is reached through metastable misfolded intermediates) [97–99] while the folding of the smaller *Azoarcus* group I intron can be approximated with a two-state reaction, where the native state is reached through a compact native-like intermediate [95,97]. This means that in the latter case the neutralization of the negative charge of the phosphate backbone and the formation of tertiary contacts take place almost simultaneously. In general, for both RNAs, the collapse to the native state becomes more cooperative with the increase of the charge density of the counterions. But, due to different folding mechanisms, the collapse of the *Azoarcus* group I intron is less sensitive to the differences among the various metal ions used, while the second transition to the native state of the *Tetrahymena* group I intron is strongly affected by size and valence of the counterions, being favored by Mg^{2+} and Ca^{2+} more than by Sr^{2+} and Ba^{2+}. This is most likely due to Mg^{2+} being specifically bound to certain structural motifs. Nevertheless, the authors point out that even if polyelectrolyte effects represent a strong driving force in RNA compaction, there are cases in which direct coordination of metal ions within specific sites of RNA determine the formation of a compact structure [97].

The folding pathways of the *Azoarcus* group I intron was additionally examined in detail by means of SAXS [100]. It was confirmed that collapse and tertiary folding for the *Azoarcus* ribozyme occur within almost the same time-scales, but it also emerged that heterogeneity of folding kinetics strongly depends on Mg^{2+} concentration.

The data derived from X-ray structures support the evidence of the structural role played by Mg^{2+} (Figure 5). In the following we will discuss this point using the example of the *Azoarcus* bacterial group I intron in complex with its 5'- and 3'-exons [20]. The crystal structure of the *Azoarcus* group I intron was obtained in two different ways and solved at two different resolutions (3.1 and 3.4 Å, Figure 5) [24,89]. Analysis of the structures reveals the presence of eighteen specifically bound K^+ and Mg^{2+} ions. In particular, five of the observed metal ions are located within the center of the ribozyme, interacting with sugar and phosphate oxygen atoms mainly

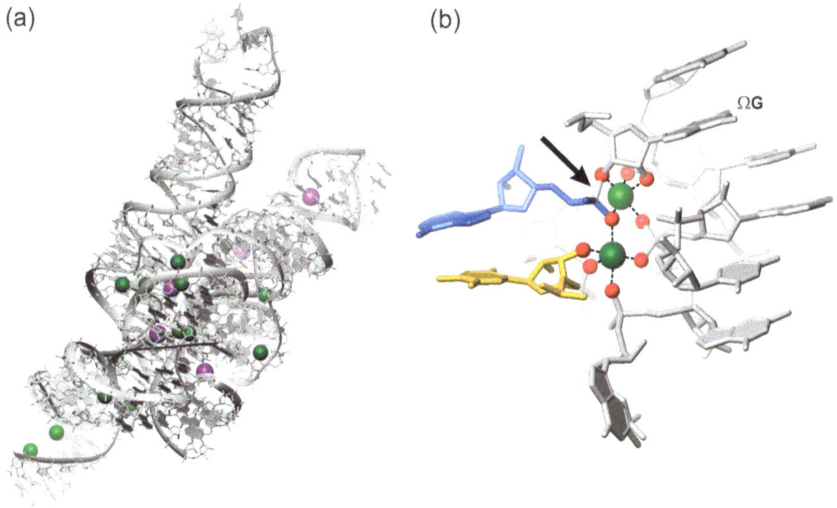

Figure 5. Crystal structure of the *Azoarcus* group I intron. (**a**) The pre-2S deoxy-ΩG structure (3.1 Å resolution) shows the intron with its exons before the second step of splicing (pre-2S) [89]. The colored spheres indicate metal ions: K^+ (purple) and Mg^{2+} (green). (**b**) Active site with two metal ions M1 and M2 in the crystal structure of the pre–2S ribo–ΩG *Azoarcus* group I intron (3.4 Å resolution). The 2′-OH of the ΩG is decisive for binding of M2. Inner sphere metal coordination is shown with dashed lines [24]. The two Mg^{2+} ions are 3.9 Å apart, suggesting a "two-metal-ion mechanism". With the exception of the catalytic center, the location of the metal ions within the two structures agrees quite well [20]. The 5′-exon is indicated in yellow, the 3′-exon in blue and the coordinating oxygens as red spheres. The arrow indicates the scissile phosphate bond. The figures were prepared with MOLMOL [187] using PDB ID 1U6B for (a) and 1ZZN for (b).

through innersphere coordination, and comprise biochemically relevant contacts. These five metals can be described as a "metal ion core" of the intron, as they lie within 12 Å of the scissile phosphate. M1 and M2, described as catalytic metal ions on the basis of biochemical data, are located close to the scissile phosphodiester and are found to be important to activate the substrate, stabilize the transition state, and form the architecture of the active site. Their catalytic activity was confirmed by the observation that the breaking of the direct contacts between these metal ions and the RNA via sulfur or amino substitutions caused a strong decrease in the splicing activity. The other three metals are not directly positioned in the active site but they have been found to be important for activity. Concerning the other metal ions, contacts to different regions of the helices were found. The coordination of the eighteen metal ions found in the structure can be summarized as follows: six of them are coordinated to two or more secondary structure elements and are responsible for tertiary contacts between distinct parts of the intron; the others are either directly coordinated to phosphate oxygens in the backbone or positioned in the negative major groove. It is also worth noting that K^+ and Mg^{2+} interact in a different manner with the RNA. Whereas K^+ mostly binds innersphere, Mg^{2+} coordinates to uncharged functional groups through a water molecule. Interestingly, the two Mg^{2+} ions within the catalytic region each show five innersphere interactions (see Figure 5). The involvement of these two metal ions in catalysis is discussed in the following section.

3.3. Metal Ions in Catalysis

Large ribozymes need divalent metal ions in order to perform splicing. Even though Mg^{2+} remains the most important cofactor for both folding and catalysis, it has been shown that the activity of the *Tetrahymena* group I intron is maintained in the presence of either Mg^{2+} or Mn^{2+} [83]. In contrast, Ca^{2+} only promotes folding, but does not support catalytic activity [98]. Even more interesting, it is possible to properly engineer *in vitro* both the *Tetrahymena* [101,102] and *Azoarcus* [103] group I introns in a way that they selectively require Ca^{2+} in splicing.

In Section 2.2, the "two-metal-ion mechanism" responsible for catalysis of protein enzymes involved in phosphoryl transfer processes has been described. The analysis of the numerous data acquired by biochemical experiments on the *Tetrahymena* group I intron confirms the overall picture of a "two-metal-ion mechanism", but adds the possibility of the involvement of a third metal in the active site [86–88]. The scheme of actions of the three metals in the active site is illustrated in Figure 6a: M_B activates the endogenous guanosine that acts as nucleophile; M_A neutralizes the charge of the leaving group while M_C stabilizes the geometry of the active site. Thio-rescue experiments on the *Tetrahymena* group I intron helped to identify the coordination sphere of each of the three metal ions. M_B coordinates to the 3′-OH nucleophile of the attacking guanosine; M_A binds the 3′-oxygen of the leaving group, as well as the *pro*-S_p non-bridging oxygen of the scissile phosphate; finally, M_C coordinates to the 2′-OH of the guanosine, as well as the *pro*-S_p non-bridging oxygen of the scissile phosphate (Figure 6a). In contrast, the X-ray structural data suggest that the active site contains less than three metal ions [24,92] (Figure 5). In this regard, it is worth noting that the catalytic core of the first X-ray structures published of the group I introns from *Tetrahymena* [90], *Twort* [91], and *deoxy-ΩG Azoarcus* [89] contained only one of the two catalytic Mg^{2+} ions [82].

In both the *Tetrahymena* and *Twort* group I introns crystal structures the scissile phosphate is missing. Nevertheless, in the *Tetrahymena* group I

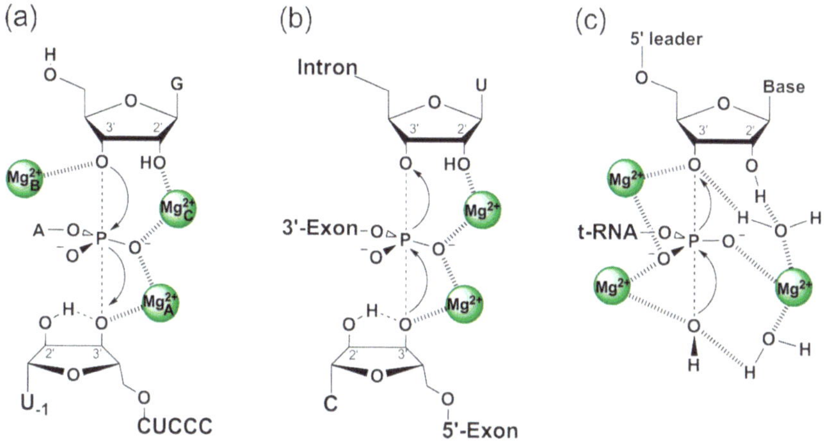

Figure 6. Transition state models for the splicing/cleavage of the three large ribozymes, as derived from biochemical studies. (**a**) Group I introns. The model is based on biochemical investigations on the *Tetrahymena* ribozyme, which employs the oligonucleotide $CCCUCU_{3'x}A$ as substrate, where X is either oxygen or sulfur, under reaction conditions that mimic the first step of splicing [87,88]. (**b**) Transition state of the second step of splicing as inferred by biochemical experiments on the yeast mitochondrial group II intron *Sc*.ai5γ [148]. (**c**) Transition state model for the hydrolysis of the scissile phosphate of pre-tRNA catalyzed by RNase P, as deduced by several biochemical studies [155,156]. Most of the Mg^{2+} interactions have not yet been confirmed.

intron the found metal ion is located very closely to both O2′ and O3′ of ΩG (conserved terminal nucleotide of the intron) in two of the four molecules of the asymmetric unit [90], suggesting that it shares the characteristics of both M_B and M_C. On the other hand, in the *Twort* group I intron the metal ion in the active site seems to be the biochemically identified M_A [91]. In the crystal structure of the *Azoarcus* group I intron published by Stahley and Strobel in 2005 [24], the presence of two Mg^{2+} in the active site strongly supports the "two-metal-ion mechanism" of catalysis. In fact, the distance between the two metal ions is 3.9 Å and their coordination sphere resembles the one observed for RNA and DNA polymerases. These two metal ions, M1 and M2 (see Figure 5), fulfill all the biochemical requirements: M1 binds the nucleophile of the second step of the splicing and can be described as the biochemically termed M_A; M2 shares characteristics of both M_B and M_C, including coordination of O3′ and O2′ of ΩG. It has also been suggested that M2 could be assigned to M_C, and possible explanations for the lack of M_B in the active site have been proposed [45]. For example, (i) the metal absent in the crystal structure could be weakly bound to the catalytic core and consequently cannot be accurately located; (ii) the arrangement of the catalytic site within the crystal structure is incompatible with the binding of the third metal ion.

Recently, Lipchock and Strobel published three new crystal structures of the *Azoarcus* group I intron [92]. One crystal structure comprises the intron in the presence of the spliced exons, while the other two structures shed light on the role of metal ions in the active site during the second step of splicing. They observed that a relaxed active site is obtained after ligation, in which metal ion coordination is lost but tertiary interactions between the exon and the intron are maintained. Interestingly, in all three structures they observed the presence of M1 and M2 in the active site. The detailed analysis of the crystal data, supported by additional biochemical experiments, lead the authors to suggest the presence of a change in the substrate–metal coordination preceding the release of ligated exons. These findings create the foundation for a revised model of substrate binding and docking, giving new insights into the complicated and intriguing landscape of group I introns splicing mechanism.

In general, structural and biochemical data on group I introns correlate quite well. As shown above, the localization of metal ions in the crystal structures of group I intron ribozymes confirmed their expected role in both folding and catalysis of these complex molecular machines.

Due to the increasing number of crystal structures of different group I introns from different organisms, the interest in defining more accurately the catalytic core of these ribozymes continues, resulting in new findings. Recently, Piccirilli, Herschlag, and coworkers analyzed in detail the nature of the active site of the *Tetrahymena* group I intron by means of double-mutant cycles [104,105]. The different crystal structures available show a different architecture around the 2′-OH group of the conserved residue adenosine A261 (numbering in *Tetrahymena*), whose involvement in catalysis has been demonstrated in the past [106]. With the use of the double-mutant cycles the authors could prove which of the interactions involving the 2′-OH group of A261 and the guanosine nucleophile are involved in catalysis. The hydrogen bond interaction between the 2′-OH group of A261 and the NH_2 group of the guanosine, that is formed not immediately after guanosine binding but during the catalytic cycle, seems to be of particular importance. Finally, they proved its role in the structural organization of the active site of the ribozyme before catalysis.

4. GROUP II INTRON RIBOZYMES

4.1. General Remarks on Group II Introns

Group II intron ribozymes are the second class of large self-splicing molecules. They are second in size only to ribosomal RNA and were identified as an independent class in 1982 [107]. Their length varies from 600 to about 2500 nucleotides, and they are found primarily in organellar genes of plants, fungi, bacteria, and lower eukaryotes [108,109]. Recently, a group II intron was found in an animal mitochondria genome [110]. Based on the conserved features of their secondary structure, most group II introns can be further classified into two subclasses, IIA and IIB, which can be divided again into smaller groups [73,111,112]. More recently, a new class was identified, the bacterial class IIC [112]. Generally, group II introns of this latter class are smaller and thought to be the most primitive group II introns [46]. Moreover, they exhibit differences with respect to the other two classes in terms of exon/intron binding site interaction and the self-splicing pathway *in vitro* [113]. In general, these group II introns can efficiently self-splice *in vitro* in the presence of high concentrations of metal ions [108], which are needed for the folding process and for the stabilization of the tertiary structure [108]. However, recently it was shown that only low Mg^{2+} concentrations are sufficient to fold the yeast mitochondrial group II intron *Sc*.ai5γ (from *Saccharomyces cerevisiae*) if the natural protein cofactor Mss116 is present *in vitro* [114]. On the other hand, *in vivo* group II introns require protein cofactors for both the splicing [74,108] and the stabilization of the tertiary structure [115].

Group II introns are phylogenetically related to a large part of the human genome [116] and are considered to be direct ancestors of the eukaryotic spliceosome [47,117], which is responsible for the splicing of nuclear pre-mRNA. The role of metal ions in folding and activity of this fascinating molecular machine is reviewed in Chapter 8 of this volume.

The primary sequence of group II intron ribozymes is not well conserved, but biochemical and mutational studies show that these ribozymes are formed by a conserved set of six domains, that contain well-defined secondary structure elements [109,111,112,118] (Figure 1b). These six domains fold independently [119]: when the individual domains are added together *in trans* (i.e., not covalently bound) in the presence of the right amount of metal ions they fold into the active tertiary structure [120]. Any of the six domains has its own function within the molecule [121]. Domain 1 (D1) is the largest one; it is an independent folding unit and provides the scaffold for docking of the other domains. Its folding is the rate determining step for the folding of the entire ribozyme. Moreover, D1 recognizes the correct splice sites. This domain is normally divided into four subdomains and it contains regions forming tertiary contacts important for catalysis. The docking between D1 and domain 5 (D5) represents the minimal structure necessary for catalytic activity. D5 is the most conserved domain: it adopts the form of a stable hairpin capped with a well known GNRA tetraloop motif [122]. The interaction between D5 and the central core of D1 through the ζ–ζ' and κ–κ' contacts is essential for catalytic activity. Domains 2 (D2) and 3 (D3) play a secondary role with respect to D1 and D5 in both folding and catalysis. Nevertheless, D2 stabilizes the folding of the entire ribozyme and D3 increases the rate of catalysis. Domain 4 (D4) does not contribute to splicing but can contain an open reading frame that can encode a maturase protein, possibly useful for splicing and intron mobility *in vivo* [75]. Finally, domain 6 (D6) contains the conserved bulged adenosine that acts as the nucleophile during the first step of splicing.

Like group I introns, also group II introns perform self-splicing by two sequential trans-esterification reactions. But, in contrast to group I introns, they use the 2′-OH of an internal adenosine as the nucleophile in the first step of the splicing, the so-called branch point adenosine. This 2′-OH attacks the scissile phosphate bond at the 5′-splice site in a S_N2-type reaction. It has been shown that this reversible step represents the rate determining step of the self-splicing reaction [123]. The released 5′-exon remains in contact with the intron through base pairs involving a different region of D1. In the second step of splicing, the 3′-OH terminus of the 5′-exon attacks the phosphate at the 3′-splice site. In this way, the intron is released in the form of a lariat, resembling the mechanism of splicing of the eukaryotic spliceosome (Figure 7). When the branch adenosine is missing, a water molecule can act as nucleophile [117]. This hydrolytic pathway has been observed to work also *in vivo*. The splicing reaction is fully reversible, making these molecules mobile genetic elements [74,75]. This is also reflected by their

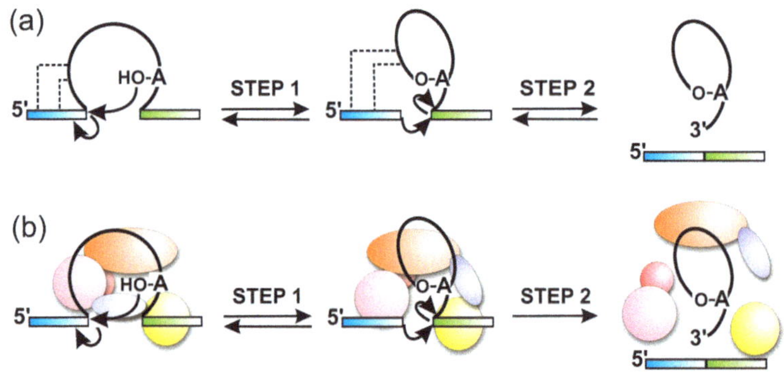

Figure 7. (**a**) Self-splicing mechanism of group II intron ribozymes. In the first step, the internal branch adenosine acts as a nucleophile attacking the scissile phosphate bond at the 5′-splice site via a S_N2 mechanism. In the second step, the two exons are ligated and the intron is excised as a lariat. (**b**) This splicing mechanism resembles the one performed by the eukaryotic spliceosome, in which the excision of the introns and the ligation of the exons are reached by the cooperative work of five RNAs and several proteins (adapted from [22]).

relationship with the spliceosome, various components of telomerase and the LINE-element form of the transposons [25,74,116].

4.2. Metal Ions in Folding

The folding process of group II introns is strongly dependent on metal ions: monovalent (usually Na^+ and K^+) as well as divalent metal ions (Mg^{2+}) are intrinsically tied to RNA for charge compensation, while specifically interacting metal ions have been proposed to play a direct role in catalysis [46]. Noteworthy, different group II introns originating from several organisms require different amounts of metal ions *in vitro* for proper folding and catalysis, despite their conserved secondary structure: up to 2 M K^+ and between 0.1 and 100 mM Mg^{2+} (Table 1).

One of the best examined group II intron constructs is the so called D135 ribozyme, a shortened construct of the yeast mitochondrial intron *Sc*.ai5γ, comprising only domains 1, 3, and 5 [62]. It was shown that the folding of the model D135 ribozyme is rapid at 100 mM Mg^{2+} concentration while it is very slow at physiological Mg^{2+} concentration (2–3 mM) [124,125]. The D135 ribozyme folds into a discrete tertiary structure via a direct pathway without kinetic traps [126]. The collapse of D1 is the rate determining step in this folding process [127], and strongly depends on metal ion concentration. Biochemical studies allowed to identify the crucial tertiary contacts responsible for folding and to locate metal ions in specific positions. In particular, it was found that the κ–ζ element is a divalent ion binding pocket that behaves as a Mg^{2+}-dependent switch responsible for the initiation of D135 folding [126,128]. It has also been shown that the amount of Mg^{2+} needed for folding is not affected by the presence of 0.5 M KCl. It seems that monovalent ions do not compete with Mg^{2+} binding sites, suggesting a high level of specificity of the metal ion binding sites of D135 [127]. The folding of D135 is reversible: i.e., the urea-denaturated state can be refolded in the presence of the right amount of Mg^{2+} [125]. These findings are in line with a two-state folding pathway [125] that does not include kinetic traps. This is intriguing considering the complexity of these large molecules and contrasts with the rugged folding pathways followed by the other large ribozymes.

A recent review [54] analyzes in detail the folding landscape of group II introns comparing the folding pathways in non-physiological and at near-physiological conditions, as well as describing the role of proteins that facilitate folding *in vivo*. The authors finally propose the folding strategy used by group II introns as a model for the folding of large RNAs.

Recently Sigel, Rueda, and coworkers [129] used single molecule Förster resonance energy transfer (smFRET) to study the influence of Mg^{2+} in the folding of the D135 ribozyme. They confirm that the folding takes place

Table 1. Concentration of monovalent and divalent metal ions needed by group II intron ribozymes originating from different organisms in order to reach the best activity *in vitro* (Table adapted from [22]).

Group II intron	Subgroup	Organism	$[K^+/NH_4^+]$ [M]	$[Mg^{2+}]$[mM]	References
Sc.aI1[a]	IIA	*Saccharomyces cerevisiae*	1	100	[180]
Sc.aI2[a]	IIA	*Saccharomyces cerevisiae*	2	100	[181]
Sc.ai5γ[a]	IIB	*Saccharomyces cerevisiae*	0.5	50–100	[62,182]
Sc.bI1[b]	IIB	*Saccharomyces cerevisiae*	1.25	10	[132]
Pl.lsu/1[c]	IIB	*Pylaiella littoralis*	1	100	[183]
Pl.lsu/2[c]	IIB	*Pylaiella littoralis*	1(0.1–1)	5(0.1–100)	[183]
Ll.ltrB[d]	IIA	*Lactococcus lactis*	0.5–1.5	50	[184,185]
Av.hsp60[e]	IIB	*Azotobacter vinelandii*	0.5	100	[186]
pOi5[f]	IIC	*Oceanobacillus iheyensis*	0.5	125	[46]

Gene location of the ribozymes listed in the table: [a]cox1, cytochrome oxidase 1; [b]cob, apocytochrome B; [c]lsu rRNA, large subunit of ribosomal RNA; [d]ltrB, relaxase gene of conjugative element pRS01; [e]hsp60, heat shock protein 60kD; [f]modified construct used in [46].

without kinetic traps and identify previously not characterized on-pathway folding intermediates. D135 folds to a more stable near-native state, from which the active structure is reached by substrate binding. The folding of D135 strongly depends on Mg^{2+} concentration and the increase of Mg^{2+} concentration causes structural dynamics that allows the establishment of the folded active state. In a successive work, the same authors studied with the same technique the influence of Ca^{2+} in the folding pathway of D135 [130]. It was found that Ca^{2+} perturbs the global 3D architecture of this molecule, besides creating two distinctly different subpopulations that cannot switch between each other [130].

4.3. Metal Ion Binding Sites

Specifically interacting metal ions can both stabilize the tertiary structure and directly participate in the self-splicing reaction. By means of Tb^{3+}-mediated cleavage (see Chapter 2) metal ion binding sites could be located in domains 1, 2, 5, and 6 of the yeast mitochondrial group II intron *Sc*.ai5γ [131]. The majority of the binding sites is located within the core of the active tertiary structure of the intron, including the bulge of the catalytic D5. Some other metals are assumed to play a structural role, as they appear in regions critical for the achievement of the tertiary structure. Successive work performed on the *bI1* group II intron showed the existence of a Mg^{2+} binding site close to the branching adenosine of D6, responsible for the first step of splicing [132].

Further on, interesting insights on the localization of metal ions in the structure of single domains coming from NMR technique were published recently. In the last few years NMR has established itself as a powerful method to determine RNA structures [133–135]. Also the detection of metal ion binding sites is possible by using NMR-active isotopes [136], mimicking ions [137] or paramagnetic ions [138].

Sigel and coworkers recently described the solution structure of domain 6 of the yeast mitochondrial group II intron *Sc*.ai5γ [139]. They studied the truncated but catalytically active construct derived from the wild-type D6 sequence, in which the internal loop is missing (Figure 8). The splicing activity with this shortened D6-27 was only slightly reduced, confirming the recent hypothesis on the branching mechanism [140]. Based on the NMR solution structure at least four or five Mg^{2+} were localized by Mg^{2+} titrations studies. These ions are specifically bound at different positions in D6-27: one at the 5′-end of the hairpin, one at the tandem C–G base pair in helix 1, one in the branch region, and one or two in the tetraloop (Figure 8). Interestingly, both NMR and fluorescence studies show that the local structure of the branch point does not change upon metal binding. They also determined the intrinsic affinities of the different Mg^{2+} ions coordinated to

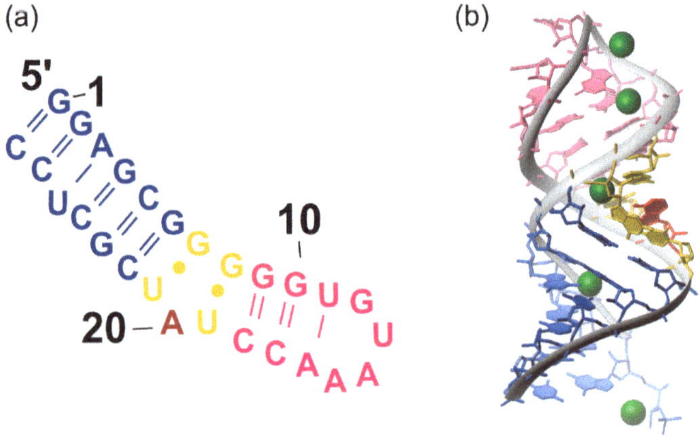

Figure 8. (**a**) Secondary structure of the shortened domain 6 of the yeast mitochondrial group II intron *Sc*.ai5γ (D6-27) used for the experiments described in [139,141,142]. The branch adenosine is indicated in red. (**b**) Tertiary structure of D6-27 with the Mg^{2+} ions in green as emerged from NMR investigations. This figure was prepared with MOLMOL [187] using PDB ID 2AHT.

D6-27 [141]. In a successive recent publication they used ^{15}N NMR to define the coordination sphere of the metal ions coordinated in D6-27 [142]. Despite the intrinsic lability of Mg^{2+} coordination, that hampers solution studies, they found a direct innersphere interaction between Mg^{2+} and N7 of a purine in the tetraloop. Concerning the critical Mg^{2+} located in the adenosine branch point, the authors suggest a specific outersphere contact between $[Mg(H_2O)_6]^{2+}$ and the flanking G–U wobble close to the branch adenosine, but no direct interaction between Mg^{2+} and the adenine N7. As a consequence, the branch adenosine is more free to act as a nucleophile, the long-range electrostatic interaction with the close Mg^{2+} possibly helping to activate it [41].

A similar approach, based on NMR, was used to prove that micromolar amounts of Mg^{2+} or Cd^{2+} promote the formation of the 5′-splice site recognition complex. It was shown that, irrespective of the nature of the ions, divalent metal ions are needed to promote the formation of the EBS1 (exon binding site 1)–IBS1 (intron binding site 1) complex. Circular dichroism spectroscopy was used to prove the efficiency of monovalent metal ions in the EBS1–IBS1 docking [143].

4.4. Metal Ion-Promoted Activity

The presence of Mg^{2+} is crucial for the activity of group II introns, and no other divalent metal ion has shown a better performance in catalysis of these ribozymes [42]. Along the same line as for the folding process, the requirement of Mg^{2+} in catalysis varies among different group II introns (Table 1). Just to cite an example, although being encoded on the same RNA strand, the *Pl*.lsu/2 is active at very low concentration of Mg^{2+} (about 0.1 mM), while *Pl*.lsu/1 needs at least 100 mM Mg^{2+} to effectively carry out catalysis *in vitro* (Table 1 in Section 4.2). Although Mg^{2+} is the natural cofactor for ribozyme catalysis, it can be partially substituted by Mn^{2+} [123]. However, in general, no enhancement in catalysis has been observed by replacing Mg^{2+} with other divalent metal ions. A recent publication dealing with the splicing of the yeast mitochondrial group II intron *Sc*.ai5γ showed that Ca^{2+} is a good inhibitor of the first step of splicing: for example, the reaction rate is decreased to half in the presence of Ca^{2+} and a 100-fold excess of Mg^{2+} [144]. This strong inhibition was not expected, as Ca^{2+} and Mg^{2+} show similar coordinating properties, though the intrinsic affinity of Ca^{2+} towards nucleic acids and derivatives is known to be lower [17,145]. Corresponding experiments indicate that the presence of 10 mM Ca^{2+}, Mn^{2+}, Ni^{2+}, Zn^{2+}, Cd^{2+}, Pb^{2+} or $[Co(NH_3)_6]^{3+}$ with 90 mM Mg^{2+} causes a serious decrease in the rate of the first step of splicing and, in some cases, provokes the degradation of the RNA [144]. The authors point out that the effect of the different metal ions in folding and catalysis of group II introns does not fit to the Irving–Williams series [146], but suggest the existence of specific binding pockets in group II introns to which different metal ions can bind more or less specifically.

Concerning the specific way of catalysis, in the past it was suggested that one Mg^{2+} is involved in the catalytic process by stabilizing the 3′-oxoanion leaving group [147]. More recent studies suggest the direct involvement of two metal ions in the catalytic activity of group II introns [148,149] (Figure 6b). Moreover, two recent crystal structures of a group IIC intron solved by Pyle and coworkers in 2008 show the presence of two metal ions in close proximity to the catalytic center [46,150]. Subsequent to the publication of these crystal structures several papers appeared, with the aim to better analyze the newest findings and describe their biological implications. It has been suggested that the basic tertiary structures of different group II introns are similar and quite conserved. The analysis of all the data available allowed to depict a meticulous picture of the architecture of these large ribozymes, with consequent implications for biological systems [25,47–50].

Both crystal structures of the intact self-spliced group IIC intron ribozyme from *Oceanobacillus iheyensis* [46,150] confirmed many of the biochemically predicted tertiary contacts [151]. In general, the structures show the catalytic domain 5 (D5) surrounded and stabilized by the other domains of the intron (Figure 9), and confirm the presence of metal ion binding sites within the catalytic core [135,149]. The high negative charge accumulated in the proximity of the backbone of the catalytic triad and the bulge within D5 is screened by the presence of two specifically coordinated metal ions, M1 and M2. Soaking experiments with Yb^{3+} suggested that these ions could be Mg^{2+}. Within the crystal structure containing also the ligated exons, M1 is bound to the intron by three innersphere contacts, whereas M2 shows two innersphere contacts to the intron and directly coordinates O3′ of the exonic U4. Both M1 and M2 show short contacts with the scissile phosphate (Figure 9). This is consistent with biochemical studies [148]. The observed arrangement predestines the two ions for catalysis and fits well with the polarity and ionic position expected for a two-ion mechanism. The distance between the two metal ions is about 3.7 Å, consistent with the distance usually found within the two metal ions involved in catalysis in other phosphodiesterase protein enzymes and in group I introns [24]. They reside on the surface of D5 and they are accessible to both 3′- and 5′-splice sites. All these evidences suggest that the self-splicing of the group II introns occurs also via a "two-metal-ion mechanism". However, this evidence is not yet

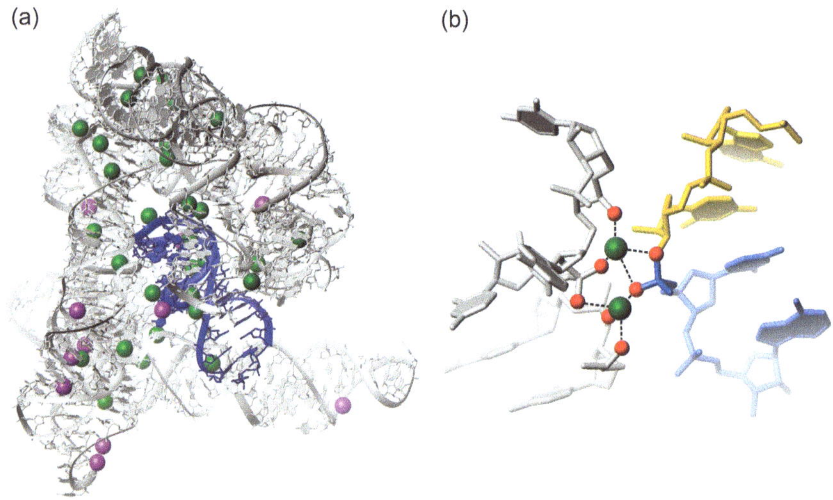

Figure 9. (**a**) Crystal structure of the group IIC intron from *Oceanobacillus iheyensis*, in which the intron has been co-crystallized with ligated exons. The structure depicts a state in which the exons have been ligated but not yet released [150]. The catalytic domain 5 (in blue) represents the core of this compact structure, surrounded by the other domains. The colored spheres indicate metal ions: K^+ (purple) and Mg^{2+} (green). (**b**) A detailed view on the coordination sphere of the catalytic metal ions M1 and M2 (green spheres). The 5′-exon is indicated in yellow, the 3′-exon in blue and the coordinating oxygens as red spheres. Innersphere metal coordination is shown with dashed lines. Both figures were prepared with MOLMOL [187] using PDB ID 3EOG.

conclusive. In fact, electron density for D6 that contains the branch-adenosine is missing in the crystal structure. This is somehow surprising as the intron was crystallized from a full length construct that included D6. Why is there no electron density for D6? There may be two reasons: D6 is disordered in the crystal or it is absent, excised after splicing *in vitro*. It is known that D6 is very mobile, which is important for its function. In order to answer this question, Pyle and coworkers performed further biochemical studies on the *Oceanobacillus iheyensis* construct before and after crystallization [151]. They realized that after splicing *in vitro* this ribozyme causes the excision of D6, and suggested that the encoded maturase protein (without which group IIC introns cannot perform the branching *in vitro*) may help in avoiding this event [25]. Nevertheless, the careful analysis of the D1–D5 scaffold obtained from the X-ray analysis in conjunction with numerous previous biochemical evidences can help in localizing D6 [25]. It should dock to a specific part of D1, including the κ region, and the branch adenosine should be in close proximity to the scissile phosphodiester linkage at the 5′-splice site, where the metal ion M1 has been suggested to be located.

5. RNASE P

5.1. General Remarks on RNase P

Nuclear ribonuclease P (RNase P) is a RNA–protein complex that contains a catalytically active part formed by a RNA molecule (200–400 nucleotides) and one or more protein subunits (120–1000 amino acids). The protein subunit is essential *in vivo* for substrate recognition [152], but the findings that some bacterial [153] and archaeal [154] RNase P are active *in vitro* at high salt concentration without the presence of the protein, suggest that the catalytic component is located within the RNA part, thus confirming the nature of this ribonuclease as a large ribozyme. This ribozyme is largely conserved over all kingdoms of life and catalyzes the hydrolysis of a phosphodiester bond of precursor RNA substrates (e.g., tRNA, 5S rRNA) that leads to the formation of mature RNAs [2,152,155,156]. In bacteria, RNase P comprises a large RNA subunit (400 nucleotides) and a small protein. Bacterial RNase P RNAs have been widely studied and many of the findings have been used to analyze their more complex archaeal and eukaryotic counterparts [155]. They present some differences in sequence, length, and secondary structure (Figure 1c), but contain five sequence-conserved regions and several helical elements constituting a conserved structural core [157]. The RNA subunit comprises two independently folding domains with distinct functions: one domain contacts the T-stem loop of the pre-tRNA substrate and the other is responsible for the catalytic activity [152,156]. Within the latter domain resides helix P4, the most conserved structural element in all RNase P RNAs that constitutes the catalytic core of the ribozyme [158]. There are two families of bacterial RNase P RNAs, A (ancestral) and B (bacillus). They show some different structural elements that surround the analogous conserved structure and fold into a similar tertiary structure [155].

These special ribozymes need metal ions for three reasons: (i) stabilization of RNA folding, (ii) ligand binding, and (iii) catalysis [152,156]. For example, it was demonstrated that more than 100 Mg^{2+} ions are non-specifically bound to the RNA subunit of RNase P, to screen the negative charge of the phosphate backbone and allow the folding to the tertiary structure, while several $Mg^{2+}$are specifically bound. Many roles are played by specifically coordinated Mg^{2+} ions (either inner- or outersphere): they promote and stabilize the binding of RNaseP and the pre-tRNA substrate, they stabilize the transition state for cleavage and enhance catalysis, and they increase the reactivity [159,160]. In general, RNase P strictly requires divalent metal ions for catalysis, mostly Mg^{2+} and Mn^{2+}, although some cases, in which other metal ions are employed for catalysis, have been reported [155].

Recent analysis of crystal structures (see below) suggest that the metal ions found in helix P4, a highly conserved motif of this large ribozyme, have a double function, i.e., structural and catalytic [21,161]. The presence of two or three metal ions involved in the trans-esterification reaction responsible of the formation of mature RNA substrate has been suggested by different chemical and biochemical studies [162,163]. The involvement of the three metal ions in catalysis was reviewed by Kazantsev and Pace in 2006 [155], and is in line with the observed biochemical behavior of *Tetrahymena* group I intron. One metal ion binds the attacking nucleophile as well as the *pro*-R_p non-bridging phosphate oxygen, thus enhancing the nucleophilicity of the attacking group. The second metal ion binds the 3′ bridging and the *pro*-R_p non-bridging oxygens, while the third binds the *pro*-S_p non-bridging oxygen and the 2′-OH of the cleavage site (Figure 6c). The three metal ions can in principle be described as the biochemically termed M_A, M_B and M_C in *Tetrahymena* group I intron (see above), but most of the predicted interactions need to be further confirmed.

The presence of the substrate that binds to this large ribozyme and the necessity of a conformational change of this ribozyme–substrate complex in order to perform catalysis, make this class of ribozymes very complicated and difficult to study. So far, no structural data for the complete RNase P, including RNA and protein, or for the complex between RNase P and its substrate are available. Moreover, no extensive and conclusive picture for the transition state of the RNase P at atomic level has been provided so far. Nevertheless, several biochemical studies and phylogenetic comparative analysis gave insight on structure and function of RNase P, as well as the enzyme–substrate complex [156]. Some of these findings have been confirmed by recent spectroscopic data, as will be discussed in the following section.

5.2. Metal Ion Binding Sites: Structural and Catalytic Role of Metal Ions

In the last few years several crystal structures of RNase P were published, comprising either the RNA subunit or the protein subunit, from different organisms (*Bacillus stearothermophilus*, *Thermotoga maritima*, *Bacillus subtilis,* and *Staphylococcus aureus*) [21,161,164–167]. Unfortunately, no high resolution structural data for the RNA–protein complex or for the complex comprising the pre-tRNA is available so far. For this reason, many of the considerations regarding conformational changes and catalysis are still based only on biochemical and biophysical studies [155,168,169].

However, it is interesting to analyze the crystal structure of the B-type RNase P RNA from *Bacillus stearothermophilus*, published in 2005 [161]. The 417-nt long molecule (solved at 3.3 Å resolution) forms six coaxially stacked helical domains stabilized by long-range tertiary contacts. The chemically active core is located in the conserved fraction of the ribozyme. In general, the structure appears unusual, as it is flat and quite extended. First attempts were undertaken to localize metal ion binding sites within the structure, by analyzing the anomalous signals of the heavy atom derivatives, mainly obtained with Os(III) hexammine. It was found that most of the metal ion binding sites lie in the major groove of double stranded helices. But, most interesting, some of them were localized within P4, the conserved catalytic core of this ribozyme. Three sites were found in the major groove of P4 and two lie in proximity to the catalytic center. Nevertheless, the authors propose that the primary role of metal ion binding sites in helix P4 could be the screening of the negative charge accumulated within.

Unfortunately, this full length construct was not found well suited for crystallographic analysis of metal binding sites because of its poor crystallization properties. For this reason, Pace and coworkers prepared a shorter construct which contains the catalytic domain, whose crystallization properties are better suited for soaking experiments with different metal ions, in order to locate the metal ion binding sites within the ribozyme (Figure 10). In this way they could depict a more detailed map of the metal binding sites in the catalytic core of the RNase P RNA [21]. By anomalous scattering individual binding sites within the catalytic region were identified and

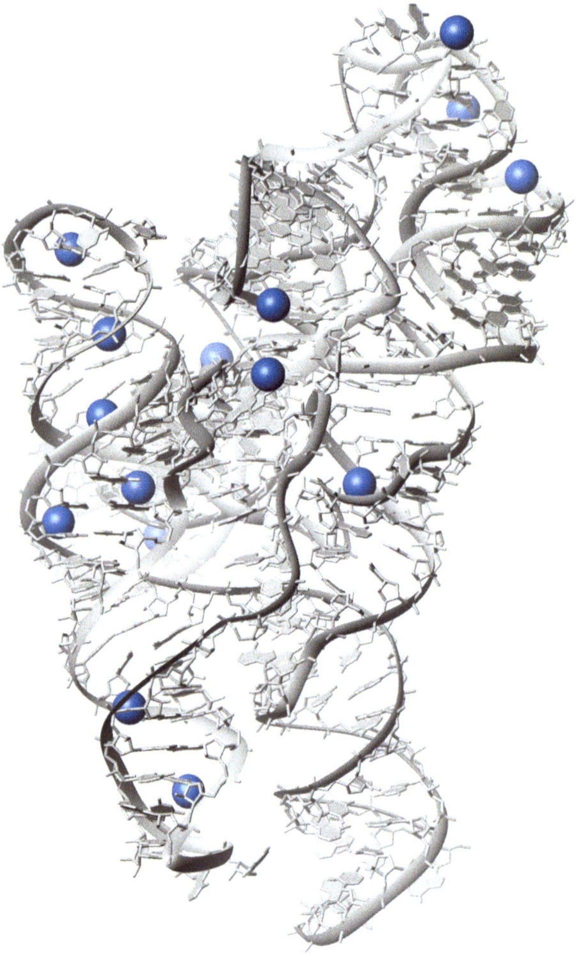

Figure 10. Crystal structure of the short construct comprising the independently folding catalytic domain of the *Bacillus stearothermophilus* RNase P RNA, used in the soaking experiments that lead to the localization of metal ion binding sites [21]. The figure was prepared with MOLMOL [187] using PDB ID 3DHS.

examined for their biological meaning. To identify possible regions of metal binding, they used a variety of metal ions, namely Mn^{2+}, Co^{2+}, Ni^{2+}, Cu^{2+}, Au^{3+}, Pb^{2+}, La^{3+}, Pr^{3+}, Sm^{3+}, Gd^{3+}, Dy^{3+}, Yb^{3+}, Lu^{3+} as well as Os(III) hexammine, but only with some of them they could find the right conditions to obtain useful derivatives. It is clear that each of these ions has its own coordinating properties, they differ in size and charge, and they can behave differently from Mg^{2+}.

Nonetheless, identifying the metal binding sites within the structure helps in defining the regions in which positive charge is potentially needed for structural or catalytic reasons.

A careful examination of the results ended up with the identification of thirteen well defined metal binding sites in the catalytic core, some of which are formed by conserved structural elements. Indeed, one of them is in line with the only one identified (due to the low resolution) in the crystal structure of *Thermotoga maritima* RNase P RNA [164]. Three metal ion binding sites, M10–M12, that partially overlap, are found in close proximity to the area identified as catalytic center by various biochemical methods [170,171]. Special attention must be dedicated to the metal ion binding site M6: it is the only site to which Mg^{2+} binding could be unambiguously attributed. Nevertheless, the biochemically predicted active site resides 12 Å far away from site M6 found in this crystal structure, suggesting that its function is not catalytic [21]. The two metal ion binding sites that are the best candidates to be considered the ones directly involved in catalysis are M10 and M11, even if they are associated only diffusely to Mg^{2+} in the absence of substrate. Finally, the authors do not exclude that the docking of the substrate leads to a conformational change allowing Mg^{2+} ions to specifically coordinate at the positions required for catalysis by a "two-metal-ion mechanism", as proposed by previous biochemical studies [172,173]. Hence, all these findings are not yet conclusive, and further experimental data need to be collected to fully elucidate the multiple roles of metal ions in the ribozyme–substrate complex.

Recently different spectroscopic techniques were applied to cast light on the role of metal ions in these intriguing ribozymes. Fierke and coworkers showed in 2010 for the RNase P from *Bacillus subtilis* in the presence of pre-tRNA, the importance of an innersphere coordinated metal ion for the stabilization of the ribozyme–substrate complex just before the trans-esterification reaction, by using fluorescence techniques [174]. In fact, once formed, the ribozyme–substrate complex undergoes a structural modification that allows it to effectively perform catalysis. It has been suggested that this conformational change could help to unwind the pre-tRNA, in order to make it ready for catalysis [175], besides protecting the RNA subunit from degradation. Briefly, after preforming the ribozyme–substrate complex in the presence of Co(III) hexammine, Mg^{2+} and Ca^{2+} were added in order to favor the stabilization of the active conformer. Kinetic and thermodynamic fluorescent measurements showed that both metal ions stabilize the active conformation, and micromolar apparent affinity constants were found. For the bond cleavage to be effective, a second Mg^{2+} ion is needed, that seems to interact with lower affinity and whose existence confirms the different roles played by catalytic active metal ions [159,160]. Moreover, time-resolved Förster resonance energy transfer (trFRET) experiments, performed in the presence of increasing amount of Ca^{2+}, allowed to follow the conformational change undergone by the ribozyme–substrate complex before the cleavage [174].

Very recently an interesting new method to localize metal ions within the active site of RNase P was published [176]. The combined data coming from NMR and XAS experiments made possible the localization of an innersphere bound metal ion in the helix P4. The data are complementary, as NMR allows to precisely identify the residues involved in the binding whereas XAS gives information about the nature of the interaction. It is known that the highly conserved P4 helix may contain catalytic and cocatalytic metal ions [177,178], but a precise picture at atomic level of this region has not yet been painted. The authors studied a stem-loop RNA that mimics the *Bacillus subtilis* RNase P helix P4 and adopts a conformation similar to that described in the crystal structure of the *Bacillus stearothermophilus* ribozyme [21,161]. Two different XAS techniques were used: X-ray absorption near edge structure (XANES) and extended X-ray absorption fine structure (EXAFS). It was proved that Zn^{2+} binds to helical P4 in a six-coordination fashion, preferentially to oxygen/nitrogen atoms with an average distance of 2.08 Å. NMR paramagnetic line broadening studies, performed by adding Mn^{2+} to the P4 mimic stem-loop, as well as the observation of chemical shift changes induced by addition of Zn^{2+} and $[Co(NH_3)_6]^{3+}$, allowed to depict a quite accurate coordination pattern around the metal ion, that resulted to be innersphere coordinated to the tandem purines, G22 and G23. These nucleotides have been already identified as possible partners for metal coordination [179], but this work allowed to distinguish between inner- and outersphere coordination. The authors highlight the potential of joint NMR and XAS techniques in order to assign and characterize metal ion binding sites in RNAs.

6. CONCLUDING REMARKS

The need of metal ions in both folding and catalysis of large ribozymes is out of doubt: metal ions help ribozymes in neutralizing the negative charge intrinsically correlated to the phosphate backbone, they allow the ribozymes to fold into the active tertiary structure and they participate actively in catalysis.

For many years several research groups around the world tried to rationalize the behavior of such fascinating molecules. Many biochemical, biophysical, phylogenetic, and genetic studies have been performed and allowed to obtain a quite solid picture of the involvement of metal ions in structure and function of these molecules. Moreover, these findings are now supported by crystallographic data. X-ray structures of many group I introns have been published in the last few years, and crystallographic data on RNase P are increasing, even if crystal structures of the complete enzyme (comprising RNA and protein subunits) and the enzyme interacting with the pre-tRNA are still missing. Obtaining a crystal structure of the group II intron was extremely helpful because of the parallel nature to the eukaryotic spliceosome. A few years ago a small group IIC intron was crystallized, and its structure was solved. This finally opened the doors to a deeper understanding of the complex eukaryotic world.

Even if the involvement of metal ions in structure and catalysis is undisputed, the differentiation between the two roles remains challenging. Many chemical, biochemical and spectroscopic techniques have been developed with the aim of identifying and characterizing the metal ion binding sites, but much work is still needed to better distinguish metal ions with structural and catalytic roles.

For all three types of ribozymes discussed a "two-metal-ion mechanism" for catalysis seems to be supported by the experimental data. Nonetheless, strong evidences has been found only for group I introns: careful biochemical studies allowed to depict an atomic-level model of the transition state, that have been partially confirmed by crystal structures obtained at different stages of the splicing pathway.

Further studies are needed in order to fill the gaps existing. Clearly, many secrets are still left to be revealed. Moreover, the numerous papers published by different research groups in the last few years confirm these fascinating molecules to be of wide interest.

ACKNOWLEDGMENTS

D. D. thanks Roland K. O. Sigel for introducing her to the world of these fascinating molecules, for stimulating discussions and for critical reading of the manuscript. Financial support by the 7th European Community Framework Programme (Marie-Curie Post-doctoral Fellowship to D.D.), the Swiss State Secretariat for Education and Research within the COST Action D39 (J.S.), the Swiss National Science Foundation (Project 200021-124834/1 to R.K.O.S.) and the University of Zürich is gratefully acknowledged.

ABBREVIATIONS

EBS1	exon binding site 1
EXAFS	extended X-ray absorption fine structure

IBS1	intron binding site 1
K_a	acidity constant
NMR	nuclear magnetic resonance
RNase P	ribonuclease P
SAXS	small-angle X-ray scattering
smFRET	single molecule Förster resonance energy transfer
trFRET	time-resolved Förster resonance energy transfer
XAS	X-ray absorption spectroscopy
XANES	X-ray absorption near edge structure

REFERENCES

1. W. G. Scott, *Curr. Opin. Struct. Biol.*, 2007, **17**, 280–286.
2. J. A. Doudna and T. R. Cech, *Nature*, 2002, **418**, 222–228.
3. A. Serganov and D. J. Patel, *Nat. Rev. Genet.*, 2007, **8**, 776–790.
4. A. M. Pyle, *Met. Ions Biol. Syst.*, 1996, **32**, 479–519.
5. N. Ban, P. Nissen, J. Hansen, P. B. Moore and T. A. Steitz, *Science*, 2000, **289**, 905–920.
6. G. F. Joyce, *Pure Appl. Chem.*, 1993, **65**, 1205–1212.
7. I. Shcherbakova, S. Mitra, A. Laederach and M. Brenowitz, *Curr. Opin. Chem. Biol.*, 2008, **12**, 655–666.
8. S. A. Strobel and J. C. Cochrane, *Curr. Opin. Chem. Biol.*, 2007, **11**, 636–643.
9. R. Hanna and J. A. Doudna, *Curr. Opin. Chem. Biol.*, 2000, **4**, 166–170.
10. M. J. Fedor, *Curr. Opin. Struct. Biol.*, 2002, **12**, 289–295.
11. V. J. DeRose, *Curr. Opin. Struct. Biol.*, 2003, **13**, 317–324.
12. D. M. J. Lilley, *Trends Biochem. Sci.*, 2003, **29**, 495–501.
13. A. M. Pyle, *J. Biol. Inorg. Chem.*, 2002, **7**, 679–690.
14. S. A. Woodson, *Curr. Opin. Chem. Biol.*, 2005, **9**, 104–109.
15. V. B. Chu, Y. Bai, J. Lipfert, D. Herschlag and S. Doniach, *Curr. Opin. Chem. Biol.*, 2008, **12**, 619–625.
16. D. E. Draper, *Biophys. J.*, 2008, **95**, 5489–5495.
17. E. Freisinger and R. K. O. Sigel, *Coord. Chem. Rev.*, 2007, **251**, 1834–1851.
18. J. H. Cate and J. A. Doudna, *Structure*, 1996, **4**, 1221–1229.
19. J. H. Cate, R. L. Hanna and J. A. Doudna, *Nat. Struct. Biol.*, 1997, **4**, 553–558.
20. M. R. Stahley, P. L. Adams, J. Wang and S. A. Strobel, *J. Mol. Biol.*, 2007, **372**, 89–102.
21. A. V. Kazantsev, A. A. Krivenko and N. R. Pace, *RNA*, 2009, **15**, 266–276.
22. R. K. O. Sigel, *Eur. J. Inorg. Chem.*, 2005, 2281–2292.
23. S. Basu, R. P. Rambo, J. Strauss-Soukup, J. H. Cate, A. R. Ferré-D'Amaré, S. A. Strobel and J. A. Doudna, *Nat. Struct. Biol.*, 1998, **5**, 986–992.
24. M. R. Stahley and S. A. Strobel, *Science*, 2005, **309**, 1587–1590.
25. A. M. Pyle, *Crit. Rev. Biochem. Mol.*, 2010, **45**, 215–232.
26. B. Lippert, *Chem. Biodivers.*, 2008, **5**, 1455–1474.

27. M. D. Smith, R. Mehdizadeh, J. E. Olive and R. A. Collins, *RNA*, 2008, **14**, 1942–1949.
28. B. Lippert, *Prog. Inorg. Chem.*, 2005, **54**, 385–447.
29. R. K. O. Sigel and B. Lippert, *Chem. Commun.*, 1999, 2167–2168.
30. R. K. O. Sigel, E. Freisinger and B. Lippert, *J. Biol. Inorg. Chem.*, 2000, **5**, 287–299.
31. B. Knobloch, R. K. O. Sigel, B. Lippert and H. Sigel, *Angew. Chem. Int. Ed.*, 2004, **43**, 3793–3795.
32. E. Freisinger, A. P. Grollman, H. Miller and C. Kisker, *EMBO J.*, 2004, **23**, 1494–1505.
33. Y. W. Yin and T. A. Steitz, *Cell*, 2004, **116**, 393–404.
34. T. A. Steitz, *Curr. Opin. Struct. Biol.*, 2004, **14**, 4–9.
35. L. W. Tari, A. Matte, H. Goldie and L. T. J. Delbaere, *Nat. Struct. Biol.*, 1997, **4**, 990–994.
36. H. Sigel, *Coord. Chem. Rev.*, 1990, **100**, 453–539.
37. T. A. Steitz and J. A. Steitz, *Proc. Natl. Acad. Sci. USA*, 1993, **90**, 6498–6502.
38. A. M. Pyle, *Science*, 1993, **261**, 709–714.
39. A. T. Perrotta and M. D. Been, *Biochemistry*, 2006, **45**, 11357–11365.
40. J. L. O'Rear, S. Wang, A. L. Feig, L. Beigelman, O. C. Uhlenbeck and D. Herschlag, *RNA*, 2001, **7**, 537–545.
41. R. K. O. Sigel and A. M. Pyle, *Chem. Rev.*, 2007, **107**, 97–113.
42. J. Schnabl and R. K. O. Sigel, *Curr. Opin. Chem. Biol.*, 2010, **14**, 269–275.
43. J. K. Frederiksen and J. A. Piccirilli, *Methods*, 2009, **49**, 148–166.
44. V. J. DeRose, in *Nucleic Acid–Metal Ion Interactions*, Ed. N. V. Hud, Royal Society of Chemistry, Cambridge, UK, 2009, pp. 154–179.
45. J. K. Frederiksen, R. Fong and J. A. Piccirilli, in *Nucleic Acid–Metal Ion Interactions*, Ed. N. V. Hud, Royal Society of Chemistry, Cambridge, UK, 2009, pp. 260–306.
46. N. Toor, K. S. Keating, S. D. Taylor and A. M. Pyle, *Science*, 2008, **320**, 77–82.
47. K. S. Keating, N. Toor, P. S. Perlman and A. M. Pyle, *RNA*, 2010, **16**, 1–9.
48. N. Toor, K. S. Keating and A. M. Pyle, *Curr. Opin. Struct. Biol.*, 2009, **19**, 260–266.
49. F. Michel, M. Costa and E. Westhof, *Trends Biochem. Sci.*, 2009, **34**, 189–199.
50. K. T. Dayie and R. A. Padgett, *RNA*, 2008, **14**, 1697–1703.
51. S. E. Butcher, *Proc. Natl. Acad. Sci. USA*, 2009, **106**, 12211–12212.
52. S. Mitra and M. Brenowitz, in *Nucleic Acid–Metal Ion Interactions*, Ed. N. V. Hud, Royal Society of Chemistry, Cambridge, UK, 2009, pp. 221–259.
53. D. P. Giedroc and N. E. Grossoehme, in *Nucleic Acid–Metal Ion Interactions*, Ed. N. V. Hud, Royal Society of Chemistry, Cambridge, UK, 2009, pp. 180–220.
54. A. M. Pyle, O. Federova and C. Waldsich, *Trends Biochem. Sci.*, 2007, **32**, 138–145.
55. S. A. Woodson, *Annu. Rev. Biophys.*, 2010, **39**, 61–77.
56. I. J. Tinoco and C. Bustamante, *J. Mol. Biol.*, 1999, **293**, 271–281.
57. R. T. Batey, R. P. Rambo and J. A. Doudna, *Angew. Chem. Int. Ed.*, 1999, **38**, 2326–2343.
58. M. H. Bailor, X. Sun and H. M. Al-Hashimi, *Science*, 2010, **327**, 202–206.

59. J. H. Davis, T. R. Foster, M. Tonelli and S. E. Butcher, *RNA*, 2007, **13**, 76–86.
60. D. K. Treiber and J. R. Williamson, *Curr. Opin. Struct. Biol.*, 1999, **9**, 339–345.
61. X. W. Fang, T. Pan and T. R. Sosnick, *Nat. Struct. Biol.*, 1999, **6**, 1091–1095.
62. J. F. Swisher, L. J. Su, M. Brenowitz, V. E. Anderson and A. M. Pyle, *J. Mol. Biol.*, 2002, **315**, 297–310.
63. R. Shiman and D. E. Draper, *J. Mol. Biol.*, 2000, **302**, 79–91.
64. J. R. Lorsch and J. W. Szostak, *Biochemistry*, 1994, **33**, 973–982.
65. G. S. Manning, *Q. Rev. Biophys.*, 1978, **11**, 179–246.
66. V. K. Misra, R. Shiman and D. E. Draper, *Biopolymers*, 2003, **69**, 118–136.
67. I. Rouzina and V. A. Bloomfield, *J. Phys. Chem.*, 1996, **100**, 4305–4313.
68. K. Chin, K. Sharp, B. Honig and A. M. Pyle, *Nat. Struct. Biol.*, 1999, **6**, 1055–1061.
69. C. C. Correll, B. Freeborn, P. B. Moore and T. A. Steitz, *Cell*, 1997, **91**, 705–712.
70. E. L. Christian and M. Yarus, *Biochemistry*, 1993, **32**, 4475–4480.
71. P. Haugen, D. M. Simon and D. Bhattacharya, *Trends Genet.*, 2005, **21**, 111–119.
72. L. Dai and S. Zimmerly, *RNA*, 2002, **8**, 1294–1307.
73. D. M. Simon, N. A. C. Clarke, B. A. McNeil, I. Johnson, D. Pantuso, L. Dai, D. Chai and S. Zimmerly, *RNA*, 2008, **14**, 1704–1713.
74. A. M. Lambowitz and S. Zimmerly, *Annu. Rev. Genet.*, 2004, **38**, 1–35.
75. A. M. Lambowitz and M. Belfort, *Annu. Rev. Biochem.*, 1993, **62**, 587–622.
76. B. A. Sullenger and T. R. Cech, *Nature*, 1994, **371**, 619–622.
77. W. Gilbert, *Nature*, 1986, **319**, 618.
78. K. Kruger, P. J. Grabowski, A. J. Zaug, J. Sands, D. E. Gottschling and T. R. Cech, *Cell*, 1982, **31**, 147–157.
79. A. J. Zaug and T. R. Cech, *Science*, 1986, **231**, 470–475.
80. Q. Vicens and T. R. Cech, *Trends Biochem. Sci.*, 2006, **31**, 41–51.
81. S. A. Woodson, *Curr. Opin. Struct. Biol.*, 2005, **15**, 324–330.
82. M. R. Stahley and S. A. Strobel, *Curr. Opin. Struct. Biol.*, 2006, **16**, 319–326.
83. C. A. Grosshans and T. R. Cech, *Biochemistry*, 1989, **28**, 6888–6894.
84. P. Rangan and S. A. Woodson, *J. Mol. Biol.*, 2003, **329**, 229–238.
85. E. Koculi, C. Hyeon, D. Thirumalai and S. A. Woodson, *J. Am. Chem. Soc.*, 2007, **129**, 2676–2682.
86. A. Yoshida, S. Sun and J. A. Piccirilli, *Nat. Struct. Biol.*, 1999, **6**, 318–321.
87. S. Shan, A. Yoshida, S. Sun, J. A. Piccirilli and D. Herschlag, *Proc. Natl. Acad. Sci. USA*, 1999, **96**, 12299–12304.
88. S. Shan, A. V. Kravchuk, J. A. Piccirilli and D. Herschlag, *Biochemistry*, 2001, **40**, 5161–5171.
89. P. L. Adams, M. R. Stahley, A. B. Kosek, J. Wang and S. A. Strobel, *Nature*, 2004, **430**, 45–50.
90. F. Guo, A. R. Gooding and T. R. Cech, *Mol. Cell.*, 2004, **16**, 351–362.
91. B. L. Golden, H. Kim and E. Chase, *Nat. Struct. Mol. Biol.*, 2005, **12**, 82–89.
92. S. V. Lipchock and S. A. Strobel, *Proc. Natl. Acad. Sci. USA*, 2008, **105**, 5699–5704.

93. J. H. Cate, A. R. Gooding, E. Podell, K. Zhou, B. L. Golden, C. E. Kundrot, T. R. Cech and J. A. Doudna, *Science*, 1996, **273**, 1678–1685.
94. I. Shcherbakova, S. Gupta, M. R. Chance and M. Brenowitz, *J. Mol. Biol.*, 2004, **342**, 1431–1442.
95. P. Rangan, B. Masquida, E. Westhof and S. A. Woodson, *Proc. Natl. Acad. Sci. USA*, 2003, **100**, 1574–1579.
96. E. Koculi, N. K. Lee, D. Thirumalai and S. A. Woodson, *J. Mol. Biol*, 2004, **341**, 27–36.
97. S. Moghaddam, G. Caliskan, S. Chauhan, C. Hyeon, R. M. Briber, D. Thirumalai and S. A. Woodson, *J. Mol. Biol.*, 2009, **393**, 753–764.
98. D. W. Celander and T. R. Cech, *Science*, 1991, **251**, 401–407.
99. J. Pan and S. A. Woodson, *J. Mol. Biol*, 1998, **280**, 597–609.
100. J. Ho Roh, L. Guo, J. Duncan Kilburn, R. M. Briber, T. Irving and S. A. Woodson, *J. Am. Chem. Soc.*, 2010, **132**, 10148–10154.
101. N. Lehman and G. F. Joyce, *Nature*, 1993, **361**, 182–185.
102. N. Lehman and G. F. Joyce, *Curr. Biol.*, 1993, **3**, 723–734.
103. A. S. Burton and N. Lehman, *Biochimie*, 2006, **88**, 819–825.
104. M. Forconi, R. N. Sengupta, M. C. Liu, A. C. Sartorelli, J. A. Piccirilli and D. Herschlag, *Angew. Chem. Int. Ed.*, 2009, **48**, 7171–7175.
105. M. Forconi, R. N. Sengupta, J. A. Piccirilli and D. Herschlag, *Biochemistry*, 2010, **49**, 2753–2762.
106. L. Ortoleva-Donnelly, A. A. Szewczak, R. R. Gutell and S. A. Strobel, *RNA*, 1998, **4**, 498–519.
107. F. Michel, A. Jacquier and B. Dujon, *Biochimie*, 1982, **64**, 867–881.
108. K. Lehmann and U. Schmidt, *Crit. Rev. Biochem. Mol.*, 2003, **38**, 249–303.
109. F. Michel and J. L. Ferat, *Annu. Rev. Biochem.*, 1995, **64**, 435–461.
110. Y. Vallès, K. M. Halanych and J. L. Boore, *PLoS ONE*, 2008, e1488.
111. F. Michel, K. Umesono and H. Ozeki, *Gene*, 1989, **82**, 5–30.
112. N. Toor, G. Hausner and S. Zimmerly, *RNA*, 2001, **7**, 1142–1152.
113. N. Toor, A. R. Robart, J. Christianson and S. Zimmerly, *Nucleic Acids Res.*, 2006, **34**, 6461–6471.
114. A. Solem, N. Zingler and A. M. Pyle, *Mol. Cell.*, 2006, **24**, 611–617.
115. R. Schroeder, A. Barta and K. Semrad, *Nat. Rev. Mol. Cell Biol.*, 2004, **5**, 908–919.
116. J. D. Boeke, *Genome Res.*, 2003, **13**, 1975–1983.
117. A. M. Pyle and A. M. Lambowitz, *The RNA World*, Eds. R. Gesterland, T. R. Cech and J. K. Atkins, Cold Spring Harbor Laboratory Press, Cold Spring Harbor, New York, 2006, pp. 469–505.
118. P. Z. Qin and A. M. Pyle, *Curr. Opin. Struct. Biol.*, 1998, **8**, 301–308.
119. P. Z. Qin and A. M. Pyle, *Biochemistry*, 1997, **36**, 4718–4730.
120. M. Boudvillain, A. de Lancastre and A. M. Pyle, *Nature*, 2000, **406**, 315–318.
121. O. Federova and N. Zingler, *Biol. Chem.*, 2007, **388**, 665–678.
122. F. M. Jucker, H. A. Heus, P. F. Yip, E. H. M. Moors and A. Pardi, *J. Mol. Biol.*, 1996, **264**, 968–980.
123. K. Chin and A. M. Pyle, *RNA*, 1995, **1**, 391–406.
124. O. Federova, C. Waldsich and A. M. Pyle, *J. Mol. Biol.*, 2007, **366**, 1099–1114.

125. L. J. Su, M. Brenowitz and A. M. Pyle, *J. Mol. Biol.*, 2003, **334**, 639–652.
126. C. Waldsich and A. M. Pyle, *J. Mol. Biol.*, 2008, **375**, 572–580.
127. L. J. Su, C. Waldsich and A. M. Pyle, *Nucleic Acids Res.*, 2005, **33**, 6674–6687.
128. C. Waldsich and A. M. Pyle, *Nature Struct. Mol. Biol.*, 2007, **14**, 37–44.
129. M. Steiner, K. S. Karunatilaka, R. K. O. Sigel and D. Rueda, *Proc. Natl. Acad. Sci. USA*, 2008, **105**, 13853–13858.
130. M. Steiner, D. Rueda and R. K. O. Sigel, *Angew. Chem. Int. Ed.*, 2009, **48**, 9739–9742.
131. R. K. O. Sigel, A. Vaidya and A. M. Pyle, *Nat. Struct. Biol.*, 2000, **7**, 1111–1116.
132. M. Hertweck and M. W. Mueller, *Eur. J. Biochem.*, 2001, **268**, 4610–4620.
133. G. Varani, F. Aboul-ela and F. H. T. Allain, *Prog. Nucl. Mag. Res. Sp.*, 1996, **29**, 51–127.
134. M. P. Latham, D. J. Brown, S. A. McCallum and A. Pardi, *ChemBioChem*, 2005, **6**, 1492–1505.
135. R. K. O. Sigel, D. G. Sashital, D. L. Abramovitz, A. G. Palmer III, S. E. Butcher and A. M. Pyle, *Nat. Struct. Mol. Biol.*, 2004, **11**, 187–192.
136. M. Vogtherr and S. Limmer, *FEBS Lett.*, 1998, **433**, 301–306.
137. N. V. Hud, P. Schultze and J. Feigon, *J. Am. Chem. Soc.*, 1998, **120**, 6403–6404.
138. R. L. Gonzalez, Jr. and I. Tinoco, Jr., *Method Enzimol.*, 2002, **338**, 421–443.
139. M. C. Erat, O. Zerbe, T. Fox and R. K. O. Sigel, *ChemBioChem*, 2007, **8**, 306–314.
140. V. T. Chu, C. Adamidi, Q. Liu, P. S. Perlman and A. M. Pyle, *EMBO J.*, 2001, **20**, 6866–6876.
141. M. C. Erat and R. K. O. Sigel, *Inorg. Chem.*, 2007, **46**, 11224–11234.
142. M. C. Erat, H. Kovacs and R. K. O. Sigel, *J. Inorg. Biochem.*, 2010, **104**, 611–613.
143. D. Kruschel and R. K. O. Sigel, *J. Inorg. Biochem.*, 2008, **102**, 2147–2154.
144. M. C. Erat and R. K. O. Sigel, *J. Biol. Inorg. Chem.*, 2008, **13**, 1025–1036.
145. R. K. O. Sigel, B. Song and H. Sigel, *J. Am. Chem. Soc.*, 1997, **119**, 744–755.
146. H. M. Irving and R. J. P. Williams, *J. Chem. Soc.*, 1953, 3192–3210.
147. E. J. Sontheimer, P. M. Gordon and J. A. Piccirilli, *Genes Dev.*, 1999, **13**, 1729–1741.
148. P. M. Gordon, R. Fong and J. A. Piccirilli, *Chem. Biol.*, 2007, **14**, 607–612.
149. P. M. Gordon and J. A. Piccirilli, *Nat. Struct. Biol.*, 2001, **8**, 893–898.
150. N. Toor, K. Rajashankar, K. S. Keating and A. M. Pyle, *Nat. Struct. Mol. Biol.*, 2008, **15**, 1221–1222.
151. N. Toor, K. S. Keating, O. Federova, K. Rajashankar, J. Wang and A. M. Pyle, *RNA*, 2010, **16**, 57–69.
152. J. C. Kurz and C. A. Fierke, *Curr. Opin. Chem. Biol.*, 2000, **4**, 553–558.
153. C. Guerrier-Takada, K. Gardiner, T. Marsh, N. Pace and S. Altman, *Cell*, 1983, **35**, 849–857.
154. J. A. Pannucci, E. S. Haas, T. A. Hall, J. K. Harris and J. W. Brown, *Proc. Natl. Acad. Sci. USA*, 1999, **96**, 7803–7808.
155. A. V. Kazantsev and N. R. Pace, *Nat. Rev. Microbiol.*, 2006, **4**, 729–740.
156. M. E. Harris and E. L. Christian, *Curr. Opin. Struct. Biol.*, 2003, **13**, 325–333.
157. J. L. Chen and N. R. Pace, *RNA*, 1997, **3**, 557–560.

158. M. E. Harris and N. R. Pace, *RNA*, 1995, **1**, 210–218.
159. J. A. Beebe, J. C. Kurz and C. A. Fierke, *Biochemistry*, 1996, **35**, 10493–10505.
160. J. C. Kurz and C. A. Fierke, *Biochemistry*, 2002, **41**, 9545–9558.
161. A. V. Kazantsev, A. A. Krivenko, D. J. Harrington, S. R. Holbrook, P. D. Adams and N. R. Pace, *Proc. Natl. Acad. Sci. USA*, 2005, **102**, 13392–13397.
162. D. Smith and N. R. Pace, *Biochemistry*, 1993, **32**, 5273–5281.
163. T. Persson, S. Cuzic and R. K. Hartmann, *J. Biol. Chem.*, 2003, **278**, 43394–43401.
164. A. Torres-Larios, K. K. Swinger, A. S. Krasilnikov, T. Pan and A. Mondragon, *Nature*, 2005, **437**, 584–587.
165. A. V. Kazantsev, A. A. Krivenko, D. J. Harrington, R. J. Carter, S. R. Holbrook, P. D. Adams and N. R. Pace, *Proc. Natl. Acad. Sci. USA*, 2003, **100**, 7497–7502.
166. T. Stams, S. Niranjanakumari, C. A. Fierke and D. W. Christianson, *Science*, 1998, **280**, 752–755.
167. C. Spitzfaden, N. Nicholson, J. J. Jones, S. Guth, R. Lehr, C. D. Prescott, L. A. Hegg and D. S. Eggleston, *J. Mol. Biol*, 2000, **295**, 105–115.
168. A. H. Buck, A. V. Kazantsev, A. B. Dalby and N. R. Pace, *Nat. Struct. Mol. Biol.*, 2005, **12**, 958–964.
169. S. Niranjanakumari, J. J. Day-Storms, M. Ahmed, J. Hsieh, N. H. Zahler, R. A. Venters and C. A. Fierke, *RNA*, 2007, **13**, 521–535.
170. A. B. Burgin and N. R. Pace, *EMBO J.*, 1990, **9**, 4111–4118.
171. N. H. Zahler, E. L. Christian and M. E. Harris, *RNA*, 2003, **9**, 734–745.
172. L. Sun and M. E. Harris, *RNA*, 2007, **13**, 1505–1515.
173. J. M. Warnecke, J. P. Fürste, W. D. Hardt, V. A. Erdmann and R. K. Hartmann, *Proc. Natl. Acad. Sci. USA*, 1996, **93**, 8924–8928.
174. J. Hsieh, K. S. Koutmou, D. Rueda, M. Koutmos, N. G. Walter and C. A. Fierke, *J. Mol. Biol.*, 2010, **400**, 38–51.
175. D. A. Pomeranz Krummel, O. Kent, A. M. MacMillan and S. Altman, *J. Mol. Biol.*, 2000, **295**, 1113–1118.
176. K. S. Koutmou, A. Casiano-Negroni, M. M. Getz, S. Pazicni, A. J. Andrews, J. E. Penner-Hahn, H. M. Al-Hashimi and C. A. Fierke, *Proc. Natl. Acad. Sci. USA*, 2010, **107**, 2479–2484.
177. E. L. Christian, N. M. Kaye and M. E. Harris, *RNA*, 2000, **6**, 511–519.
178. E. L. Christian, K. M. J. Smith, N. Perera and M. E. Harris, *RNA*, 2006, **12**, 1463–1467.
179. M. M. Getz, A. J. Andrews, C. A. Fierke and H. M. Al-Hashimi, *RNA*, 2007, **13**, 251–266.
180. S. K. Hebbar, S. M. Belcher and P. S. Perlman, *Nucleic Acids Res.*, 1992, **20**, 1747–1754.
181. H. R. Huang, M. Y. Chao, B. Armstrong, Y. Wang, A. M. Lambowitz and P. S. Perlman, *Mol. Cell. Biol.*, 2003, **23**, 8809–8819.
182. K. A. Jarrell, C. L. Peebles, R. C. Dietrich, S. L. Romiti and P. S. Perlman, *J. Biol. Chem.*, 1988, **263**, 3432–3439.
183. M. Costa, J. M. Fontaine, S. Loiseaux-de Goër and F. Michel, *J. Mol. Biol.*, 1997, **274**, 353–364.

184. R. Saldanha, B. Chen, H. Wank, M. Matsuura, J. Edwards and A. M. Lambowitz, *Biochemistry*, 1999, **38**, 9069–9083.
185. M. Matsuura, R. Saldanha, H. Ma, H. Wank, J. Yang, G. Mohr, S. Cavanagh, G. M. Dunny, M. Belfort and A. M. Lambowitz, *Genes Dev.*, 1997, **11**, 2910–2924.
186. C. Adamidi, O. Federova and A. M. Pyle, *Biochemistry*, 2003, **42**, 3409–3418.
187. R. Koradi, M. Billeter and K. Wüthrich, *J. Mol. Graphics*, 1996, **14**, 29–32 & 51–55.
188. H. Sigel and R. Griesser, *Chem. Soc. Rev.*, 2005, **34**, 875–900.
189. (a) H. Pelletier, M. R. Sawaya, A. Kumar, S. H. Wilson and J. Kraut, *Science*, 1994, **264**, 1891–1903; (b) H. Pelletier, *Science*, 1994, **266**, 2025–2026; (c) H. Pelletier, M. R. Sawaya, W. Wolfle, S. H. Wilson and J. Kraut, *Biochemistry*, 1996, **35**, 12762–12777.
190. (a) T. A. Steitz, *Nature*, 1998, **391**, 231–232; (b) T. A. Steitz, *J. Biol. Chem.*, 1999, **274**, 17395–17398.

Met. Ions Life Sci. **2011**, *9*, 235–251

8

The Spliceosome and Its Metal Ions

Samuel E. Butcher
Department of Biochemistry, University of Wisconsin-Madison,
433 Babcock Drive, Madison WI 53706-1544, USA
< butcher@biochem.wisc.edu >

ABSTRACT: The spliceosome is a massive complex of 5 RNAs and many proteins that associate to catalyze precursor messenger RNA splicing. The process of splicing

Metal Ions in Life Sciences, Volume 9 Edited by Astrid Sigel, Helmut Sigel, and Roland K. O. Sigel

Published by the Royal Society of Chemistry, www.rsc.org DOI: 10.1039/978184973251200235

involves two phosphoryl transfer reactions that result in intron excision and ligation of the flanking exons. Since it is required for normal protein production in eukaryotic cells, pre-mRNA splicing is an essential step in gene expression. Although high resolution structural views of the spliceosome do not yet exist, a growing body of evidence indicates that the spliceosome is a magnesium-dependent enzyme that utilizes catalytic metal ions to stabilize both transition states during the two phosphoryl transfer steps of splicing. A wealth of data also indicate that the core of the spliceosome is comprised of RNA, and suggest that the spliceosome may be a ribozyme. This chapter presents the evidence for metal ion catalysis by the spliceosome, draws comparisons to similar RNA enzymes, and discusses the future directions for research into the mechanism of pre-mRNA splicing.

KEYWORDS: metal ion catalysis · pre-mRNA splicing · ribozymes · self-splicing introns · spliceosome

1. INTRODUCTION

Eukaryotic genes are interrupted by regions called introns that must be removed before a functional gene product can be produced. This editing step is called splicing, and occurs in the nucleus at the level of the precursor messenger RNA (pre-mRNA). Pre-mRNA splicing is the process by which internal segments of the pre-mRNA transcript, the introns, are removed and the flanking sequences, or exons, are ligated. It is often assumed that the exons contain the protein-coding information, while the intronic regions of the pre-mRNA are non-coding. However, this is an incorrect assumption since many non-coding exons are known [1], while some type of introns code for proteins [2]. The enzyme responsible for pre-mRNA splicing is the spliceosome, a large assembly of 5 RNAs and many proteins. The spliceosome catalyzes a metal-ion dependent, two-step RNA splicing mechanism, which is the subject of this chapter.

The spliceosome is a large ($\sim$60S) and dynamic supramolecular assembly of 5 RNAs and many proteins. The 5 RNAs are named U1, U2, U4, U5 and U6 RNA [3,4], and are also known as small nuclear RNAs (snRNA), or spliceosomal RNA. The snRNAs associate with proteins to form small nuclear ribonucleoprotein particles (snRNPs). In addition to the 5 snRNPs, other protein complexes also associate with the spliceosome, such as the Prp19 complex [3]. More than 150 different proteins have been found to associate with the human spliceosome [5,6], while the yeast spliceosome has approximately half as many proteins [6]. Metazoans also have a "minor" spliceosome, which is responsible for splicing a minor class of introns with different 5' and 3' boundary sequences [7]. The minor spliceosome is composed of different snRNPs (U11, U12, U4atac and U6atac; the U5 snRNP is the same in both the major and minor spliceosomes). The minor spliceosome catalyzes the same splicing chemistry as the major spliceosome and appears to be highly homologous to the major spliceosome, and therefore will not be discussed further in this chapter.

Spliceosome assembly is a highly dynamic process [3]. Large-scale rearrangements in the spliceosome occur upon assembly and activation. These large-scale rearrangements include extensive, ATP-dependent remodeling of the RNA secondary structure by the action of DExD/H-box helicase proteins [3]. The remodeling catalyzed by these helicases results in the disruption of extensive base-pairing interactions between the U4 and U6 RNAs, formation of new base pairing interactions between U2 and U6 RNAs, displacement of U1 RNA interactions with the 5' splice site in favor of interaction of U6 RNA with the 5' splice site, and dissociation of the U1 and U4 snRNPs from the complex [3]. This large-scale remodeling of the spliceosome structure results in formation of the catalytically active spliceosomal C complex, comprised of the U2, U5, and U6 snRNPs. It has been hypothesized that the ATP-dependent rearrangement of the spliceosome structure serves to ensure that the spliceosome has been properly assembled on the pre-mRNA before commitment to splice [3]. The interesting assembly steps that lead up to the formation of the catalytically active spliceosomal C complex are outside the scope of this chapter, and have been reviewed elsewhere [3].

Currently, there are no high-resolution structural data of the intact spliceosome or even of an individual snRNP. The crystal structure of the U1 snRNP has been solved at an intermediate resolution of 5.5 Å [8]. However, a growing body of X-ray crystallographic and NMR structural data has been accumulated on individual components of the spliceosome [9]. In addition, great progress has been made in defining the overall low-resolution shape of the spliceosome by electron microscopy [10]. Ultimately, a high-resolution crystal structure of the catalytically active spliceosome will be required in order to fully understand the pre-mRNA splicing mechanism. Progress towards this goal has been hindered by the dynamic nature of the spliceosome and the fact that, thus far, it has not been possible to isolate the milligram quantities of homogenous spliceosomes that would be required for high-resolution structural studies.

Fortunately, many elegant biochemical experiments over the past two decades have revealed much about the pre-mRNA splicing mechanism and its metal ion dependence. Furthermore, high-resolution crystal structures have recently been solved of a group II ribozyme that performs splicing chemistry identical to that catalyzed by the spliceosome [11,12]. This crystal structure gives valuable insights into the metallochemistry of the spliceosome. These studies are reviewed and discussed below.

2. THE PRE-mRNA SPLICING MECHANISM

2.1. Chemistry of Splicing

The first step that must occur in the chemistry of pre-mRNA splicing is the specific recognition of the pre-mRNA substrate. This is no simple task, since the average human gene is interrupted by approximately 9 introns, with the average intron being around 5400 nucleotides in length [13]. Conversely, the average human exon is relatively short, approximately 170 nucleotides long [13]. How then are splice sites accurately identified by the spliceosome amongst this background of thousands of nucleotides?

The 5' and 3' boundaries of the intron are each delineated by consensus sequences. The 5' boundary of the intron is typically marked with the consensus sequence 5'-GUAUGU-3', and the 3' end of the intron is delineated by a conserved 5'-YAG-3', where Y is a pyrimidine [3,14]. In addition to these boundary sequences, an internal sequence near the 3' end of the intron, termed the branch point, is also recognized. The branch point has the consensus sequence UACUAAC, where the underlined A contains the ribose 2'-OH group that acts as a nucleophile in the first step of splicing. In between this branch point consensus sequence and the 3' end of the intron is a conserved polypyrimidine tract, which is recognized by the protein U2AF65 [3]. The intron is recognized by the spliceosome in a highly regulated, step-wise assembly process [3]. The first step in intron recognition involves interaction of the U1 snRNP at the 5' end of the intron, which is followed by recognition of the branch site by the U2 snRNP. The U4/U5/U6 tri-snRNP then assembles onto the intron, forming the assembled but inactive spliceosomal B complex. This complex is then activated, in part by the action of the helicase protein Brr2 and the Prp19 complex, to form the catalytically active spliceosomal C complex. The C complex contains the U2, U5, and U6 snRNPs.

2.2. Biochemical Elucidation of Stereochemistry and Metal Ion Requirements

The spliceosome catalyzes two transesterification reactions (Figure 1A). The first step involves nucleophilic attack of the 5' splice site by the 2'-OH of the branch site adenosine, resulting in an unusual 2'-5' phosphodiester bond and formation of a circular "lariat" intron. The second step involves attack of the newly formed 3'-OH group at the end of the 5' exon at the 3' splice site, which releases the intron lariat and results in formation of the spliced exons (Figure 1A and B). The intron lariat is subsequently debranched and degraded by cellular enzymes.

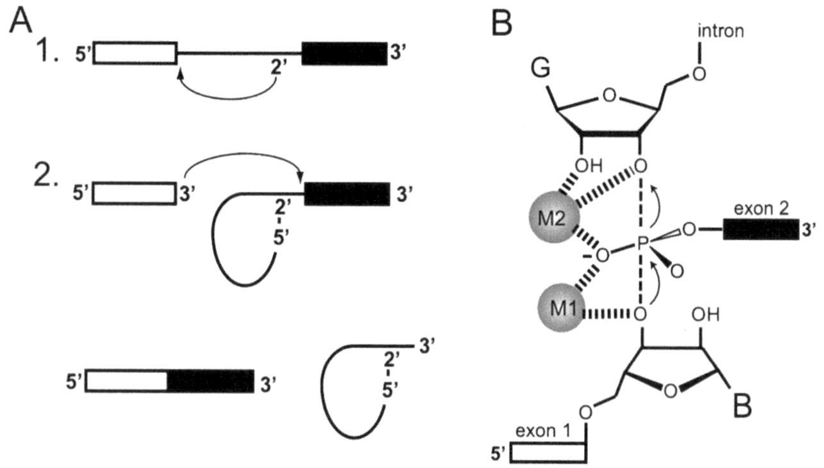

Figure 1. The pre-mRNA splicing mechanism. **A.** Schematic diagram of pre-mRNA splicing. The 5' exon is represented by a white rectangle, the intron is a black line, and the 3' exon is a black rectangle. The first step is the attack of a 2' hydroxyl group from within the intron (the branch point adenosine 2'-OH) at the 5' splice site, resulting in a 2'-5' lariat intron and a 5' exon with a free 3' hydroxyl. The second step involves attack of the 3' hydroxyl at the 3' splice site, resulting in exon ligation and release of the lariat intron. **B.** Schematic diagram of the hypothesized transition state for the second step of splicing. The last nucleotide of the intron is a guanosine and is labeled with a "G", while the last nucleotide of the exon is variable and is labeled with a "B" at the nucleobase position. The remaining parts of the 5' and 3' exons are shown as white and black rectangles, as in A. Two hypothesized metal ions stabilize the transition state and are shown as M1 and M2. Dashed lines indicate hypothesized bonds between the metal ions and oxygens. Experimental evidence for coordination of M2 to the 3' oxygen leaving group is discussed in the text.

The stereochemistry of pre-mRNA splicing was determined by Moore and Sharp [15]. By incorporating chiral phosphorothioates into the pre-mRNA substrate, they observed that both steps were inhibited by the Rp, but not the Sp phosphorothioate diastereomer. This indicates both that the pro-R oxygen of the active site phosphate is very important and possibly involved in metal ion coordination, and that the spliceosome probably shifts between two different active sites during catalysis of the two steps of splicing [15]. Since the Sp phosphorothioate-containing substrates were still active, their stereochemical configurations could be analyzed with phosphodiesterase enzyme after the splicing reaction. It was found that the phosphorothioates underwent stereochemical inversion after splicing, indicating that both steps occur as single "in-line" S_N2 nucleophilic displacement reactions. Interestingly, this mechanism is analogous to both the group I self-splicing introns [15] and the group II self-splicing introns [16]. Since the group II self-splicing introns perform the same splicing reaction with identical stereochemistry, it has been hypothesized that the two systems evolved from a common molecular ancestor [17].

Incorporation of a 3' sulfur atom in place of the 3' oxygen at the 5' splice site showed that interaction between metal ion and the 3' oxygen leaving group is important for the first step of splicing [18]. In contrast, a sulfur substitution at the 3' position of the last nucleotide of the intron, which is the leaving group in the second step of splicing, initially showed no evidence for a metal ion–leaving group interaction in the second reaction step, suggesting that the two steps of splicing occur via different catalytic mechanisms. This idea was consistent with the data from Moore and Sharp [15], which suggested that there are two distinct active sites in the spliceosome. However, when a bimolecular exon ligation assay was used to monitor the second step of splicing, it was found that the 3' oxygen leaving group at the last position of the intron does indeed coordinate a metal ion essential for the second step of splicing [19]. The simplest explanation for these data is that the second step of splicing is not rate-limiting under normal splicing conditions, and that defects in metal ion coordination at the 3' splice site can be masked by a slower, rate-limiting step. It was hypothesized that the slow, rate-limiting step is a conformational change between the first and second steps of splicing [19]. The metal ion that stabilizes the 3' oxygen leaving group during the second (exon ligation) step of splicing is shown in Figure 1B as M2.

2.3. Parallels with Self-Splicing Introns and Other Enzymes

There are both chemical and stereochemical similarities between the second step of pre-mRNA splicing catalyzed by the spliceosome and the second step of splicing catalyzed by group I self-splicing introns [15]. Both second steps result in exon ligation, both utilize the 3'-OH of the 5' exon as the nucleophile, and both contain a conserved guanosine as the last nucleotide in the intron which is attached through a 3' oxygen to the phosphate at the 3' splice site [15]. In addition, both the group I self-splicing introns and the spliceosome are inhibited by an Rp phosphorothioate at the second step, suggesting that both rely on magnesium ion coordination to the non-bridging pro-R phosphate oxygen during the exon ligation step [15]. The crystal structure of a group I intron in complex with substrate and metal ions has been solved [20,21]. Interestingly, this structure shows that the ribozyme uses a two metal ion mechanism similar to the mechanism utilized by protein phosphoryltransferases, as originally hypothesized by Steitz and Steitz [22]. The observation of two metal ions in the active site of the group I self-splicing intron crystal structure is consistent with biochemical data for 2 out of the 3 hypothesized metal ions at the active site [23]. The catalytic magnesium ions in the crystal structure are coordinated such that they are 3.9 Å apart in the active site, forming direct innersphere contacts to the bridging oxygens on either side of the scissile phosphate. This distance is identical to the metal ion spacing in *E. coli* DNA polymerase I and other phosphoryltransferase enzymes [22]. In the group I intron crystal structure that mimics the second step of splicing [21], the pro-Rp oxygen at the exon junction is coordinated by both catalytic magnesium ions during the second step of splicing. Interestingly, the equivalent pro-Rp oxygen in pre-mRNA splicing is highly sensitive to phosphorothioate substitution [15]. This evidence, along with the biochemical data indicating direct magnesium coordination to the 3' oxygen leaving group [19], suggests that a metal ion coordinates to both of these positions during pre-mRNA splicing. Therefore, it is possible to hypothesize a 2 metal ion mechanism for pre-mRNA splicing (Figure 1B) analogous to the group I intron mechanism. In this model, metal ion 2 (M2) coordinates simultaneously to the 3' oxygen leaving group and the non-bridging pro-Rp phosphate oxygen at the scissile phosphate. The coordination of M2 to the 2'-OH group and the position of metal ion 1 (M1) are hypothetical, based on the general model of Steitz and Steitz [22]. The positions of M1 and M2 in this model are consistent with the crystal structures of both the group I [21] and the group II [11,12] self-splicing introns.

Remarkable similarities exist between the spliceosome and the group II self-splicing introns. The group II introns are ribozymes that catalyze a splicing reaction identical to pre-mRNA splicing (Figure 1A), but in the absence of protein. The stereochemistry of group II intron splicing is identical to pre-mRNA splicing as well, with both steps occurring as single "in-line" S_N2 nucleophilic displacement reactions accompanied by the corresponding inversion of stereochemistry [16,19,24]. In the following sections, I will discuss how the structure and function of the group II self-splicing introns relate to the pre-mRNA splicing mechanism.

3. IS THE SPLICEOSOME A RIBOZYME?

3.1. Evolution of the Spliceosome and Group II Self-Splicing Introns

Whether or not the spliceosome is a ribozyme has been a longstanding question [25–27], particularly in light of its mechanistic identity with the group II ribozymes and the finding that the ribosome, a macromolecular ribonucleoprotein complex similar in both size and complexity to the spliceosome, is indeed a ribozyme [28]. It has been hypothesized that the spliceosome and the group II self-splicing introns share a common molecular ancestor, and that the spliceosomal RNAs arose from disarticulated group II introns [29]. This hypothesis was further extended by Cavalier-Smith, who proposed that eukaryotic introns and spliceosomal RNAs were acquired when an intron-less eukaryote was invaded by a mitochondrial symbiont carrying mobile group II introns [30]. This hypothesis is fully consistent with the prevalence of group II introns in α-proteobacteria and the mobility of group II introns in contemporary eubacteria [31]. Since group II introns are retrotransposable elements that often contain an open reading frame that codes for their own reverse transcriptase, maturase and endonuclease enzymes, it is relatively easy to envision how an organellar group II intron could have invaded a eukaryotic nuclear genome. The universal presence of spliceosomal introns in eukaryotes suggests that eukaryotic introns are as ancient as eukaryotes themselves [32], and that introns evolved so that genes could be constructed in modular fashion [33]; this is the so-called "introns-early" hypothesis [34,35]. Gilbert hypothesized that introns provided an evolutionary advantage by increasing genetic diversity through alternative splicing [36], while Martin and Koonin forwarded the hypothesis that the eukaryotic nuclear envelope evolved in order to allow the time and compartmentalization required for pre-mRNA splicing, which is slow relative to translation [31]. Analysis of the conservation of spliceosomal proteins suggests that the spliceosome has changed relatively little over the course of eukaryotic evolution [37]. The introns-early hypothesis, the mobility of group II introns, and the mechanistic identity between pre-mRNA splicing and the group II self-splicing introns are all consistent with the RNA World hypothesis [38] and the idea that the spliceosome is a ribozyme. However, in order to prove the hypothesis that the spliceosome is a ribozyme, one of two things would be required: (1) reconstitution of robust pre-mRNA splicing activity *in vitro* in the absence of proteins or (2) a high resolution crystal structure of the spliceosome showing an RNA-only active site. No one has yet succeeded in crystallizing the spliceosome, but significant progress towards reconstitution of splicing activity *in vitro* has been made.

3.2. Protein-Free Splicing Reactions as Model Systems

Valadkhan and Manley first reported splicing-related catalysis in 2001 [39]. Their approach was to use truncated versions of the human U2 and U6 RNAs, and to react these RNAs in the presence of magnesium with a substrate RNA that mimics the branch site. However, instead of finding formation of a new 2'-5' phosphodiester bond, they found formation of a phosphotriester linkage involving the 2' oxygen of the branch site adenosine and the phosphate between two conserved nucleotides in U6 RNA (A53 and G54). Furthermore, this reaction was extremely inefficient, with only 0.1%

of the substrate converted into product after 20 hours [39]. Therefore, although this activity is very intriguing and apparently dependent upon magnesium, it does not by itself seem convincing enough to unequivocally argue that the spliceosomal RNAs are catalytically active in the absence of protein. Later, the authors showed that this reaction suffers from an unfavorable equilibrium constant, but can be slightly enhanced by pseudouridylation of the U2 RNA [40], which occurs naturally in humans. Still, this reaction was very inefficient and formed an unnatural phosphotriester. By modifying the end of the U6 RNA to mimic a 5' splice site, Valadkhan et al. were able to show that a Y-shaped branched product could be isolated that more closely resembles a splicing reaction [41].

Recently, a two-step splicing reaction highly similar to a pre-mRNA splicing reaction was developed from derivatives of the human U2 and U6 spliceosomal RNA sequences [42]. This is a significant step towards proving the spliceosome is a ribozyme hypothesis, for two reasons [27]. First, the reaction produces products that are chemically identical to the exon splicing reaction, with a 5'-3' bond. Although the chemistry of this reaction is not identical to splicing because the first step is generated via hydrolysis rather than a 2'-5' branching reaction, the products are the same in both pathways. Interestingly, there is precedence for a first step hydrolysis reaction among the group II ribozymes [43]. Second, the reaction is approximately 10-fold more efficient than the previous protein-free reactions [39–41]. This is significant because in the absence of catalytic activity, one would expect the opposite: the efficiency should either decrease or stay the same, due to the additional second step. This argues that there is indeed some residual catalytic activity in the protein-free U2 and U6 spliceosomal RNAs. However, a counter-argument is that RNA sequence space may have a high degree of intrinsic reactivity [44,45], and RNA-based, non-catalytic phosphoryl transfer reactions can be observed due to the intrinsic reactivity of RNA and template driven proximity effects [27,46]. Therefore, in the future it will be interesting to test the protein-free 2-step reaction for stereochemistry and metal ion coordination, to see if it parallels the known stereochemistry of the spliceosome. For example, is this reaction inhibited by an Rp phosphorothioate substitution at the second step (Figure 1B)?

4. STRUCTURAL BIOLOGY OF THE SPLICEOSOME

4.1. Interaction of Conserved Spliceosomal RNA Domains with Metal Ions

The pre-mRNA splicing mechanism is dependent upon the coordination of magnesium ion to the 3' oxyanion leaving groups for both the first and second steps of splicing [18,19], in order to shield the build-up of electronegative charge in the transition state (Figure 1B). According to the model of Steitz and Steitz [22] and by analogy to the group I [20,21] and group II [12,47] ribozyme structures, the splicing mechanism is likely to require the coordination of two metal ions, one to activate the attacking 3'-OH nucleophile, and one to stabilize the 3' oxyanion leaving group as depicted in Figure 1B for the second step of splicing. Therefore, an outstanding question is, what additional groups in the spliceosome are responsible for coordinating these catalytic metal ions?

If a phosphoryl oxygen is important for magnesium ion coordination, a sulfur substitution at that position should result in a switch in metal ion specificity towards more thiophilic ions, such as cadmium [48]. In the case of ribozymes, careful kinetic analysis of phosphorothioate substituted RNA in the presence of magnesium versus cadmium can be used to elucidate both the number of metal ions coordinated in the transition state as well as their locations [23,49]. In the spliceosome, the U2 and U6 RNAs are the most likely candidates for ribozyme activity and hence metal ion coordination. The U6 RNA has been remarkably conserved throughout eukaryotic evolution [50], which is consistent with its known role in the active site of the spliceosome. The U2 and U6 RNAs base pair with each other, and directly hydrogen bond to the pre-mRNA substrate [26,51] (Figure 2A). U6 RNA forms base pairs with, and can be crosslinked to, the 5' splice site [52] (Figure 2A). Therefore, U6 RNA was a likely candidate for participating in metal ion coordination during splicing. Consistent with a role in metal ion coordination, substitution of U6 RNA with Rp phosphorothioates inhibits pre-mRNA splicing [53,54]. These studies identified several pro-Rp oxygens in U6 that were important for splicing, including two that are essential for the first step (G50 and U80), and one for the second step of splicing (A59) (this numbering system refers to *S. cerevisiae*). However, initial attempts to rescue these phosphorothioates with addition of manganese failed, for unknown reasons [54]. One possible explanation was that substitution with Rp phosphorothioates can sometimes perturb RNA structure [55], and the aforementioned studies tested U6 RNA with multiple phosphorothioate substitutions.

Lin and colleagues tested single phosphorothioate substitutions in U6 RNA [56]. These were technically challenging experiments, involving chemical synthesis of an oligonucleotide with a single phosphorothioate substitution, purification of the individual Rp and Sp phosphorothioate diastereomers, ligation of these RNAs to a partial transcript of U6, reconstitution of the single phosphorothioate substituted U6 RNA into active spliceosomes, and testing for splicing activity. The previously identified important phosphate at U80 was tested (the position of U80 is indicated with an asterisk in Figure 2). Rewardingly, it was found that an Sp

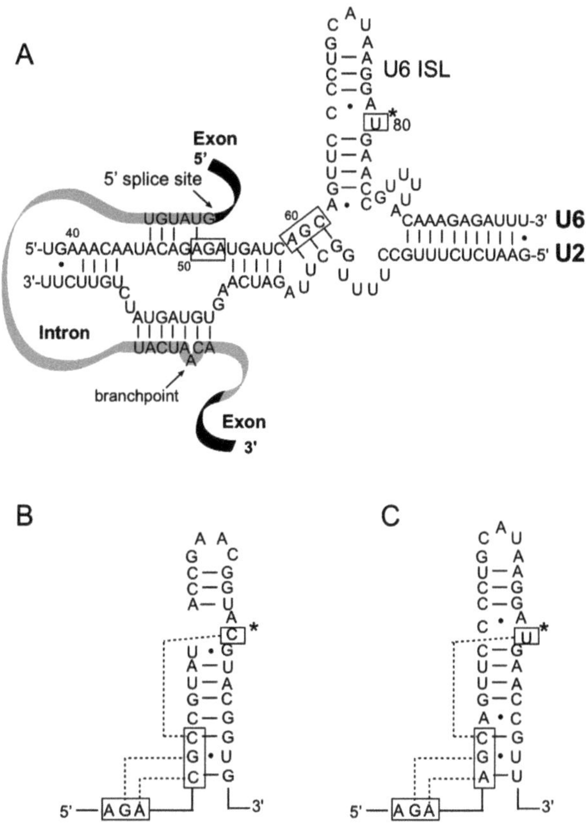

Figure 2. Secondary structure diagram of RNA interactions in the spliceosome and comparison of the U6 ISL with domain 5 from group II introns. **A.** Interactions between U2, U6 and pre-mRNA. Highly conserved and functionally important regions of U6 RNA are boxed. An asterisk indicates the site of metal ion binding at the U80 phosphate. The intron is shown as a grey ribbon, with exons in black. **B.** Secondary structure of domain 5 from the *O. iheyensis* group II self-splicing intron. Tertiary contacts seen in the crystal structure are indicated with dashed lines. Note that the boxed CGC triad is most commonly an AGC triad in other group II introns. Site of metal ion coordination is marked with an asterisk. **C.** Secondary structure of the U6 ISL with hypothesized tertiary contacts, as inferred from the group II intron crystal structure, shown as dashed lines.

phosphorothioate substitution at U80 inhibited the first step of splicing, and this inhibition could indeed be rescued by thiophilic ions [56]. However, only the first step of splicing could be rescued, and not the second. The reason for this is unknown, but it is consistent with the hypothesis that a change in active site conformation occurs between the first and second steps of splicing

[15,19]. This work demonstrated that a single atom substitution within the massive spliceosome alters its chemistry to be dependent upon a thiophilic metal ion for the first step of splicing; this in turn strongly suggests that the pro-Sp phosphate oxygen of U80 coordinates a metal ion during the first step of splicing. However, it is not yet known if the metal ion coordinated at U80 is the same metal ion that also coordinates to the 3' oxygen leaving group during the first step [18] (Figure 1B).

A role for U80 in metal ion coordination within the spliceosome active site is consistent with its high degree of conservation throughout eukaryotic evolution. Furthermore, mutations at U80 are lethal [57,58], possibly due to disruption of the U6 RNA structure [59]. The nucleobase of U80 appears to play a role in both spliceosome assembly and activation [60]. Thus, the U80 position of U6 RNA is clearly very important for spliceosome function. Interestingly, the U80 nucleotide is located within a highly conserved helix, termed the U6 intramolecular stem-loop or ISL (Figure 2).

The structure of the U6 ISL has been solved [61–63]. In addition, the metal ion binding properties and dynamic motions of this domain have been extensively investigated by NMR [64–67]. The isolated U6 ISL intrinsically binds metal ion at the U80 pro-Sp oxygen [61,64,67], consistent with the biochemical and functional studies performed on the entire spliceosome [56]. Metal ion binding to the U80 phosphate is pH dependent because the RNA undergoes a dynamic conformational change in response to the protonation state of the adjacent A79 adenine ($pK_a = 6.5$) [61,64,65,67]. Possibly, the dynamics at the U80 position are related to the conformational change that is hypothesized to occur in the spliceosomal active site as it transitions between the two steps of splicing. Interestingly, the structure of the U6 ISL is very similar to the structure of the highly conserved domain 5 from group II self-splicing introns, which resides at the catalytic center of these ribozymes [63].

4.2. Structural Clues from a Group II Intron

The crystal structure of the group IIC self-splicing intron from *Oceanobacillus iheyensis* was recently solved at 3.1–3.4 Å resolution [11,12]. This structure gives us deep insights into how RNA can catalyze splicing [47,68,69]. The structure shows the domain 5 helix at the center of the intron, which is consistent with a wide body of functional data for the group II ribozymes [11,12,47,69,70]. The conserved bulge region of domain 5 participates in an elaborate network of tertiary interactions and coordinates 2 magnesium ions approximately 3.7 Å apart, consistent with biochemical data [71] and the general two metal ion mechanism hypothesis of Steitz and Steitz [22]. In the crystal structure, a substrate analogue consisting of a

6 nucleotide mimic of the ligated exons is located precisely across from the two metal ion active site of domain 5 [11].

The similarities between domain 5 of group II introns and the U6 ISL are striking (Figure 2B and C). Both are helices of the same overall length, both have highly conserved primary and secondary structures, both are capped by a loop from the GNRA family of RNA folds, both contain a bulge located at the same position in the middle of the helix, both bulges serve to coordinate metal ions, and both bulges are equidistant from a highly conserved and catalytically important "AGC triad" sequence [63,69], although domain 5 from the *O. iheyensis* group IIC self-splicing intron is unusual among group II introns because it has a CGC triad in place of AGC (Figure 2B and C). Isolated domain 5 and U6 ISL structures have been solved and can be superimposed with a backbone root mean squared deviation of 2.6 Å [63]. Similarities even extend to the functional level, as a group II intron domain 5 can function in place of the U6 ISL in the U12-dependent (or "minor") spliceosome [72].

The tertiary structure of the *O. iheyensis* group IIC self-splicing intron can be used to make predictions for tertiary contacts involving U6 RNA [69]. Domain 5 forms several functionally important base triples in its major groove (Figure 2B, dashed lines). These include a C•C-G triple between the bulge C and the helix, a G•G-U triple, and another C•C-G triple (Figure 2B). These base triples align the catalytic core of the group II intron by bringing a distal single stranded region (J2/3) into alignment with the metal ion coordinating bulge nucleotides. The U6 ISL has analogous features, but instead of a bulged C at its metal binding region, it has a bulged U (U80, Figures 2A and C). However, the bulged U may still be able to form a similar base triple (a U•C-G instead of C•C-G), since the uracil has a ketone group in the same position as cytosine [69]. In addition, the U6 ISL has a highly conserved and functionally important "AGC triad" in a similar location as domain 5 (Figure 2B and C). In the spliceosome, the AGC triad forms a short helix with U2 RNA during the second step of splicing [73,74] (Figure 2A), but also seems to have other functions [74], one of which may be to form an alternate pairing within U6 to extend the ISL during the first step of splicing (Figure 2C) [63]. This arrangement has been observed by NMR [63] and is highly similar to the group II intron domain 5 configuration (Figure 2B and C).

5. CONCLUDING REMARKS AND FUTURE DIRECTIONS

In conclusion, it seems very likely that the spliceosomal active site will reveal at least two catalytic metal ions in an RNA-rich active site; it is also likely that this active site will resemble something similar to the model proposed in

Figure 1B for the second step of splicing. There is good evidence to support the hypothesis that the spliceosome is a ribozyme at its core, and that this massive ribonucleoprotein complex once shared a common molecular ancestor with the group II self-splicing introns. However, our current structural view of the catalytically active spliceosomal C complex comes from electron microscopy at a resolution of approximately 30 Å [75], and it seems that a high resolution crystal structure of this complex will be some years to come in the future.

Future directions for investigating the catalytic mechanism of the spliceosome will involve developing ways to overexpress and purify the spliceosome so that it can be crystallized. In the meantime, much can be learned about the structures of the individual snRNPs and how they assemble, from a variety of genetic, biochemical, and biophysical approaches. The *in vitro* RNA-only 2 step reaction from Valadkhan and co-workers [42] may also allow biochemists to probe the splicing mechanism and its dependence upon metal ions. Finally, the group II intron crystal structure makes specific predictions about how the U2 and U6 RNAs may fold and coordinate metal ions in the spliceosomal active site [11,12,47,69,70], and future directions should involve testing some of these predictions.

ACKNOWLEDGMENTS

I would like to thank Professor David A. Brow for stimulating my interest in the spliceosome and for many productive scientific discussions.

ABBREVIATIONS

atac	adenine-thymine/adenine-cytosine
ATP	adenosine 5′-triphosphate
Brr2	bad response to refrigeration 2
DExD/H-box	asp-glu-x-asp/his helicase
NMR	nuclear magnetic resonance
pre-mRNA	precursor messenger RNA
Prp19	pre-mRNA processing protein 19
Rp	R chiral configuration
SN2	bimolecular nucleophilic substitution
snRNA	small nuclear RNA
snRNP	small nuclear ribonucleoprotein particle
Sp	S chiral configuration
U6 ISL	U6 intramolecular stem-loop

REFERENCES

1. M. Q. Zhang, *Hum. Mol. Genet.*, 1998, **7**, 919–932.
2. A. Beauregard, M. J. Curcio and M. Belfort, *Annu. Rev. Genet.*, 2008, **42**, 587–617.
3. D. A. Brow, *Annu. Rev. Genet.*, 2002, **36**, 333–360.
4. M. C. Wahl, C. L. Will and R. Luhrmann, *Cell*, 2009, **136**, 701–718.
5. M. S. Jurica and M. J. Moore, *Mol. Cell*, 2003, **12**, 5–14.
6. P. Fabrizio, J. Dannenberg, P. Dube, B. Kastner, H. Stark, H. Urlaub and R. Luhrmann, *Mol. Cell*, 2009, **36**, 593–608.
7. A. A. Patel and J. A. Steitz, *Nat. Rev. Mol. Cell Biol.*, 2003, **4**, 960–970.
8. D. A. Pomeranz Krummel, C. Oubridge, A. K. Leung, J. Li and K. Nagai, *Nature*, 2009, **458**, 475–480.
9. D. B. Ritchie, M. J. Schellenberg and A. M. MacMillan, *Biochim. Biophys. Acta*, 2009, **1789**, 624–633.
10. R. Luhrmann and H. Stark, *Curr. Opin. Struct. Biol.*, 2009, **19**, 96–102.
11. N. Toor, K. Rajashankar, K. S. Keating and A. M. Pyle, *Nat. Struct. Mol. Biol.*, 2008, **15**, 1221–1222.
12. N. Toor, K. S. Keating, S. D. Taylor and A. M. Pyle, *Science*, 2008, **320**, 77–82.
13. M. K. Sakharkar, V. T. Chow and P. Kangueane, *In Silico Biol.*, 2004, **4**, 387–393.
14. S. Saxonov, I. Daizadeh, A. Fedorov and W. Gilbert, *Nucleic Acids Res.*, 2000, **28**, 185–190.
15. M. J. Moore and P. A. Sharp, *Nature*, 1993, **365**, 364–368.
16. R. A. Padgett, M. Podar, S. C. Boulanger and P. S. Perlman, *Science*, 1994, **266**, 1685–1688.
17. P. A. Sharp, *Cell*, 1985, **42**, 397–400.
18. E. J. Sontheimer, S. Sun and J. A. Piccirilli, *Nature*, 1997, **388**, 801–805.
19. P. M. Gordon, E. J. Sontheimer and J. A. Piccirilli, *RNA*, 2000, **6**, 199–205.
20. P. L. Adams, M. R. Stahley, A. B. Kosek, J. Wang and S. A. Strobel, *Nature*, 2004, **430**, 45–50.
21. M. R. Stahley and S. A. Strobel, *Science*, 2005, **309**, 1587–1590.
22. T. A. Steitz and J. A. Steitz, *Proc. Natl. Acad. Sci. USA*, 1993, **90**, 6498–6502.
23. J. L. Hougland, A. V. Kravchuk, D. Herschlag and J. A. Piccirilli, *PLoS Biol.*, 2005, **3**, e277.
24. M. J. Moore and P. A. Sharp, *Science*, 1992, **256**, 992–997.
25. C. A. Collins and C. Guthrie, *Nat. Struct. Biol.*, 2000, **7**, 850–854.
26. S. E. Butcher and D. A. Brow, *Biochem. Soc. Trans.*, 2005, **33**, 447–449.
27. S. E. Butcher, *Proc. Natl. Acad. Sci. USA*, 2009, **106**, 12211–12212.
28. T. R. Cech, *Science*, 2000, **289**, 878–879.
29. T. R. Cech, *Cell*, 1986, **44**, 207–210.
30. T. Cavalier-Smith, *Trends Genet.*, 1991, **7**, 145–148.
31. W. Martin and E. V. Koonin, *Nature*, 2006, **440**, 41–45.
32. E. V. Koonin, *Biol. Direct*, 2006, **1**, 22.
33. J. E. Darnell and W. F. Doolittle, *Proc. Natl. Acad. Sci. USA*, 1986, **83**, 1271–1275.

34. S. W. Roy and W. Gilbert, *Proc. Natl. Acad. Sci. USA*, 2005, **102**, 5773–5778.
35. I. B. Rogozin, Y. I. Wolf, A. V. Sorokin, B. G. Mirkin and E. V. Koonin, *Curr. Biol.*, 2003, **13**, 1512–1517.
36. W. Gilbert, *Nature*, 1978, **271**, 501.
37. L. Collins and D. Penny, *Mol. Biol. Evol.*, 2005, **22**, 1053–1066.
38. W. Gilbert, *Nature*, 1986, **319**, 618.
39. S. Valadkhan and J. L. Manley, *Nature*, 2001, **413**, 701–707.
40. S. Valadkhan and J. L. Manley, *RNA*, 2003, **9**, 892–904.
41. S. Valadkhan, A. Mohammadi, C. Wachtel and J. L. Manley, *RNA*, 2007, **13**, 2300–2311.
42. S. Valadkhan, A. Mohammadi, Y. Jaladat and S. Geisler, *Proc. Natl. Acad. Sci. USA*, 2009, **106**, 11901–11906.
43. M. Podar, V. T. Chu, A. M. Pyle and P. S. Perlman, *Nature*, 1998, **391**, 915–918.
44. D. J. Smith and M. M. Konarska, *RNA*, 2009, **15**, 1–3.
45. D. J. Smith and M. M. Konarska, *RNA*, 2009, **15**, 8–13.
46. D. P. Bartel and J. W. Szostak, [see comment], *Science*, 1993, **261**, 1411–1418.
47. N. Toor, K. S. Keating and A. M. Pyle, *Curr. Opin. Struct. Biol.*, 2009, **19**, 260–266.
48. R. K. O. Sigel, B. Song and H. Sigel, *J. Am. Chem. Soc.*, 1997, **119**, 744–755.
49. M. Forconi, J. Lee, J. K. Lee, J. A. Piccirilli and D. Herschlag, *Biochemistry*, 2008, **47**, 6883–6894.
50. D. A. Brow and C. Guthrie, *Nature*, 1988, **334**, 213–218.
51. T. Villa, J. A. Pleiss and C. Guthrie, *Cell*, 2002, **109**, 149–152.
52. C. H. Kim and J. Abelson, *RNA*, 1996, **2**, 995–1010.
53. P. Fabrizio and J. Abelson, *Nucleic Acids Res.*, 1992, **20**, 3659–3664.
54. Y. T. Yu, P. A. Maroney, E. Darzynkiwicz and T. W. Nilsen, *RNA*, 1995, **1**, 46–54.
55. J. S. Smith and E. P. Nikonowicz, *Biochemistry*, 2000, **39**, 5642–5652.
56. S. L. Yean, G. Wuenschell, J. Termini and R. J. Lin, *Nature*, 2000, **408**, 881–884.
57. H. D. Madhani, R. Bordonne and C. Guthrie, *Genes Dev.*, 1990, **4**, 2264–2277.
58. D. S. McPheeters, *RNA*, 1996, **2**, 1110–1123.
59. D. G. Sashital, A. M. Allmann, S. R. Van Doren and S. E. Butcher, *Biochemistry*, 2003, **42**, 1470–1477.
60. C. J. McManus, M. L. Schwartz, S. E. Butcher and D. A. Brow, *RNA*, 2007, **13**, 2252–2265.
61. A. Huppler, L. J. Nikstad, A. M. Allmann, D. A. Brow and S. E. Butcher, *Nat. Struct. Biol.*, 2002, **9**, 431–435.
62. N. J. Reiter, L. J. Nikstad, A. M. Allmann, R. J. Johnson and S. E. Butcher, *RNA*, 2003, **9**, 533–542.
63. D. G. Sashital, G. Cornilescu, C. J. McManus, D. A. Brow and S. E. Butcher, *Nat. Struct. Mol. Biol.*, 2004, **11**, 1237–1242.
64. N. J. Reiter, H. Blad, F. Abildgaard and S. E. Butcher, *Biochemistry*, 2004, **43**, 13739–13747.
65. V. Venditti, L. Clos, 2nd, N. Niccolai and S. E. Butcher, *J. Mol. Biol.*, 2009, **391**, 894–905.
66. V. Venditti, N. Niccolai and S. E. Butcher, *Nucleic Acids Res.*, 2008, **36**, e20.

67. H. Blad, N. J. Reiter, F. Abildgaard, J. L. Markley and S. E. Butcher, *J. Mol. Biol.*, 2005, **353**, 540–555.
68. K. T. Dayie and R. A. Padgett, *RNA*, 2008, **14**, 1697–1703.
69. K. S. Keating, N. Toor, P. S. Perlman and A. M. Pyle, *RNA*, 2010, **16**, 1–9.
70. N. Toor, K. S. Keating, O. Fedorova, K. Rajashankar, J. Wang and A. M. Pyle, *RNA*, 2010, **16**, 57–69.
71. P. M. Gordon, R. Fong and J. A. Piccirilli, *Chem. Biol.*, 2007, **14**, 607–612.
72. G. C. Shukla and R. A. Padgett, *Mol. Cell*, 2002, **9**, 1145–1150.
73. H. D. Madhani and C. Guthrie, *Cell*, 1992, **71**, 803–817.
74. A. K. Hilliker and J. P. Staley, *RNA*, 2004, **10**, 921–928.
75. M. S. Jurica, D. Sousa, M. J. Moore and N. Grigorieff, *Nat. Struct. Mol. Biol.*, 2004, **11**, 265–269.

Met. Ions Life Sci. **2011**, *9*, 253–275

9

The Ribosome: A Molecular Machine Powered by RNA

Krista Trappl and Norbert Polacek
Innsbruck Biocenter, Division of Genomics and RNomics, Medical University Innsbruck, Fritz-Pregl-Strasse 3, 6020 Innsbruck, Austria
<Krista.Trappl@i-med.ac.at>
<Norbert.Polacek@i-med.ac.at>

Metal Ions in Life Sciences, Volume 9 Edited by Astrid Sigel, Helmut Sigel, and Roland K. O. Sigel

Published by the Royal Society of Chemistry, www.rsc.org DOI: 10.1039/9781849732512‌00253

ABSTRACT: Metal ions are the salt in the soup of essentially every biological system. Also in the ribosome, the largest natural ribozyme that produces all proteins in every living cell, metal ions have been found contributing significantly to the highly dynamic and accurate process of translation. The ribosome is considered a molecular fossil of the 'RNA world' and it could be shown that the evolutionarily oldest parts of the particle, which build the catalytic center and surrounding domains, are densely packed with divalent metal ions. Nevertheless, metal ions do not seem to directly participate in ribosomal catalysis, their important roles in the ribosome, however, cannot be denied. It is probable that mono- and divalent metal ions primarily promote the functionally competent architecture of the ribosomal RNAs, but more direct roles in mRNA decoding and reading frame maintenance are likely. Decades of biochemical studies and the recent high resolution crystallographic structures of the ribosome strongly indicate that metal ions are involved in essentially every phase of the ribosomal elongation cycle, thus contributing significantly to the precise translation of the genetic code.

KEYWORDS: evolution · peptide bond formation · protein synthesis · ribosome · RNA world · rRNA

1. THE RIBOSOME – THE LARGEST NATURAL RIBOZYME

1.1. RNA Makes Proteins

The ribosome is one of the largest cellular machines known to date and has appeared on the scene of evolution about 4 billion years ago. Its primary task during gene expression is the translation of the genetic information encoded within messenger RNAs (mRNAs) into the amino acid sequence of polypeptides and proteins. The bacterial ribosome (on which we will focus in this chapter) with its molecular weight of 2.6–2.8 MDa sediments during gradient centrifugation at 70S and is always composed of two unequally sized subunits, namely the small 30S and the large 50S subunit (Figure 1).

It has been realized already decades ago that about 2/3 of the ribosome's mass originates form ribosomal RNA (rRNA) and 1/3 of ribosomal proteins (r-proteins) [1]. The 50S particle contains the 23S rRNA and 5S rRNA, non-protein-coding RNAs of ~2,900 and 120 nucleotides in length, respectively, and about 33 different r-proteins. The small subunit on the other hand is composed of a single rRNA chain, the ~1,500 residue long 16S rRNA, and about 20 r-proteins. The 50S subunit is the home of the catalytic heart of the ribosome where the individual amino acid monomers are chemically linked via peptide bonds to form the polypeptide chain. This active site is referred to as the peptidyl transferase center (PTC) and is located in a deep cleft on the interface side of the 50S subunit.

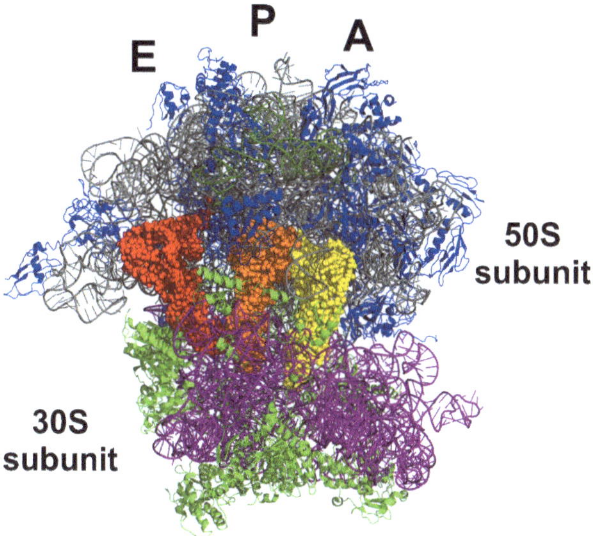

Figure 1. The bacterial ribosome. The crystallographic structure of the *Thermus thermophilus* 70S ribosome is depicted including three tRNAs bound to the A-site (yellow), P-site (orange), and E-site (red). 23S rRNA and 5S rRNA are blue and dark green, respectively, while 16S rRNA is colored magenta. Large subunit (50S subunit) r-proteins are shown in blue and small subunit (30S subunit) r-proteins in green. The structures were generated from pdb files 2WDM and 2WDN [19].

The mean by which the PTC catalyzes peptide bond formation has been a question of long and rather intense controversies over decades (reviewed in [2]). Initially it was thought that r-proteins are responsible for catalysis simply based on the fact that RNA enzymes (ribozymes) had not yet been discovered and protein enzymes were the only known biological catalysts. Nevertheless, influenced by Francis Crick's ground-breaking hypothesis in 1968 that the ancestral proto-ribosome might have consisted of RNA only [3], scientists also started to entertain the idea that rRNA might be more than merely the scaffold for the r-proteins. These ribocentric freaks among the ribosomologists received support for their theories in the beginning of the 1980ies when the first natural ribozymes were discovered [4,5]. In the following years additional RNA enzymes have been discovered and added to the list of RNA catalysts in a variety of biological systems ranging from viruses to mammals (reviewed in [6]). These ribozymes range from about 20 to several hundred nucleotides in length and can be functionally grouped into self-cleaving (e.g., the hammerhead ribozyme) and self-splicing (e.g., the group I intron ribozyme) RNA enzymes. With respect to the ribosome, the debates whether or not the ribosome is a ribozyme continued until the beginning of the new millennium when X-ray crystallography finally convincingly showed that the inner core of the PTC is an all-RNA active site with five nucleotides of domain V of 23S rRNA building the chemical center of the ribosome [2].

The ribosome is special among natural ribozymes because it is the only one that does not catalyze RNA phosphodiester bond cleavage or ligations. Instead it promotes amide bond formation. With its molecular mass of almost 3 MDa the ribosome is by far the largest natural ribozyme known to date, and it is likely that it will keep its number one chart position also in the future. From an evolutionary point of view it is remarkable that all proteins in contemporary cells are made by the ribosome, an RNA-based molecular relic 'invented' in the pre-biotic 'RNA world' about 4 billion years ago (see Section 4 below). It appears that the ribosome has successfully 'refused' to be replaced by protein enzymes during the course of evolution and thus has proven itself as a tremendously successful RNA-based catalyst that obviously outlived the transition from the 'RNA world' to present-day biology.

1.2. The Ribosomal Elongation Cycle

During protein synthesis the ribosome is engaged in the elongation cycle which can be subdivided into the initiation, elongation, termination, and the recycling stage [1]. During the initiation phase the 30S subunit binds an mRNA via base-pair interactions between the Shine-Dalgarno sequence and the complementary anti-Shine-Dalgarno sequence located in the 3'-end of 16S rRNA. The initiator tRNA (fMet-tRNA^{fMet} in bacteria) is delivered to the 30S particle as ternary complex with IF2 and GTP, where it binds to the AUG start codon located in the 30S P-site with the aid of IF1 and IF3. This is the only phase of protein synthesis when a tRNA directly binds to the ribosomal P-site.

Subsequently the 50S subunit joins which results in triggering of the GTPase activity of IF2 and subsequent dissociation of the IFs from the 70S initiation complex. During the elongation phase of translation, the next amino acid is delivered to the A-site in form of the ternary complex composed of aminoacyl-tRNA (aa-tRNA), elongation factor Tu, and GTP (aa-tRNA•EF-Tu•GTP). Accuracy of translation is maintained in the 30S decoding center primarily at the decoding step by a sophisticated mechanism relying on geometry recognition of the minor groove of the codon-anticodon mini-helix by the conformationally flexible 16S rRNA residues A1492, A1493, and G530 [7]. Only correct Watson-Crick interactions between the mRNA codon and the tRNA anticodon (primarily monitored at the first two nucleotides of the codon) are allowed and trigger, by a so far not completely solved signaling pathway, GTP hydrolysis on EF-Tu. The acceptor end of the aa-tRNA then swings into the 50S A-site, a process named accommodation. Also this step contributes to translation fidelity since it has been shown that near-cognate aa-tRNA that has successfully by-passed the codon-anticodon recognition stage can be rejected during accommodation [8].

Once the CCA terminal end of the aa-tRNA is fully accommodated, peptide bond formation occurs immediately. In bacteria about 15–20 peptide bonds are formed every second. As a result of transpeptidation, the peptidyl moiety of peptidyl-tRNA located at the P-site is transferred to the aa-tRNA in the A-site, which results in a nascent peptide chain, elongated by one amino acid. The growing peptide leaves the PTC via the peptide exit tunnel, a ~100 Å long path that spans the entire 50S subunit and has its end at the solvent side of the large ribosomal subunit. In the final step of the elongation cycle the deacylated tRNA and the pept-tRNA are translocated from the P- and A-sites to the E- and P-sites, respectively. This process is driven by the action of EF-G and GTP hydrolysis, which results in the invasion of a new codon into the decoding center of the small ribosomal subunit A-site.

Binding of EF-G • GTP to the ribosome results in a ratchet-like rotation of the 30S relative to the 50S subunit. In this translocation intermediate state the acceptor ends of two tRNAs already moved on the 50S subunit to the E- and P-sites while remaining their positions in the small subunit. These so-called hybrid state positions (P/E and A/P for deacylated and pept-tRNA, respectively) are finally resolved by the action of EF-G and GTP hydrolysis which promotes tRNA movement in the 30S subunit, thus establishing the classical E/E and P/P tRNA states of the post-translocational ribosome. Once an mRNA stop codon (UAA, UAG, or UGA) enters the A-site, the termination phase of protein synthesis is initiated.

Stop codons are usually not read by tRNA anticodons but by specific peptide motifs of class I release factors (RF1 and RF2 in bacteria). Binding of RF1/2 to the A-site of the ribosome triggers pept-tRNA hydrolysis in the PTC. This reaction is the second chemical reaction that is catalyzed in the ribosome during protein synthesis. A universally conserved GGQ tripeptide motif, which is located in a flexible loop of the RF, reaches into the PTC where it contacts and repositions conserved 23S rRNA residues, such as the crucial nucleotide A2602 [9,10]. These structural changes are thought to be key for the GGQ motif to coordinate the hydrolytic water molecule that finally cleaves the ester bond connecting the tRNA and the synthesized peptide [2].

Finally, the recycling phase of translation frees the ribosome from the mRNA, the tRNA and the RF1/2. RF1/2 are ejected with the help of the class II release factor RF3, which exchanges GDP for GTP when bound to the 70S particle. This triggers the dissociation of RF1/2 form the A-site and subsequent to GTPase activity, also RF3 leaves the post-termination ribosome. The ribosome is finally dissociated into its subunits via binding of the ribosomal recycling factor RRF and EF-G GTPase activity.

2. RIBOSOMAL BIOGENESIS

2.1. rRNA Processing

The processing of rRNAs starts generally before the transcription of a ribosomal operon is finished and is basically a two-step procedure. The first aspect is the trimming and the second is the modification of specific rRNA nucleotides. The primary processing enzymes are endonucleases like RNase III, a phosphodiesterase, which is playing a major role in the maturation of 23S rRNA and requires a bivalent metal ion to hydrolyze phosphodiester linkages. The task of RNase III is the cleavage within the spacer sequences flanking 23S and 16S rRNA [11]. In the case of bacterial RNases III, probably Mg^{2+} is mainly necessary, but several other metal ions, like Mn^{2+}, Co^{2+}, and Ni^{2+} can support catalysis. In *Escherichia coli* it was shown that a two-metal-ion catalytic mechanism is responsible for RNase III activity. The hydrolysis of RNA phosphodiester is depending on two Mg^{2+} ions [12]. For the processing of 5S rRNA RNase E is needed, which requires Na^+, K^+, or NH^{4+}, and Mg^{2+} or Mn^{2+}, to produce the mature molecule from the 9S pre-complex [13].

The second step, the nucleotide modification, is different for each rRNA. The final modification pattern of 23S rRNA is formed as an early event during assembly, meaning that the enzymes involved in the chemical modifications act on the 23S rRNA precursor. Some of these 23S rRNA modifications are essential for the fast and productive ribosome assembly [11]. On the other hand, 16S rRNA modification occurs later during ribosome biogenesis, namely at the stage of the 30S RNP particle. As a consequence, 16S rRNA modifications are not necessary for assembly *per se*. Nucleobase methylations count for most of the introduced modifications (14 out of 23) in bacterial 23S rRNA, while the remaining are pseudo-uridylations. The main part of the modifications is found in or close to the peptidyl transferase center, in domain V of the 23S rRNA, as well as at residues building the wall of the nascent peptide exit tunnel. While rRNA modifications appear to cluster near or at functional centers, the precise location of the introduced nucleotide modification does not seem to be universally conserved [11]. Thus their precise role in ribosome functioning has remained elusive despite decades of investigations.

2.2. rRNA Folding and Stabilization

Folding of rRNA is, like rRNA processing, a two-step procedure. The first step of this event is the formation of the secondary structure, which is stimulated by the presence of monovalent cations (particularly K^+), divalent cations, cationic polyamines, and basic proteins. In this stage their task is to screen for negative charge and compensate for the electrostatic penalty. The second step is the generation of the RNA tertiary structure, which is achieved by folding of the secondary structure elements into a functionally active three-dimensional interaction network. Aside of r-proteins, this formation is controlled by stringent metal ion requirements, especially divalent cations like magnesium are necessary, but potassium is also contributing. Generally there are four requirements for the formation of the tertiary structure: first of all, a certain RNA primary sequence, second, a metal ion balance (normally Mg^{2+} to Ca^{2+} and K^+ to Na^+), a certain metal ion concentration, and last, rRNA binding proteins and polyamines [14].

For the stabilization of the rRNA structure metal ions play a critical role. They can either bind diffusely via electrostatic forces, by screening charge to balance electrostatic repulsion within the rRNA backbone, or by site-specific binding. A majority of site-bound Mg^{2+} has no direct interaction with the rRNA, but the water ligands of magnesium hexahydrate are contacting the nucleobases to stabilize certain rRNA motifs. These so-called outersphere interactions are thermodynamically significant, are frequently common in the major groove edge of guanosine and can characteristically be replaced by exchange-inert mimics like cobalt(III) or osmium(III)hexammine [14].

The other option of Mg^{2+} bound to the rRNA is via direct interactions, which is achieved by substituting one to three water ligands of the hydrated magnesium with direct contacts between the metal ion and the RNA [14]. For this so-called "innersphere" contact, the main interaction partners on the rRNA are the non-bridging phosphoryl oxygen atoms of the RNA phosphodiester backbone (90% of innersphere bound Mg^{2+} within the 23S rRNA), the purine N7 heteroatoms, nucleobase keto groups and the ribose 2′-OH [15]. These interactions are thermodynamically significant and stabilize rRNA structure but do not function catalytically in the ribosome, which will be discussed later [14]. 116 innersphere bound Mg^{2+} ions could be found in the large ribosomal subunit of *Haloarcula marismortui* and according to their geometric arrangement of RNA and protein atoms, they can been distinguished [15]. They can bind different numbers of innersphere ligands, up to five ligands, but there have never been six innersphere ligands observed. The most common binding type for Mg^{2+} ions in the large ribosomal subunit, 40 examples have been found, is the one with two bound ligands, in which the position of the RNA or protein contact is in an orthogonal conformation. The second most frequently found binding conformation (37 cases) are Mg^{2+} ions with just one direct interaction partner. In both of these cases the metal ions are often bound to RNA motifs, which require a neutralization of their negative charge because of phosphate groups in close juxtaposition [15].

Metal ions can bind to certain binding motifs in the rRNA structure, for example the deep, narrow major groove in RNA duplexes, which have a distinct negative potential. One of the most common binding sites is the major groove of tandem G-U base pairs, which are creating a high negative charge, that serves as a good binding site for direct or outersphere metal ion linkage [14]. Furthermore, internal loop structures, like the loop E motif, often bind metal ions. In this case Mg^{2+} is bound as a binuclear cluster to bring nearby backbones close together. Another frequently observed metal ion binding motif is the A-platform motif, which specifically binds dehydrated potassium ions for structure stabilization [14]. Noteworthy, most of the metal ions observed in the 50S X-ray structure concentrate near the functional and evolutionarily oldest parts of the 23S rRNA, such as the PTC, where r-proteins are mainly missing (Figure 2) [15].

2.3. Ribosomal Proteins

The globular domains of r-proteins are generally located at the surface of the ribosome and they have characteristic long peptide extrusions. These tails seem to function as a connection to the inner core of the particle, which is mostly composed of rRNA. These r-protein tails are mostly devoid of secondary structure and are thought to glue together individual 23S rRNA domains probably replacing ancestral metal ions originally forming rRNA-rRNA tertiary structure connections in the primordial ribosome (see Section 4 below). Initially it was assumed that r-proteins participate directly in ribosomal catalysis, however, not a single r-protein nor a mixture of proteins could be identified with catalytic activity (reviewed in [2]).

Even though we now realize that the ribosome is essentially an RNA-driven molecular machine, a functional ribosomal particle without any r-proteins has never been observed *in vivo* or constructed in the test tube. Nowadays many functional tasks of r-proteins are known, for example the interaction with the P-site tRNA, which is known to involve proteins S9, S13, L5, and L27 [16]. In the A-site r-proteins S12, S13, L16, and L27 contribute to tRNA binding while r-proteins S7, S11, L28, L33, and L1 are involved in E-site tRNA binding. Also for the major catalytic task of the ribosome, the peptidyl transferase reaction, which is carried out just by RNA residues [17], proteins L2, L3, and L27 are additionally required for efficient amide bond synthesis [18,19]. Interestingly, proteins L2 and L3

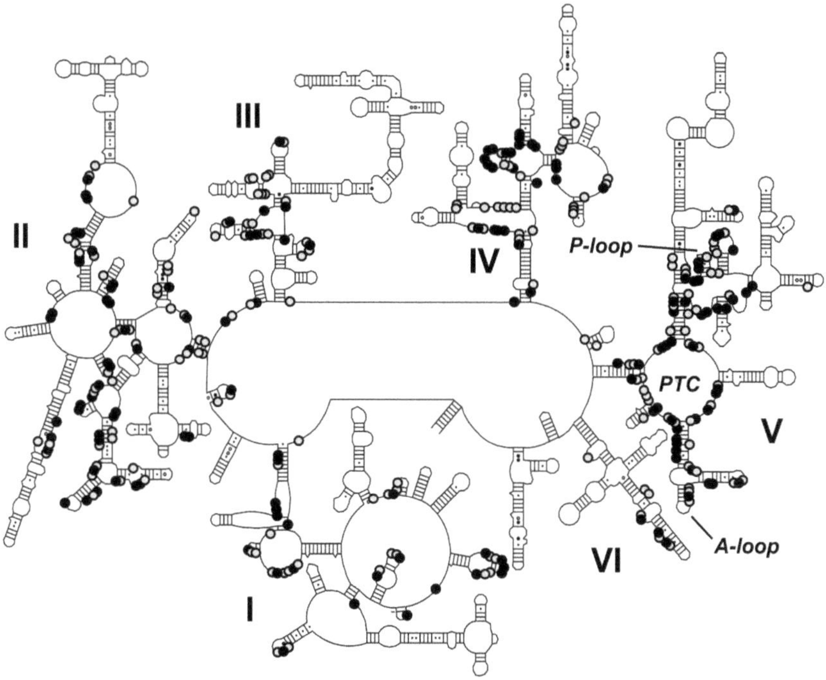

Figure 2. Magnesium ion coordination to 23S rRNA. The schematic secondary structure of a bacterial 23S rRNA is depicted. Nucleotides involved in outer- or innersphere Mg^{2+} ion coordination [15] are shown as gray or black circles, respectively. The six 23S rRNA domains are indicated by roman numerals and the peptidyl transferase center (PTC) as well as the location of the A- and P-loops of 23S rRNA are highlighted.

have the longest and most complex extensions compared to the other r-proteins [20].

Although metal ions in the ribosome are known to bind preferentially to rRNA, there are also interactions with r-proteins. L2, L3, and L22 in the large ribosomal subunit do have direct interaction via innersphere binding to six Mg^{2+} ions [15]. These divalent ions are stabilizing their tertiary and quaternary structures by connecting non-bridging phosphate oxygen atoms of the rRNA with amino acid side chains of the specific proteins. In addition, there are also outersphere contacts of magnesium and sodium ions within the 50S subunit, that are mediating the recognition of rRNA by r-proteins, by binding to their interfaces [15].

3. THE MOLECULAR ANATOMY OF FUNCTIONAL CENTERS

3.1. The Decoding Center

During protein synthesis only one amino acid misincorporation occurs per ~3,000–10,000 peptide bonds formed ([7] and references therein). The ribosome achieves this high accuracy by selecting aa-tRNAs that possess a tRNA anticodon capable of forming correct Watson-Crick base pairing with the A-site mRNA codon displayed in the 30S decoding center. It has been recognized that tRNA selection is even more accurate than expected from the strength of the codon-anti-codon interaction alone, since replacing a cognate Watson-Crick A-U pair by a mismatch G-U interaction would not significantly affect base pairing in solution, on the ribosome, however, even these 'near-cognate' aa-tRNAs are efficiently rejected from entering the A-site. With the presentation of high resolution crystallographic structures of the small 30S subunit with tRNA substrate analogs this riddle could be solved.

Upon cognate tRNA binding to the A-site during the so-called initial selection step, an induced fit of the decoding center occurs which results in conformational reorientations of three universally 16S rRNA nucleotides at positions A1492, A1493 of helix 44 and G530 from the 30S shoulder domain [21]. These residues contact the minor groove of the codon-anticodon helix by forming A-minor motif interactions with the first two base pairs of the codon-anticodon helix in a sequence-independent manner. The third codon-anticodon pair, however, is less rigorously monitored by the ribosomal decoding center since some non-Watson-Crick base pairs are allowed in this so called 'wobble' position. The signal of an induced fit triggered by cognate aa-tRNA binding results is transmitted to EF-Tu which hydrolyzes GTP and dissociates from the aa-tRNA allowing accommodation of the acceptor arm into the 50S PTC. Also this accommodation step contributes to translation fidelity since it could be shown that near-cognate aa-tRNAs that by-passed the initial selection step can be rejected at this stage as well (reviewed in [8]).

Metal ions, especially Mg^{2+}, have been shown to greatly influence the decoding process. Early observations indicated that the initial selection process as well as the ability to reject near-cognate aa-tRNAs after EF-Tu GTP hydrolysis, are negatively influenced by increased Mg^{2+} concentrations, thus increasing the error rate of translation [22,23]. In the crystallographic structures of the 30S subunit two constrained Mg^{2+} ions have been observed in 16S rRNA helix 44 likely stabilizing the flipped-out conformation of A1492 and A1493 [21]. It is therefore possible that elevated Mg^{2+} ion levels induce this flipped-out conformation, thus mimicking the helix 44 architecture usually established only in the presence of cognate tRNAs, which can explain the observed decrease in decoding fidelity. Furthermore, the ribose 2′-OH of the 3rd codon interacts with C518 and Pro48 of S12 via a Mg^{2+}-mediated interaction.

Additionally, in order to allow simultaneous codon-anticodon interactions with both A- and P-site tRNA, a 45° kink in the mRNA between the A- and P-site codon has been observed that seems to be stabilized by a Mg^{2+} ion [24,25]. This Mg^{2+} ion likely also plays a role in fixing the frame of the mRNA relative to the 30S subunit. Thus, these metal ions appear to shape the ribosomal decoding center in its functionally competent conformation. Beside ribosome-coordinated Mg^{2+} ions, also tRNA-bound Mg^{2+} seems to be involved in A-site tRNA binding. Up to five Mg^{2+} ions bound to the anticodon loop of tRNA have been predicted to contribute. It has been suggested that Mg^{2+} stabilizes the stacking interaction of the purine base at position 37 of the tRNA anticodon loop, which is most often modified or hypermodified, with the first codon-anticodon base pair [26]. A very recent X-ray study provided unprecedented high-resolution insight into the role of the modified purine base at $tRNA^{Phe}$ position 37 and its coordinated Mg^{2+} ion for codon-anticodon stabilization. It has been shown that the sulfur atom of the methylthio group of the hypermodified nucleoside 2-methylthio-N6 isopentenyl adenosine forms a cross-strand stacking interaction with the base at the first nucleotide in the mRNA codon not only in the A-site, but also in the P- and E-sites [27]. Interestingly, in the P-site, the methylthio group coordinates a fully hydrated Mg^{2+} ion, which results in an intricate network of interactions between the first nucleotide of the P-site codon and the third nucleotide of the E-site codon (Figure 3). Cumulatively, these stabilizations of the codon-anticodon interactions in all three tRNA binding sites of the 30S subunit likely contribute to translation fidelity by preventing frame-shifting.

3.2. The Peptidyl Transferase Center

The PTC is the catalytic heart of the ribosome and its chemical center is composed of five universally conserved 23S RNA residues at positions C2063, A2451, U2506, U2585, and A2602, which are forming the inner core. Their possible functional implications in catalyzing the two chemical reactions of protein synthesis, peptide bond formation, and pept-tRNA hydrolysis, is indicated by the fact that mutations at any of those sites confers lethal phenotypes in *E. coli* [28,29]. Biochemical and more recent X-ray crystallographic structures highlighted A2451 as the key residue for peptide bond synthesis and A2602 for pept-tRNA hydrolysis (reviewed in [2,30]). The tRNA substrates for ribosomal catalysis are positioned and held in

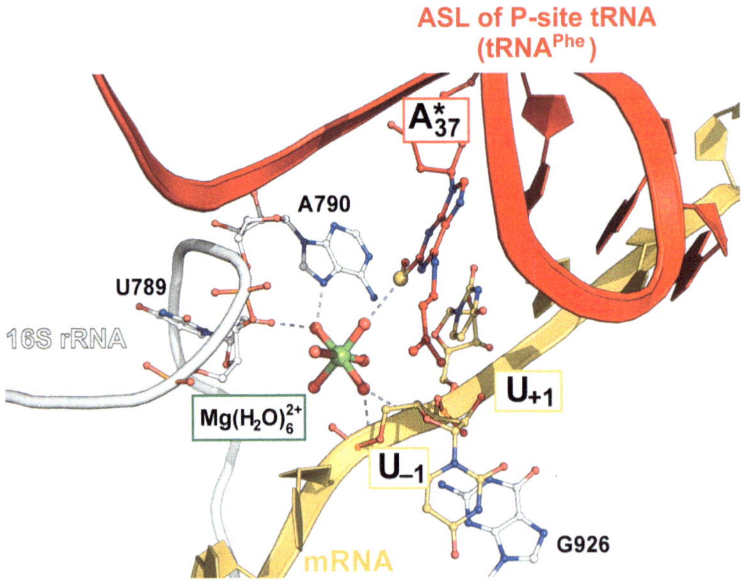

Figure 3. A magnesium ion contributes to anchoring of the tRNA in the 30S subunit P-site. A fully hydrated magnesium ion is coordinated by the sulfur atom of the 2-methylthio-N6 isopentenyl modification at tRNA position A37 [27]. U_{+1} and U_{-1} depict the first P-site and the last E-site mRNA codon position (yellow), respectively. The anticodon-stem loop (ASL) of tRNA^{Phe} is shown in red, and the 16S rRNA residues involved in P-site tRNA anchoring are shown in gray.

place in the active site by Watson-Crick interactions of their 3′-CCA ends with 23S rRNA PTC residues of the second layer. In the P-site, tRNA nucleotides C76 and C75 pair to G2252 and G2251 located in the 23S rRNA P-loop, while C75 of A-site tRNA is positioned by base-pairing to G2553 of the A-loop (see Figures 2 and 4). The terminal tRNA position A76 is stabilized in both A- and P-sites by A-minor interactions with 23S rRNA base pairs U2506-G2583 and A2450-C2501, respectively [31]. These tRNA-rRNA interactions are critical to optimally position the tRNA substrates for efficient peptide bond formation and pept-tRNA hydrolysis [19,32,33]. While it has been firmly established that the PTC is a ribozyme [34], one r-protein has been seen to approach the inner core pretty closely, namely the N-terminal tail of L27 [19]. L27 appears to contact and stabilize both A- and P-tRNA CCA ends. Direct contribution to catalysis, however, is unlikely, since L27 is dispensable for cell survival (reviewed in [35]).

Synthesis of a peptide bond involves aminolysis by the nucleophilic α-amino group of the A-site aa-tRNA on the ester bond that carries the nascent peptide at the C3′ position of the terminal ribose of pept-tRNA in

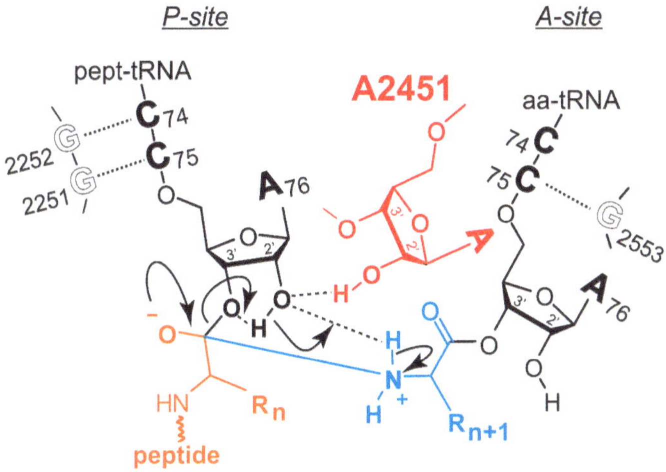

Figure 4. Reaction mechanism for peptide bond synthesis proposing a role for A2451. A2451 of 23S rRNA (red), peptidyl moiety of pept-tRNA in the P-site (orange), and aminoacyl moiety of A-site bound aa-tRNA (blue) are shown as well as the interactions of both CCA ends with 23S rRNA residues. The hydrogen bonding interaction between the A2451 2′-OH and the pept-tRNA A76 2′-O bond assists in positioning of the P-site tRNA A76 ribose in its functionally competent conformation [17]. In this model the nucleophilic attack of the α-amino group on the ester carbonyl carbon is accompanied by a concomitant acceptance of a proton from the α-amino group by the A76 2′-O of the pept-tRNA which simultaneously donates its proton to the vicinal 3′-O. Black arrows indicate pair-wise electron movement for proton shuttling after the attack of the α-amino nucleophile has established the tetrahedral intermediate.

the P-site. This leads to the transfer of the peptidyl moiety from the P- to the A-site tRNA and concomitantly results in the growth of the nascent peptide chain by one amino acid. During the termination phase of protein synthesis, water is the nucleophile and attacks the carbonyl carbon ester of P-site bound pept-tRNA, which results in hydrolysis of the ester bond and the release of the fully synthesized polypeptide. How do the PTC residues contribute to these chemical reactions? Accumulating evidence suggests that RNA enzymes exploit the same catalytic strategies as protein catalysis to achieve the observed rate enhancements of $\sim 10^6$ to 10^{11} compared to uncatalyzed reactions [36–38]. This includes decreasing entropic costs via substrate positioning, stabilizing transition state structures and facilitating the movement of protons. To achieve this, ribozymes utilize coordinated metal ions, functional groups on nucleobases, ribose hydroxyls, the phosphate backbone, and precisely constrained water molecules (reviewed in

[38–40]. Many parameters and requirements for peptide bond formation that have been revealed in the past, such as the absolute requirement for Mg^{2+}, Mn^{2+}, or Ca^{2+} and the pH optimum of ~8.5–9.5 with progressive inhibition of catalysis at lower pH values at least, using the so-called puromycin reaction as model reaction for peptidyl transfer (reviewed in [41]), are compatible with the notion that the PTC is a metallo ribozyme.

Catalytic models have been entertained, suggesting direct participation of Mg^{2+} ions [41]. However, no electron density for a potential catalytic divalent metal ion in close enough vicinity of peptide bond formation has been seen in any of the crystal structures [15,42] even though the PTC and surrounding 23S rRNA parts of domain V showed an increased Mg^{2+} ion density compared to other 23S rRNA domains (see Figures 2 and 5) [15]. This suggests that metal ions play a structural rather than a direct catalytic role during ribosomal catalysis. Some of the Mg^{2+} ions that surround the PTC have the common feature of two paired Mg^{2+} ions that are connected by a common bridging phosphate [43]. This particular Mg^{2+} binding motif has been named magnesium microcluster and turned out to be universally conserved. Three of these magnesium microclusters surrounding the PTC are thought to provide the framework for positioning crucial groups of inner core PTC residues. A fourth magnesium microcluster has been identified close to the exit site of the nascent polypeptide exit tunnel (see below). The sole metal ion observed in the inner PT cavity was a potassium ion that seems to help coordinating the orientation of the active site residues A2451, G2447, and G2061 [34]. Since mutations at A2451 and G2447, however, turned out not to interfere fundamentally with peptide bond formation, its contribution to catalysis remains unclear [44–46].

How is peptide bond formation catalyzed? Initially the nucleobase of the central PTC nucleotide A2451 has been suspected to directly participate in transpeptidation in a general acid-base mechanism. Subsequent standard as well as atomic mutagenesis, however, did not seem to support a crucial role for the nucleobase at A2451 for catalysis (reviewed in [2]). Furthermore, evidence was presented demonstrating that general acid-base chemistry is unlikely to be used by the PTC to synthesize peptide bonds [47]. A decisive next step towards a molecular understanding of peptide bond formation was achieved by using chemically modified P-site tRNA substrates [48,49] as well as chemically engineered ribosomes that carried non-natural nucleotide analogs in the PTC in transpeptidation assays [17,50,51]. In these studies ribose 2′-OH groups have been identified to be crucial for efficient amide bond formation. One of these critical 2′-OH groups is located at A76 of P-site bound pept-tRNA and the other at the 23S rRNA residue A2451. It was proposed that the A2451 2′-OH donates its proton to form a direct hydrogen bond interaction with A76 2′-OH of the pept-tRNA. This tRNA-rRNA interaction stabilizes the productive sugar conformation of A76,

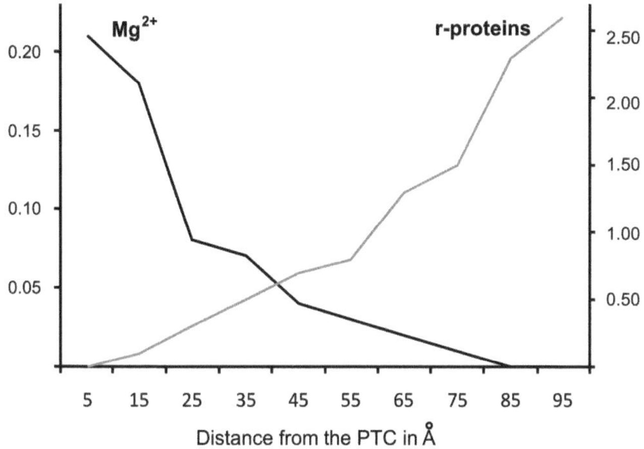

Figure 5. Magnesium ion concentration (black) and r-protein concentration (grey) in dependence on the distance in Å from the PTC. Numbers on the y-axis denote Mg^{2+} ion density (left) or r-protein density (right) normalized to the numbers of rRNA nucleotides in the corresponding PTC distance [68]. The number of magnesium ions is decreasing with distance from the PTC, while the number of r-proteins relatively to the number of rRNA residues is increasing towards the surface of the ribosomal particle.

which allows the establishment of a 6-membered ring system required for proton shuttling (cf. Figure 4). After the attack of the α-amino nucleophile when the tetrahedral intermediate is formed, the A76 2′-OH simultaneously donates its proton to the lone pair of A76 3′O leaving group and receives the proton from the positively charged nitrogen by its own lone pair.

3.3. The Nascent Peptide Exit Tunnel

As a consequence of the recurring addition of an amino acid during every elongation cycle, the newly polymerized peptide leaves the catalytic center via progressing through the nascent peptide exit tunnel. This exit tunnel starts at the PTC and goes all the way through the large ribosomal subunit to end at the 50S solvent side [52]. The tunnel is ~100 Å long and 10–25 Å wide with the walls mainly composed of 23S rRNA residues. The sole protein contributions for the tunnel architecture are via the loops of r-proteins L4, L22, and L23 [53]. Residues of L4 and L22 form the narrowest constriction along the tunnel, which is also referred to as the 'discriminator gate' [54].

The tunnel dimensions allow the first co-translational folding of the nascent peptide secondary structure to occur, namely the formation of α-helices. The functional relevance of the peptide exit tunnel is mirrored by the fact that members of the macrolide family of antibiotics (e.g., erythromycin) bind to the tunnel close to the discriminator gate mainly interacting with 23S rRNA residues A2058 and A2059 [55,56]. Initially it was assumed that the exit tunnel behaves neutral in respect to the nascent polypeptide sequence, but accumulating evidence indicates that certain nascent peptides can sequence-specifically interact with the tunnel wall. These specific peptide-tunnel interactions can yield to ribosome stalling with concomitant translation arrest which opens up a time window for gene regulation processes to occur [53]. The molecular events how the ribosome reads the sequence of the nascent peptide and submits a signal back to the PTC in order to pause amino acid polymerization, are not fully understood. Evidence has been presented highlighting a special role of the highly mobile 23S rRNA nucleotide A2062 which seems to be in an ideal position to interact and read the amino acid sequence at the entrance to the tunnel [57,58]. Very recent data support this hypothesis showing A2062 to be part of a putative 'information network' capable of transmitting signals from the tunnel back to the PTC. Beside the nucleobase of A2062, this interaction network involves a coordinated monovalent metal ion as well as the universally conserved PTC wobble base pair A2450-C2063 and A76 of P-site tRNA [75].

3.4. The Intersubunit Bridges

During the elongation cycle, the ribosome is engaged in dynamic conformational rearrangements. The largest observed structural dynamics involve the 30S head region, the L1 and L7/L12 stalks of the 50S subunit as well as the counter clockwise ratcheting movement of the entire 30S subunit relative to the 50S subunit (reviewed in [59]). The latter movement has been implicated in the tRNA translocation process and has been shown to be triggered by binding of EF-G · GTP to the elongation factor binding site near the A-site region of the ribosome. For ratcheting (and hence tRNA translocation) to occur in a productive and coordinated manner, the dynamic formation and breaking of intersubunit bridges is essential [60]. A dozen bridges have been identified that are either formed by rRNA-rRNA (bridges B2a, B2b, B2c, B3, B7a), by rRNA-protein (B1a, B4, B5, B6, B7b, B8), or by protein-protein interactions (B1b) [25].

It is notable that the rRNA-rRNA bridges primarily involve domain IV of 23S rRNA that locates close to the 50S PTC and interacts mainly with helix 44 (which also contains the decoding center) of 16S rRNA. These central rRNA bridges are in an optimal position to contact the tRNAs and have thus been suspected to be also involved in tRNA translocation through the intersubunit space. In contrast, bridges that contain r-protein components locate more to the periphery of the intersubunit region. The bridge B2a is formed by H69 of 23S rRNA and h44 of 16S rRNA is of particular interest because it is at the geometric center of the subunit interface and marks the border between A- and P-sites. Since H69 was found to be a mobile element and its tip has been observed to contact the D-stems of both A- and P-site tRNA, it was proposed that bridge B2a assists in tRNA translocation [61]. Another strong indication that bridge B2a is indeed a true functional center of the ribosome came from studies demonstrating that antibiotics (such as viomycin, capreomycin, neomycin, paromomycin, or gentamicin) target this intersubunit bridge [62–64].

It has long been know that Mg^{2+} ions are crucial for ribosomal subunit association. Reducing the Mg^{2+} ion concentration below ~2 mM results in the dissociation of *E. coli* ribosomes into its subunits. Now in the post-crystallographic era of ribosome research, the contribution of metal ions to subunit association can be understood in more detail. For forming bridge B2a two Mg^{2+} ions are involved in gluing together 23S rRNA residue A1913 of H69 and the 16S rRNA 'decoding' nucleotides A1492 and A1493 of h44 on the one hand and U38 of A-site bound tRNA on the other hand [24]. An example of a Mg^{2+}-mediated intersubunit bridge could be seen between 23S rRNA residues 1832/1833 of H67 and 16S rRNA helixes 27 (C899) and h24 (C770) forming bridge B2c. Also bridges involving r-protein components have been observed to depend on coordinated Mg^{2+} ions (e.g., B5, B8). These Mg^{2+} ions are coordinated between rRNA residues and r-proteins, thereby additionally helping establishing stable bridging interactions between the subunits [24].

4. METAL IONS AND THE EVOLUTION OF THE RIBOSOME

The ribosome is one of the most ancient molecular machines known and it probably already existed before LUCA (last universal common ancestor) appeared on the scene. The two major tasks of the ribosome, the selection of the correct amino acids and the peptide bond formation are catalyzed or promoted by RNA, whereas the r-proteins are primarily playing an ancillary role. The finding that the ribosome is a ribozyme provides the strongest argument in favor of the assumption that the ribosome is originating from the so-called 'RNA world' [65]. In this primordial era all biochemically relevant reactions were performed by nucleic acids (e.g., RNA or a precursor polymer thereof) since no protein catalysts have evolved yet. It was proposed before that, when the proto-ribosome evolved, it first started with the formation of non-templated synthesis of peptides. It is highly improbable that the first formed peptides themselves had any significant enzymatic activities, but via binding to RNA they could have enlarged the structural and, hence, functional repertoire of the RNA world nucleic acids including the proto-ribosome itself [66].

It has been amply demonstrated that the interaction of low-molecular-weight molecules with RNA can lead to the establishment of specific RNA folds that are not attainable without the ligand. Julian Davies speculated that some antibiotics, or other primordial "low-molecular-weight effectors", might be the functional remnants of the first rRNA-binding peptides that helped the all-RNA-based proto-ribosome to adopt its functional competent architecture [67]. Therefore, the first translation system that produced "functional" peptides did not evolve to pave an exit path out of the RNA world, but rather "aimed at" improving the properties of RNA molecules in the pre-protein world. Thus, it is likely that from these initial polypeptides, the first r-proteins have been recruited. By taking a closer look on the r-proteins nearest to the PTC, one recognizes that they are special in several regards. In fact, they are bigger, in terms of their residue number and molecular weight, but less structured than the ones further away. This could also be a hint, that they are more related to ancient proteins, which are as well supposed to show a low amount of α-helices and β-sheets [68,69].

Presumably over the time more and more proteins and new elements of the 23S rRNA have been added to the proto-ribosome. This proto-ribosome might have consisted of nothing more than the domain V part of the modern 23S rRNA, which is considered to be the oldest fragment of the ribosome. Domain IV and large parts of domain II are structurally highly connected with this ancient part of the ribosome and it is therefore most likely that they have a long-time co-history in terms of evolution. Whereas domain III, for example, has just minor connections to the PTC and seems therefore to be "younger" than the other parts of the present-day ribosome [70].

Most probably, there has been a co-evolution of rRNA and r-proteins, which makes it difficult nowadays to imagine that the rRNA has done the job alone in earlier times. However, as it is possible to retain peptidyl transferase activity in ribosomal particles that have been stripped of the majority of the r-proteins [71,72], argues in favor of the principal existence of an all-RNA ribosome in the RNA world. By applying another approach, based on the analysis of A-minor motifs and their structural dependencies regarding the involved elements within the 23S rRNA, one comes to a similar conclusion [73]. Namely that starting from an inner core, the PTC, the evolution of the ribosome has begun roughly four billion years ago. By the stepwise addition of more and more rRNA elements and later on also proteins, the present-day ribosome was "constructed", which can be traced back, as Bokov and Steinberg [73] concluded, by the structural dependence of the newer rRNA elements on the older ones. Based on their rRNA-evolution model they assume that it could be possible to fix the point in time when the RNA world started to change to a protein-based way of living [73]. In addition, their analysis showed that the double helices of the A-minor motifs are mainly found in regions of the 23S rRNA, which are assumed to be the oldest. Such double helices are normally stabilized by metal ions, which also seem to be supported by the observation that Mg^{2+} ions, which are "innersphere" interacting with phosphate oxygen of the rRNA in the large ribosomal subunit, are concentrated near the PTC (cf. Figure 2). This observation is correlating with the fact that the number of r-proteins is gradually increasing with the distance to the PTC (see Figure 5) [68].

Taking into account that the distance from the PTC might correlate with the evolutionary age of a ribosome domain, it seems that r-proteins have replaced the metal ion interaction with the rRNA during evolution. This matches also with the surveillance that the PTC domain shows an over-represented accumulation of single stranded rRNA regions, the structural integrity of which is known to require metal ion coordination [74]. The above mentioned conclusions are in good agreement with data obtained by comparative sequence analysis of rRNA from *T. thermophilus* (for the small subunit) and *H. marismortui* (for the large subunit) with the rRNA of the *C. elegans* mitochondria, which is known as one of the most reduced ribosomes, in terms of rRNA content. Mears and coworkers could align the majority of the nucleotides of *C. elegans* mitochondrial rRNA to the prokaryotic sequence and remarkably to those regions supposed to be the most ancient ones by the papers cited here already before [74].

In summary, we would like to conclude that there are parts of the rRNA, which can be considered to be older than others in evolutionary terms. These are the regions near the PTC in the 50S subunit and the fragments strongest connected to them. This holds true also in terms of architectural similarities and metal ion occurrence. Additionally, it could be established that some of the data is true for all domains of life, which boosts the theory that there has been a proto-ribosome, probably composed only of RNA as proposed by Francis Crick [3], that appeared on the scene of evolution before the phylogenetic tree started to branch. Thus, the ribosome can be viewed as an appropriate chronometer to trace back the evolution of life on earth.

5. CONCLUDING REMARKS

Metal ions, in particular Mg^{2+}, have always been closely connected with the evolution and function of the central players in biology. Also for the ribosome, the ~3 MDa large RNP complex responsible for protein synthesis in all living cells, metal ions play a fundamentally important part. Since the ribosome is a gigantic molecular machine that is mainly composed of rRNA, mono- and divalent metal ions are pivotal for RNA folding and stabilization.

Decades of biochemical and recent high resolution crystallographic studies revealed the ribosome as RNA enzyme. Unlike other larger RNA enzymes, such as the self-splicing group I and group II introns, the ribosome does not seem to utilize coordinated Mg^{2+} ions directly to promote catalysis. Instead, it appears that Mg^{2+} ions play a structural rather than a catalytic role in ribosome functioning. The fact that Mg^{2+} ions are mainly found in the evolutionarily oldest parts of the ribosome (such as the peptidyl transferase center) argues for a decisive role of metal ions during the course of the evolution of the ribosome. Even in the contemporary ribosome, metal ions cluster around the functionally crucial sites and likely contribute to protein synthesis at essentially all steps of the ribosomal elongation cycle.

ACKNOWLEDGMENTS

We would like to thank Nina Clementi and Matthias Erlacher for helpful comments on the manuscript. Our thanks are extended to Marat Yusupov and Lasse Jenner for providing Figure 3. Research in our lab is funded by the Austrian Science Foundation (FWF) grant Y315 to N.P.; K.T. is supported by an MFI grant (9440 to N.P) from the Medical University Innsbruck.

ABBREVIATIONS

A-site	acceptor site
aa-tRNA	aminoacyl-transfer RNA
E-site	exit site
EF Tu	elongations factor Tu
GDP	guanosine 5′-diphosphate
GTP	guanosine 5′-triphosphate
IF	initiation factor
LUCA	last universal common ancestor
Met	methionine
mRNA	messenger RNA
P-site	peptidyl site
pept-tRNA	peptidyl transfer RNA
Phe	phenylalanine
PTC	peptidyl transferase center

r-proteins	ribosomal proteins
RF	release factor
RNP	ribonucleoprotein
rRNA	ribosomal RNA

REFERENCES

1. D. N. Wilson, G. Blaha, S. R. Connell, P. V. Ivanov, H. Jenke, U. Stelzl, Y. Teraoka and K. H. Nierhaus, *Curr. Protein Pept. Sci.*, 2002, **3**, 1–53.
2. M. D. Erlacher and N. Polacek, *RNA Biol.*, 2008, **5**, 5–12.
3. F. H. Crick, *J. Mol. Biol.*, 1968, **38**, 367–379.
4. K. Kruger, P. J. Grabowski, A. J. Zaug, J. Sands, D. E. Gottschling and T. R. Cech, *Cell*, 1982, **31**, 147–157.
5. C. Guerrier-Takada, K. Gardiner, T. Marsh, N. Pace and S. Altman, *Cell*, 1983, **35**, 849–857.
6. A. Serganov and D. J. Patel, *Nat. Rev. Genet.*, 2007, **8**, 776–790.
7. J. M. Ogle and V. Ramakrishnan, *Annu. Rev. Biochem.*, 2005, **74**, 129–177.
8. M. V. Rodnina, K. B. Gromadski, U. Kothe and H. J. Wieden, *FEBS Lett.*, 2005, **579**, 938–942.
9. M. Amort, B. Wotzel, K. Bakowska-Zywicka, M. D. Erlacher, R. Micura and N. Polacek, *Nucleic Acids Res.*, 2007, **35**, 5130–5140.
10. N. Polacek, M. G. Gomez, K. Ito, Y. Nakamura and A. S. Mankin, *Mol. Cell*, 2003, **11**, 103–112.
11. K. H. Nierhaus, in *Protein Synthesis and Ribosome Structure*, Ed. K. H. Nierhaus and D. E. Wilson, Wiley-VCH Verlag GmbH & Co. KGaA, Weinheim, 2004, pp. 85–107.
12. W. Sun, A. Pertzev and A. W. Nicholson, *Nucleic Acids Res.*, 2005, **33**, 807–815.
13. T. K. Misra and D. Apirion, *J. Biol. Chem.*, 1979, **254**, 11154–11159.
14. A. M. Pyle, *J. Biol. Inorg. Chem.*, 2002, **7**, 679–690.
15. D. J. Klein, P. B. Moore and T. A. Steitz, *RNA*, 2004, **10**, 1366–1379.
16. P. Khade and S. Joseph, *FEBS Lett.*, 2010, **584**, 420–426.
17. K. Lang, M. D. Erlacher, D. N. Wilson, R. Micura and N. Polacek, *Chem. Biol.*, 2008, **15**, 485–492.
18. D. E. Brodersen and P. Nissen, *FEBS J.*, 2005, **272**, 2098–2108.
19. R. M. Voorhees, A. Weixlbaumer, D. Loakes, A. C. Kelley and V. Ramakrishnan, *Nat. Struct. Mol. Biol.*, 2009, **16**, 528–533.
20. D. N. Wilson and K. H. Nierhaus, *Crit. Rev. Biochem. Mol. Biol.*, 2005, **40**, 243–267.
21. J. M. Ogle, D. E. Brodersen, W. M. Clemons, Jr., M. J. Tarry, A. P. Carter and V. Ramakrishnan, *Science*, 2001, **292**, 897–902.
22. R. C. Thompson, D. B. Dix, R. B. Gerson and A. M. Karim, *J. Biol. Chem.*, 1981, **256**, 6676–6681.
23. J. Davies, W. Gilbert and L. Gorini, *Proc. Natl. Acad. Sci.*, 1964, **51**, 883–890.
24. M. Selmer, C. M. Dunham, F. Vt Murphy, A. Weixlbaumer, S. Petry, A. C. Kelley, J. R. Weir and V. Ramakrishnan, *Science*, 2006, **313**, 1935–1942.

25. M. M. Yusupov, G. Z. Yusupova, A. Baucom, K. Lieberman, T. N. Earnest, J. H. Cate and H. F. Noller, *Science*, 2001, **292**, 883–896.
26. A. L. Konevega, N. G. Soboleva, V. I. Makhno, Y. P. Semenkov, W. Wintermeyer, M. V. Rodnina and V. I. Katunin, *RNA*, 2004, **10**, 90–101.
27. L. B. Jenner, N. Demeshkina, G. Z. Yusupova and M. M. Yusupov, *Nat. Struct. Mol. Biol.*, 2010, **17**, 555–560.
28. M. A. Bayfield, J. Thompson and A. E. Dahlberg, *Nucleic Acids Res.*, 2004, **32**, 5512–5518.
29. E. M. Youngman, J. L. Brunelle, A. B. Kochaniak and R. Green, *Cell*, 2004, **117**, 589–599.
30. M. D. Erlacher and N. Polacek, *Biotechnologia*, 2009, **1**, 65–80.
31. P. Nissen, J. A. Ippolito, N. Ban, P. B. Moore and T. A. Steitz, *Proc. Natl. Acad. Sci.*, 2001, **98**, 4899–4903.
32. T. M. Schmeing, K. S. Huang, D. E. Kitchen, S. A. Strobel and T. A. Steitz, *Mol. Cell*, 2005, **20**, 437–448.
33. T. M. Schmeing, K. S. Huang, S. A. Strobel and T. A. Steitz, *Nature*, 2005, **438**, 520–524.
34. P. Nissen, J. Hansen, N. Ban, P. B. Moore and T. A. Steitz, *Science*, 2000, **289**, 920–930.
35. N. Polacek and A. S. Mankin, *Crit. Rev. Biochem. Mol.*, 2005, **40**, 285–311.
36. G. M. Emilsson, S. Nakamura, A. Roth and R. R. Breaker, *RNA*, 2003, **9**, 907–918.
37. T. K. Stage-Zimmermann and O. C. Uhlenbeck, *RNA*, 1998, **4**, 875–889.
38. T. J. Wilson and D. M. Lilley, *Science*, 2009, **323**, 1436–1438.
39. N. G. Walter, *Mol. Cell*, 2007, **28**, 923–929.
40. M. J. Fedor and J. R. Williamson, *Nat. Rev. Mol. Cell Biol.*, 2005, **6**, 399–412.
41. A. Barta and I. Halama, in *Ribosomal RNA and Group I Introns*. Ed. R. Green and R. Schroeder, R.G. Landes Company, Chapman & Hall, New York, 1996, pp. 35–54.
42. T. A. Steitz, *FEBS Lett.*, 2005, **579**, 955–958.
43. C. Hsiao and L. D. Williams, *Nucleic Acids Res.*, 2009, **37**, 3134–3142.
44. M. Beringer, S. Adio, W. Wintermeyer and M. Rodnina, *RNA*, 2003, **9**, 919–922.
45. M. Beringer, C. Bruell, L. Xiong, P. Pfister, P. Bieling, V. I. Katunin, A. S. Mankin, E. C. Bottger and M. V. Rodnina, *J. Biol. Chem.*, 2005, **280**, 36065–36072.
46. N. Polacek, M. Gaynor, A. Yassin and A. S. Mankin, *Nature*, 2001, **411**, 498–501.
47. P. Bieling, M. Beringer, S. Adio and M. V. Rodnina, *Nat. Struct. Mol. Biol.*, 2006, **13**, 423–428.
48. S. Dorner S, C. Panuschka, W. Schmid and A. Barta, *Nucleic Acids Res.*, 2003, **31**, 6536–6542.
49. J. S. Weinger, K. M. Parnell, S. Dorner, R. Green and S. A. Strobel, *Nat. Struct. Mol. Biol.*, 2004, **11**, 1101–1106.
50. M. D. Erlacher, K. Lang, N. Shankaran, B. Wotzel, A. Huttenhofer, R. Micura, A. S. Mankin and N. Polacek, *Nucleic Acids Res.*, 2005, **33**, 1618–1627.
51. M. D. Erlacher, K. Lang, B. Wotzel, R. Rieder, R. Micura and N. Polacek, *J. Am. Chem. Soc.*, 2006, **128**, 4453–4459.

52. N. Ban, P. Nissen, J. Hansen, P. B. Moore and T. A. Steitz, *Science*, 2000, **289**, 905–920.
53. H. Ramu, A. Mankin and N. Vazquez-Laslop, *Mol. Microbiol.*, 2009, **71**, 811–824.
54. H. Nakatogawa and K. Ito, *Cell*, 2002, **108**, 629–636.
55. F. Schlunzen, R. Zarivach, J. Harms, A. Bashan, A. Tocilj, R. Albrecht, A. Yonath and F. Franceschi, *Nature*, 2001, **413**, 814–821.
56. D. Tu, G. Blaha, P. B. Moore and T. A. Steitz, *Cell*, 2005, **121**, 257–270.
57. B. Seidelt, C. A. Innis, D. N. Wilson, M. Gartmann, J. P. Armache, E. Villa, L. G. Trabuco, T. Becker, T. Mielke, K. Schulten, T. A. Steitz and R. Beckmann, *Science*, 2009, **326**, 1412–1415.
58. N. Vazquez-Laslop, C. Thum and A. S. Mankin, *Mol. Cell*, 2008, **30**, 190–202.
59. A. Korostelev, D. N. Ermolenko and H. F. Noller, *Curr. Opin. Chem. Biol.*, 2008, **12**, 674–683.
60. L. H. Horan and H. F. Noller, *Proc. Natl. Acad. Sci.*, 2007, **104**, 4881–4885.
61. A. Bashan, I. Agmon, R. Zarivach, F. Schluenzen, J. Harms, R. Berisio, H. Bartels, F. Franceschi, T. Auerbach, H. A. Hansen, E. Kossoy, M. Kessler and A. Yonath, *Mol. Cell*, 2003, **11**, 91–102.
62. M. A. Borovinskaya, R. D. Pai, W. Zhang, B. S. Schuwirth, J. M. Holton, G. Hirokawa, H. Kaji, A. Kaji and J. H. Cate, *Nat. Struct. Mol. Biol.*, 2007, **14**, 727–732.
63. S. K. Johansen, C. E. Maus, B. B. Plikaytis and S. Douthwaite, *Mol. Cell*, 2006, **23**, 173–182.
64. R. E. Stanley, G. Blaha, R. L. Grodzicki, M. D. Strickler and T. A. Steitz, *Nat. Struct. Mol. Biol.*, 2010, **17**, 289–293.
65. T. A. Steitz and P. B. Moore, *Trends Biochem. Sci.*, 2003, **28**, 411–418.
66. H. F. Noller, *RNA*, 2004, **10**, 1833–1837.
67. J. Davies, *Mol. Microbiol.*, 1990, **4**, 1227–1232.
68. C. Hsiao, S. Mohan, B. K. Kalahar and L. D. Williams, *Mol. Biol. Evol.*, 2009, **26**, 2415–2425.
69. L. Giri, W. E. Hill, H. G. Wittmann and B. Wittmann-Liebold, *Adv. Protein Chem.*, 1984, **36**, 1–78.
70. J. Hury, U. Nagaswamy, M. Larios-Sanz and G. E. Fox, *Orig. Life Evol. Biosph.*, 2006, **36**, 421–429.
71. H. F. Noller, V. Hoffarth and L. Zimniak, *Science*, 1992, **256**, 1416–1419.
72. P. Khaitovich, A. S. Mankin, R. Green, L. Lancaster and H. F. Noller, *Proc. Natl. Acad. Sci.*, 1999, **96**, 85–90.
73. K. Bokov and S. V. Steinberg, *Nature*, 2009, **457**, 977–980.
74. J. A. Mears, J. J. Cannone, S. M. Stagg, R. R. Gutell, R. K. Agrawal and S. C. Harvey, *J. Mol. Biol.*, 2002, **321**, 215–234.
75. A. Chirkova, M. D. Erlacher, N. Clementi, M. Zywicki, M. Aigner and N. Polacek, *Nucleic Acids Res.*, 2010, **38**, 4844–4855.

Met. Ions Life Sci. **2011**, *9*, 277–297

10

Metal Ion Requirements in Artificial Ribozymes that Catalyze Aminoacylation and Redox Reactions

Hiroaki Suga, [1, 2, 3] ***Kazuki Futai*** [3] ***and Koichiro Jin*** [2]

[1]Department of Chemistry, School of Science, The University of Tokyo, 7-3-1 Hongo, Bunkyo-ku, 113-0033, Tokyo, Japan

[2]Department of Chemistry and Biotechnology, School of Engineering, The University of Tokyo, 7-3-1 Hongo, Bunkyo-ku, 113-0033, Tokyo, Japan

[3]Research Center for Advanced Science and Technology, The University of Tokyo, 4-6-1 Komaba, Meguro-ku, 153-8904, Tokyo, Japan

<hsuga@chem.s.u-tokyo.ac.jp>

Metal Ions in Life Sciences, Volume 9 Edited by Astrid Sigel, Helmut Sigel, and Roland K. O. Sigel

Published by the Royal Society of Chemistry, www.rsc.org DOI: 10.1039/9781849732512 00277

ABSTRACT: The means of *in vitro* selection has yielded a number of artificial ribozymes with functions that have not been discovered as yet in modern biological systems. Like naturally occurring ribozymes, most artificial ribozymes also use metal ions for the support of catalysis. Here we choose two such ribozymes, flexizyme and ribox, that exhibit specific activities of tRNA aminoacylation and redox chemistry, respectively, and comprehensively summarize the roles of metal ions in conjunction with their structure and function.

KEYWORDS: aminoacylation · cobalt hexammine · redox · ribozyme · terbium cleavage · transfer RNA · translation

1. INTRODUCTION

Divalent and monovalent metal ions in naturally occurring ribozymes play crucial roles in various ribozyme functions. In some cases the metal ions directly participate in catalysis, but in other cases they play a role in structural stabilization and in forming the catalytic core, i.e., they play an indirect role [1,2]. Such specific roles in naturally occurring ribozymes have been elucidated by an enormous number of elegant biochemical as well as structural studies and many of these are summarized in various chapters in this book.

To date, several dozens of artificial ribozymes have been discovered by means of *in vitro* selection or SELEX [3]. Many of these are able to catalyze cleavage of the phosphodiester bond like naturally occurring ribozymes while some of these can execute chemistry other than cleavage of the phosphodiester bond, e.g., acyl transfer [4–7], Michael [8], Diels-Alder [9,10], Claisen [11], aldol [12], and redox reactions [13,14]. Because the selection generally includes metal ions such as K^+ or Mg^{2+}, and occasionally also other divalent metal ions, the function of these generated ribozymes often relies on the presence of metal ions. For instance, the leadzyme, the simplest ribozyme known to date, was selected in the presence of Pb^{2+} for the self-cleavage reaction, and therefore, it functionally relies on Pb^{2+} [15]. Despite the fact that almost all artificial ribozymes require metal ions, only a limited number of studies regarding the roles of metal ions have been conducted.

In this chapter, we introduce two *in vitro* evolved ribozymes, the flexizymes and the redox ribozymes (see Figures1b and 6b in Sections 2 and 4, respectively), generated in our laboratory. Flexizymes catalyze the aminoacylation of tRNA yielding aminoacyl-tRNAs, which are the essential components in protein synthesis. This family of ribozymes evolved in three selection steps in the presence of Mg^{2+} and K^+, to reveal the essential catalytic residues [7,16–18]. Extensive biochemical studies on the roles of metal ions were also performed to predict the Mg^{2+} coordination geometries and its binding sites [19].

More recently, the X-ray structure of the second generation, referred to as Fx3 [20], became available, and therefore, it is an excellent case to review the roles of metal ions in conjunction with biochemical data. Another ribozyme, capable of catalyzing redox reactions and referred to as ribox, was selected in the presence of Zn^{2+} and Mg^{2+}, and so it uniquely relies on both metal ions albeit their functional roles are clearly different [13,14]. To this end, we hope that these two unique artificial ribozymes illuminate not only the functional roles of metal ions found in these ribozymes but also the importance of studies on metal ions in other artificial ribozymes.

2. FLEXIZYMES

2.1. Evolutionary History and Utility

Aminoacylation of tRNA is the reaction to generate aminoacyl-tRNA molecules that act as the essential amino acid carriers in the translation system and synthesis of proteins. This event could have been executed solely by ribozymes, and the evolution of such ribozymes along with a ribosome-like ribozyme could have made possible the transition from the 'pure' RNA world to the RNA-peptide (protein) world. Regardless of the evolutionary importance, this chemical reaction is quite challenging for RNA to achieve the following functions: (1) aminoacylation must take place specifically at the 3′-end of the hydroxyl group over other internal 2′-hydroxyl groups in tRNA or the ribozyme itself; (2) the ribozyme must recognize the amino acid of interest and simultaneously tRNA; and (3) the amino acid substrate must be activated by a certain leaving group by the ribozyme itself or other mechanisms. Although protein aminoacyl-tRNA synthetases activate amino acids by coupling of the amino acid with ATP generating aminoacyl-AMP, the acyl moiety could be pre-activated by a certain ester group (e.g., thioester) via prebiotic mechanisms. Hence, our initial goal was to achieve insights into points (1) and (2). Prior to our work, Yarus et al. reported an elegant experiment for the isolation of self-aminoacylating ribozymes, but because their ribozymes were only capable of self-charging phenylalanine (Phe) onto the non-CCA-3′ end in the presence of pre-activated Phe-AMP, neither point (1) nor (2) was achieved at that time [21–25]. A ribozyme that fulfilled points (1) and (2) was generated by the following selection strategy with a unique RNA library design published in 2001 [7].

The RNA library was designed on the basis of modern tRNA processing events in prokaryotes, where precursor tRNAs are processed by RNase P, a naturally occurring ribozyme, to yield mature tRNAs. In the modern biological world, the 5′-leader sequences appear to be variable depending on the kinds of tRNA as well as the species of prokaryotes but no function has been discovered yet. Our hypothesis was that in the RNA world the 5′-leader sequence could have had a function to catalyze the self-aminoacylation of the 3′-end of tRNA or a tRNA-like molecule and after its aminoacylation a RNase P-like ribozyme cleaved it at the correct site to generate aminoacyl-tRNA. Therefore, we placed a random sequence in the 5′-leader region to build the RNA library and selected an active population against a *N*-biotinylated L-Phe cyanomethyl ester (biotin-Phe-CME). To make a long story short, the selection in the presence of Mg^{2+} and K^+ yielded a single clone, referred to as pre-24, and the catalytic domain alone was referred to as r24 (Figure 1a). The r24 domain was further truncated by removing unnecessary regions of the P1 and P2 stems, and the construct was referred to as r24mini (Figure 1b) [18]. It should be noted that the r24 ribozyme was extensively studied for its metal ion dependence, which is discussed later in detail. Importantly, the ribozyme domain attached to the 5′-end of tRNA could be cleaved away from tRNA by the catalysis of RNase P, and the ribozyme domain itself is able to transfer biotin-Phe onto tRNA. Moreover, our biochemical study on r24mini has revealed that it charges Phe onto the 3′-hydroxyl group, not the 2′-hydroxyl group, at the 3′-end of tRNA [26], and also recognizes the aromatic ring of Phe derivatives, not the CME leaving group (i.e., the leaving group can be altered to a thioester or AMP without loss of activity) [7].

This r24mini was further transformed into a much shorter construct consisting of only 45 nucleotides, referred to as Fx3, the original flexizyme [16]. This construct was later used for crystallization in order to solve the X-ray three-dimensional structure (*vide infra*) [20]. Moreover, to reset the specificity of Fx3 toward the aromatic amino acids (Phe or Phe analogs), a library varying the putative amino acid binding region was used to select two types of variants, eFx and dFx (Figure 1c), which are capable to recognize amino acid substrates activated with *p*-chlorobenzyl thioester (CBT) and 3,5-dinitrobenzyl ester (DBE), respectively [17]. This alteration of the recognition elements of Fx from the aromatic side chain to the aromatic leaving group librated the flexizymes from the restrictions of the kind of side chain and thus expanded their utility to different amino acids including those with non-proteinogenic side chains. Most recently, the flexizymes became an extraordinary tool for biochemical studies of the ARS and ribosome as well as developing biotechnologies to express non-standard peptides [27–35].

2.2. Roles of Mg^{2+} in Flexizymes

All flexizymes evolved to date require Mg^{2+} for catalytic activity (note that K^+ is also included for the buffer component, but varying its concentration

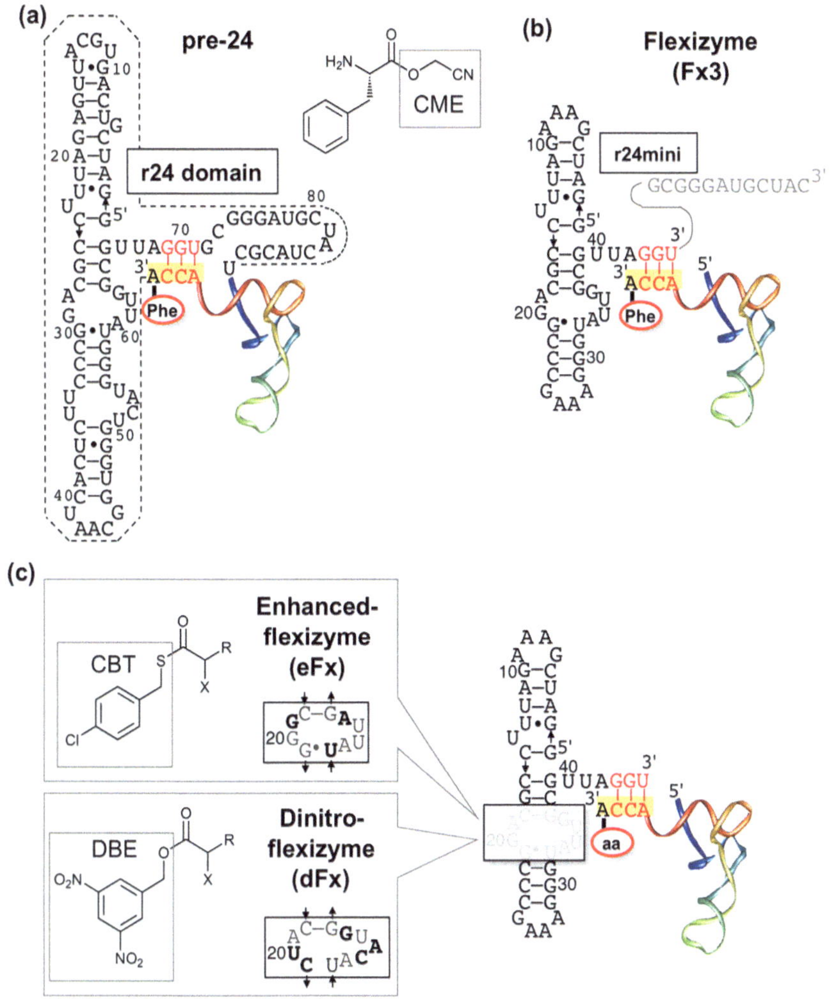

Figure 1. The secondary structures of flexizymes. (**a**) The secondary structure of pre-24. Thick lines and dots indicate predicted Watson-Crick and G-U Wobble base pairing, respectively. The broken line indicates the core region of r24. Bases involved in the base pair interaction between Fx3 and tRNA are highlighted in red. CME denotes the cyanomethyl ester leaving group. (**b**) The secondary structure of r24mini and flexizyme (Fx3). The r24mini has a catalytically unnecessary region at the 3′-end highlighted by gray bases and removal of this region yielded Fx3 [16]. (**c**) The secondary structures of the enhanced- and dinitro-flexizyme. The boxes indicate the sequences of the catalytic core for each flexizyme. CBT and DBE denote 4-chlorobenzyl thioester and 3,5-dinitrobenzyl ester, respectively.

at low levels, <250 mM, does not affect the catalytic activity). This is simply because they were evolved in the presence of Mg^{2+}. The metal ion-dependent kinetic studies have been performed in pre-24, the early version of the flexizyme conjugated with tRNA, accumulating knowledge about how critical the concentrations of Mg^{2+} are as well as about the divalent metal ion species responsible for ribozyme activity [19].

The study on the dependence of the metal ion species on pre-24 showed that Mg^{2+} is by far the best among other divalent metal ions tested, followed by Ca^{2+} with half of the activity and Mn^{2+} with a marginal one. Other divalent metal ions did not show any activity. The titration plot ([Mg^{2+}] versus k_{obs}) displayed a biphasic behavior, where the activity rapidly increases up to 20 mM Mg^{2+}, then slowly increases beyond 20 mM (Figure 2a). The addition of spermidine did not change the biphasic behavior, indicating that non-specific structural Mg^{2+} does not contribute to this behavior. Moreover, Hill analysis of the activity in the first phase gave a coefficient of 0.97, suggesting that a single metal ion is most likely responsible for the activity in this region. Because the second phase showed a nearly linear increase in activity, this region could not be analyzed by a Hill plot. However, the observed gentle slope indicates the possibility that two or more metal ions are involved in the second phase. This suggests that at least two Mg^{2+} ions are probably involved in catalysis, where one of them can bind to a high affinity metal-binding site and the other(s) can bind to (a) low affinity site(s).

A notable outcome was obtained from the substitution experiment of Mg^{2+} with $Co(NH_3)_6^{3+}$ that supposedly acts as a mimic of the hexahydrated form of Mg^{2+}. Unlike $Mg(OH)_6^{2+}$, $Co(NH_3)_6^{3+}$ is inert against ligand exchange, and thus even in the presence of EDTA it stays as it is [36–39]. $Co(NH_3)_6^{3+}$ in the presence of EDTA stimulated the catalysis with an initial rate similar to the Mg^{2+}-dependent profile below 2.5 mM, whereas at higher concentrations (20–120 mM) the activity becomes nearly plateaus (Figure 2a). This observation might be an indication that the high affinity site could be occupied with a single hydrated Mg^{2+} (i.e., in an outersphere coordination mode), while the other low affinity site(s) could be with non-hydrated Mg^{2+} ions (i.e., in an innersphere coordination mode).

Substitution of Mg^{2+} with monovalent metal ions has been used to investigate a structural role of metal ions in ribozymes [36,40–44]. A critical difference between Mg^{2+} and monovalent ions is that Mg^{2+} can lower the pK_a of a coordinating OH group and facilitate the generation of a hydroxide or alkoxide ion, whereas monovalent ions can coordinate to the OH group, but they cannot lower its pK_a. It should be noted that $Co(NH_3)_6^{3+}$ is another kind of metal ion that is unable to generate a hydroxide or alkoxide ion. However, because of its strong electropositivity, it can still bind strongly to the electronegative environment of the metal binding site(s) [45,46]. Thus, when $Co(NH_3)_6^{3+}$ stimulates ribozyme catalysis, it is possible that the

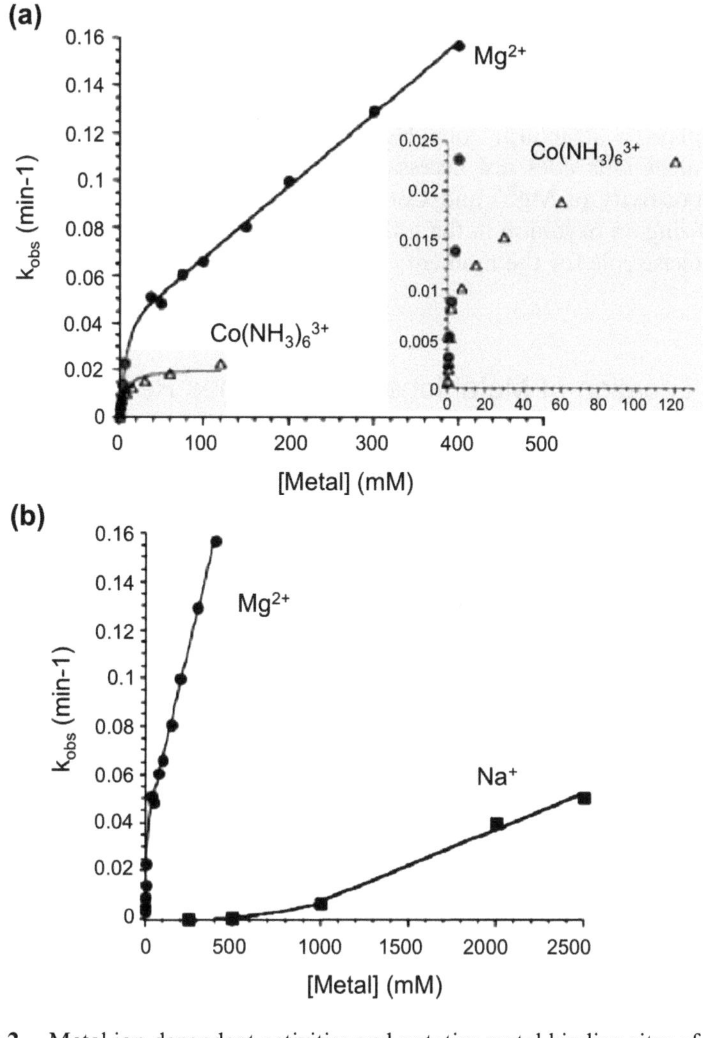

Figure 2. Metal ion-dependent activities and putative metal binding sites of pre-24. (**a**) Catalytic activity of pre-24 as a function of Mg^{2+} (filled circle) or $Co(NH_3)_6^{3+}$ (open triangle) concentration. Data were fitted to smooth interpolations. The insert shows activity below 120 mM $Co(NH_3)_6^{3+}$ and 7 mM Mg^{2+}. (**b**) Observed catalytic constant (k_{obs}) as a function of Na^+ (filled square) or Mg^{2+} (filled circle) concentration. Data were fitted to smooth interpolations.

essential metal ions do not generate such a hydroxide or alkoxide ion, but play other roles, such as stabilizing the catalytically critical structure or neutralizing an oxyanion in the transition state [47]. Because the catalytic activity of pre-24 was assisted by $Co(NH_3)_6^{3+}$, it was worthy to test whether monovalent ions could also support ribozyme catalysis. At an extreme concentration of monovalent ions (Li^+ and Na^+) as high as 2.5 M, the catalysis was in fact supported at a similar level to that with 100 mM Mg^{2+} (Figure 2b), indicating that the metal ion occupying the high affinity site likely plays a structural role. However, the observed ineffectiveness of monovalent ions does not necessarily rule out the possibility of a strong electropositivity of Mg^{2+} and $Co(NH_3)_6^{3+}$ that may play a critical role in neutralizing an oxyanion in the transition state, leaving an open question of their precise role for the moment.

2.3. Location of Metal Ions in Flexizymes Revealed by Biochemical and X-Ray Structural Studies

Hexacoordinate terbium ion (Tb^{3+}) has an ionic radius (0.92 Å) that is close to Mg^{2+} (0.72 Å), and it has the same preference for coordination to oxygen ligands [48,49]. Despite their similarities, Tb^{3+} is capable of cleaving RNA at neutral pH whereas Mg^{2+} generally is incapable to do so. This activity is attributed to a property of Tb^{3+} that generates the 2′-alkoxide ion in the binding sites, followed by an in-line attack on the 3′-bridging phosphate resulting in RNA cleavage. Thus, Tb^{3+} cleavage sites can approximate the number of Mg^{2+}-binding sites, and hence, Tb^{3+}-dependent hydrolytic mapping has been used to assess Mg^{2+} binding sites [42,50–53]. Moreover, when Tb^{3+} is titrated with Mg^{2+} or other catalytically functional but non-hydrolytic metal ions, the cleavage of the RNA strand near its binding site can be competitively inhibited [47]. This approach might allow us to postulate the location of metal binding site(s).

At a pre-determined concentration of Tb^{3+} of 50 μM which induced a clean profile of specific cleavages of pre-24 at certain sites, increasing concentrations of Mg^{2+}, varying from 1 mM to 50 mM, were able to titrate the competitive inhibitory sites against Tb^{3+}-induced cleavage. A similar experiment was also performed with $Co(NH_3)_6^{3+}$ to identify virtually identical inhibitory sites as in the above Mg^{2+} experiment. The identified sites are localized in two regions, <u>A</u>UU and <u>G</u>UU at the positions of 60–62 and 66–68, respectively (the underlined ribose was pronouncedly cleaved) (Figures 3a and 3b). It should be noted that our experimental data published in 2002 used a partly incorrect secondary structure assignment because pre-24 contained a catalytically non-essential region at the 3′-end, but later we redefined the secondary structure shown in Figure 1b. Although this series of experiments published in 2002 gave only low-resolution information regarding the metal binding site(s), the X-ray structural data solved in 2008 surprised us with the remarkable consistency.

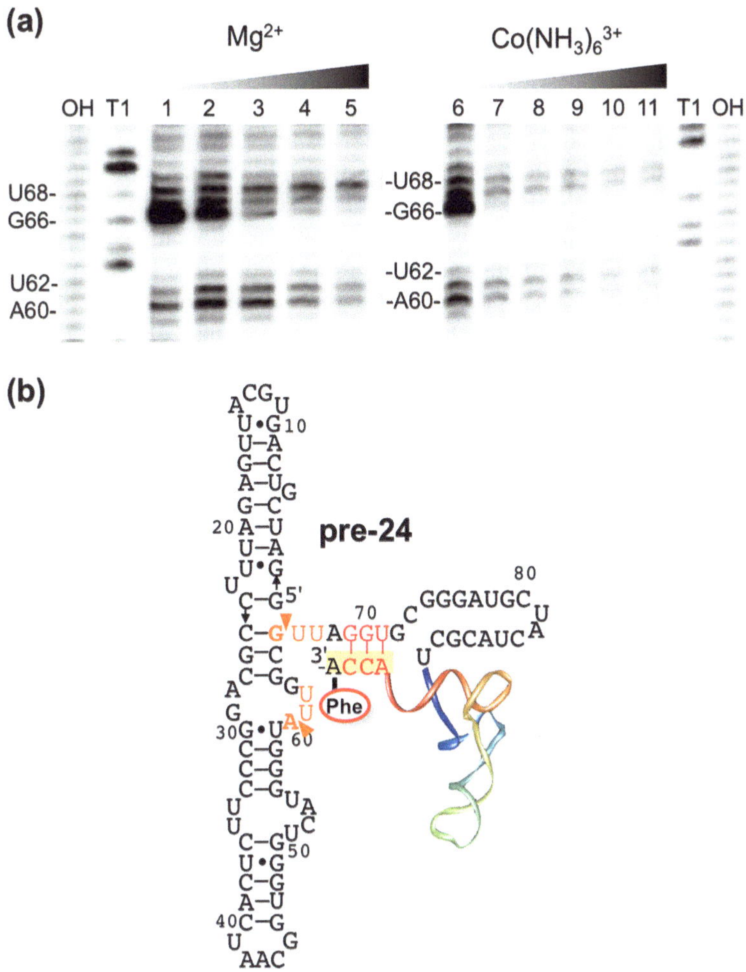

Figure 3. (**a**) Tb^{3+}-induced ribozyme cleavage in competition with metals. Tb^{3+} cleavage of 5′-labeled pre-24 at various Mg^{2+} concentrations (left). Samples in lanes 1–5 were incubated with 50 μM Tb^{3+} and Mg^{2+} concentrations of 0, 1, 10, 20, and 50 mM, respectively. OH and T1 represent alkaline hydrolysis and T1 endonuclease ladders, respectively. Tb^{3+} cleavage of 5′-labeled pre-24 at various $Co(NH_3)_6^{3+}$ concentrations (right). Samples in lanes 6–11 were incubated with 100 μM Tb^{3+} and $Co(NH_3)_6^{3+}$ concentrations of 0 0.5, 1, 5, 10, and 25 mM, respectively. (**b**) The secondary structure of the pre-24 and the Tb^{3+}-induced cleavage site. Bold orange letters indicate bases whose ribose is cleaved by Tb^{3+} and triangles indicate the cleavage site that is pronounced in each cleaved region.

Ferré-D'Amaré teamed up with us to design a construct of flexizyme (Fx3) conjugated with a mini-helix RNA, a tRNA analog, via the 3′-5′ direct linkage. In this construct, the tetra-loop in the P2 region was substituted with a U1A-binding loop, so that the complex with U1A facilitated crystallization as well as refinement of the structure. Moreover, the construct was co-crystallized with Phe ethyl ester, which is an analog of Phe-CME but hydrolytically more stable (i.e., an inhibitor), and thus the three-dimensional structure of Fx3 should represent its active form complexed with both substrates, the tRNA analog and the amino acid [20].

Putting detailed descriptions of the three-dimensional structure of the flexizyme aside, the structure has enabled us to visualize two unique Mg^{2+} binding sites (Figure 4). The first site is located at the non-canonical base pair region, AGG and UAUUG at positions 23–25 and 44–48, respectively, where the underlined bases are bulged. Mg^{2+} is in a hydrated form whose water molecules interact with the 6-oxo and the N7 of G25 and two phosphate oxygens of C22-p-A23-p-G24. Coordination of the hydrated Mg^{2+} to these residues makes the non-canonical pairs (A23 · G48, G24 · U47 and G25 · U44) and the two unpaired nucleotides (A45 and U46) open (or 'stretch'), constituting the Phe binding site. Thus, Mg^{2+} plays a crucial structural role in creating the pocket for the aromatic ring of Phe upon its binding to the major groove of the amino acid binding site. It should be noted that our Tb^{3+}/Mg^{2+} probing identified the site of AUU at positions 60–62 in pre-24, which corresponds to AUU at positions 45–47 in the Fx3–mini-helix construct (cf. Figures 3d and 4a). Thus, the biochemical data had correctly identified the Mg^{2+} binding site.

The second metal binding site is located at the 5′-end of Fx3 where two non-hydrated Mg^{2+} ions strongly coordinate to the 5′-triphosphate. Although the functional role of these Mg^{2+} ions is not so obvious from the X-ray structure, the Mg^{2+} titration experiment using pre-24 disclosed earlier has defined (a) low affinity metal binding site(s) occupied by more than one Mg^{2+}. Note that $Co(NH_3)_6^{3+}$ would not be able to bind to the 5′-phosphate as seen in the X-ray structure due to its intertness in ligand exchange. This is consistent with the biochemical data that $Co(NH_3)_6^{3+}$ is unable to support the second phase of metal-dependent catalysis. It should be noted that the Tb^{3+}-probing studies also suggested another cleavage site, 51G–53U in the Fx3–mini-helix RNA construct. Despite the fact that this cleavage site is not the same as the one which is occupied with unhydrated Mg^{2+} binding at the 5′-triphosphate, they are in close proximity in the three-dimensional space. The observed cleavage site, 51G–53U with the pronounced cleavage at 51G, consists of a constrained hairpin turn motif in the active form. Therefore, Tb^{3+} embedded in the 5′-triphosphate probably reaches to the 51G region and induces the specific cleavage as observed (G66 in Figure 3b). In conclusion, the Mg^{2+} binding sites as well as modes predicted by the

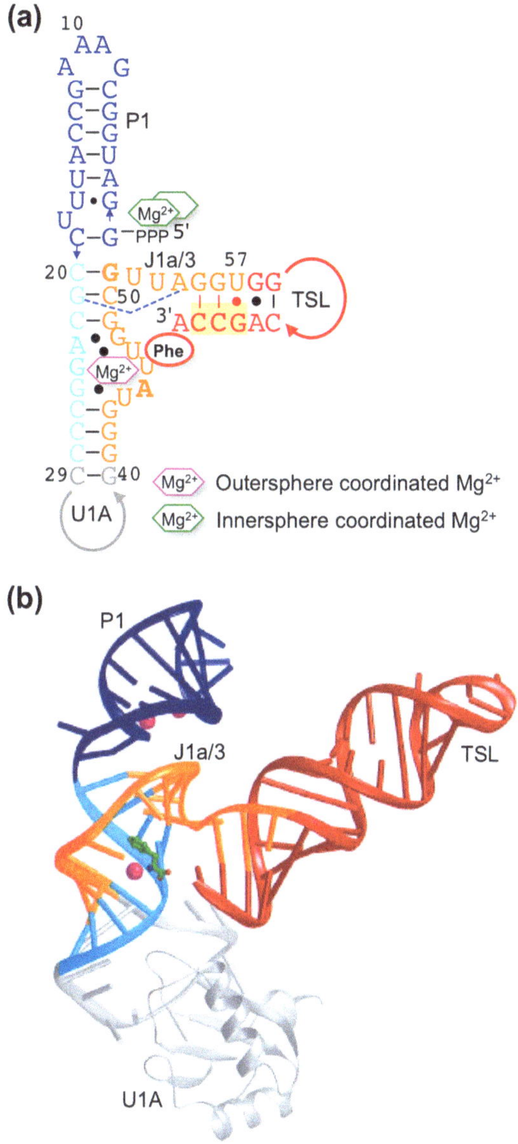

Figure 4. Overall structure of the flexizyme-minihelix RNA. (**a**) Sequence and secondary structure of the crystallization construct. Binding sites of well-ordered Mg^{2+} ions and of phenylalanine are indicated. The blue broken line indicates an A-minor interaction. TSL and U1A denote T stem-loop and U1A protein-binding loop. (**b**) Diagram of the three-dimensional structure. Phenylalanine is depicted in green, and Mg^{2+} ions in magenta. The regions with colors are consistent with the secondary structure shown in Figure 4a.

biochemical data are surprisingly consistent with those visualized in the X-ray structure.

3. REDOX RIBOZYMES

3.1. Significance of Redox Ribozymes in the RNA World

Oxidation-reduction (redox) reactions have a central role in metabolic chemistry in living organisms [54]. To the best of our knowledge, only protein enzymes are responsible for catalyzing redox reactions in modern metabolic processes, but many of such redox enzymes use ribonucleotide cofactors, such as nicotinamide adenine dinucleotide (NAD^+/NADH) and flavin adenine dinucleotide (FAD/$FADH_2$), to carry out hydride and electron transfer processes [55]. It has been postulated that these nucleotide cofactors could be molecular remnants of ancient life in the RNA world, which argues for the possibility of RNAs being responsible for the redox chemistry before the advent of protein enzymes [56,57]. Thus, the discovery of redox ribozymes would add not only a critical function to the catalytic repertoire of RNA but also give credence to the RNA world hypothesis. Despite the fact that there were a number of experiments for RNA aptamers that bind to the nucleotide redox cofactors [58], no redox ribozyme was known before our report. In fact, we reported for the first time a ribozyme that uses NAD^+/NADH to oxidize/reduce an alcohol/aldehyde. This ribozyme is referred to as ribox (ribozyme catalyzing redox reactions) [13,14].

Alcohol dehydrogenases (ADHs) are a family of the well-studied redox enzymes that catalyze the oxidation of alcohol to aldehyde. Representative ADHs use Zn^{2+}as a metal ion cofactor assisted by the coordination of two cysteines and one histidine in the active site (Figure 5) [59]. When the alcohol substrate binds to the enzyme active site, Zn^{2+} coordinates to the hydroxyl group that presumably lowers the pK_a value of the α-hydrogen (adjacent to the hydroxy group) and facilitates the hydride transfer reaction to NAD^+. Clearly, the precise positioning of the respective cofactors, Zn^{2+} and NAD^+, to the alcohol substrate is critical in order to promote the redox reaction.

3.2. Selection of Redox Ribozymes

One can imagine that an ancient redox ribozyme, if it had existed in the RNA world, used a similar or maybe exactly the same mechanism. In the context of the means of selection for redox ribozymes, two major challenges existed: (1) How could we design a RNA pool that was able to react with

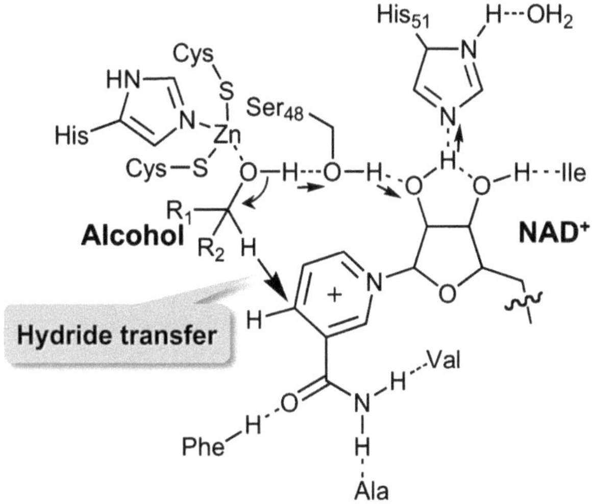

Figure 5. Schematic illustration of the catalytic site of alcohol dehydrogenase with a alcohol substrate, Zn^{2+} and NAD^+. The black arrows indicate the hydride transfer promoted by three proton relays.

three molecules, alcohol substrate, NAD^+, and Zn^{2+}? (2) How could we design the selection procedure to isolate active ribozymes that catalyze hydride transfer? Because in most selections a strategy was used to select self-modifying species, a reacting substrate was designed with an affinity tag and thus, active sequences could be fished out directly by means of an appropriate affinity purification. On the other hand, the hydride itself that is transferred by redox catalysis would not be directly taggable, and the change in molecular mass between the substrate and product is only 2 Da. Therefore, the challenge was how to distinguish the product, i.e., the self-modifying ribozyme, from inactive species in the pool.

We achieved the selection of redox ribozymes by a procedure involving two unique strategies. First, an alcohol substrate, a benzyl alcohol, was chosen in our study and attached by chemical ligation between its 2-bromoacetoamide moiety and a RNA pool that had a 5′-phosphothioate. Second, when the redox catalysis took place from the 5′-benzyl alcohol to benzaldehyde, the aldehyde moiety was tagged by biotin-hydrazide to yield a covalent hydrazone bond (Figure 6a). Our preliminary experiments showed that both steps went nearly quantitatively, so that the procedure was suitable to select ribozymes. We thus performed a selection of redox ribozymes in the presence of 0.5 mM Zn^{2+}, 100 mM Mg^{2+}, and 100 mM NAD^+ as cofactors. We found several active sequences, indicating that RNAs are capable of catalyzing redox chemistry. One of the clones showing the highest activity

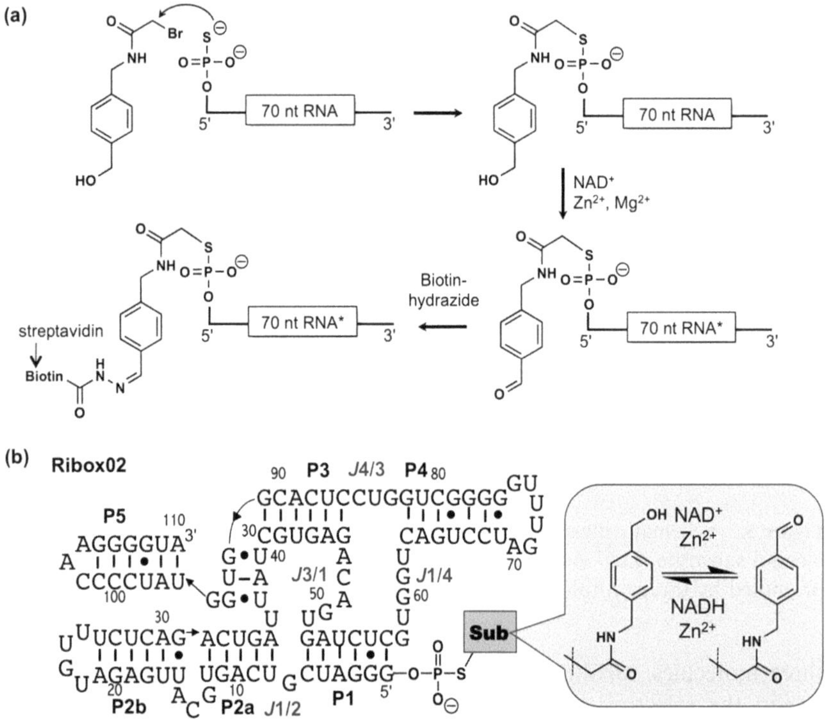

Figure 6. Selection and secondary structure of redox ribozymes. (**a**) Selection strategy for the isolation of redox ribozymes. The pool RNA consists of a random region of 70 nucleotide (nt) flanked by constant regions of 20 nt at the 5′- and 3′-ends. The pool RNA is synthesized with guanosine-5′-monophosphorothioate (GMPS) at the 5′-end, and then reacted with a benzyl alcohol derivative. After incubating this pool with NAD^+, Zn^{2+}, and Mg^{2+}, active RNAs are chemoselectively modified with biotin hydrazide. Subsequently, the biotin-tagged RNAs are captured on resin-immobilized streptavidin. (**b**) The proposed secondary structure of ribox02.

and representing the most abundant clone, was referred to as ribox02 (Figure 6b) and chosen for further study. Ribox02 gives the rate acceleration over the background with at least seven orders of magnitude, indicating that it is one of the most prominent artificial ribozymes evolved *in vitro* to date.

A series of biochemical studies, including nuclease and chemical mapping along with compensatory mutations, on the stems in ribox02 determined its secondary structure to be a pseudoknot (Figure 6b). Two catalytically unnecessary stem regions, P2b and P5, were predicted and substituted with tetraloops to yield a simplified ribox02, consisting of 75 nucleotides, without any loss of the parental ribozyme activity (Figure 7a). More recently, a series

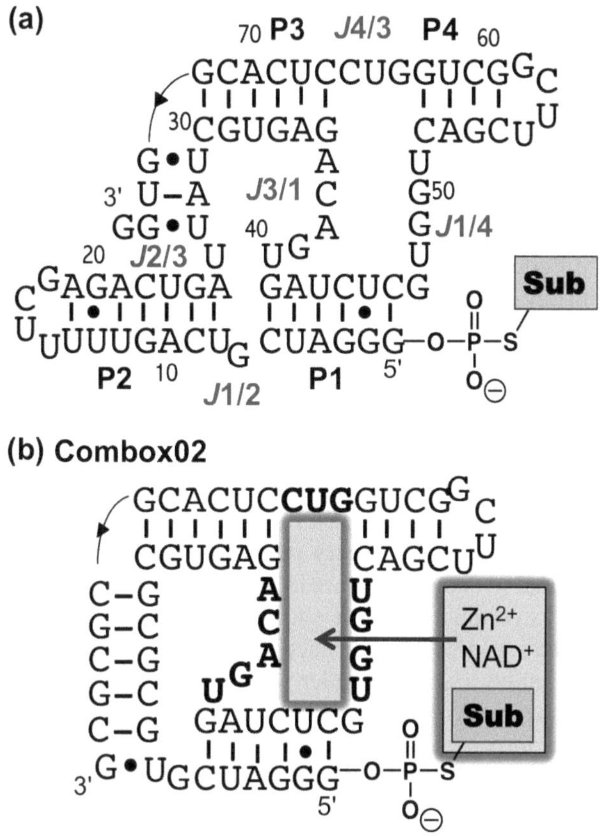

Figure 7. (**a**) The proposed secondary structure of simplified ribox02. (**b**) The proposed secondary structure of combox02. The gray box encompassed by bold bases indicates the catalytic domain of combox02 that presumably interacts with the alcohol substrate and cofactors.

of mutations and structural minimization was performed to yield a very compact ribox02, referred to as combox02 (Figure 7b). This simple ribozyme essentially gives the same catalytic activity as the parental ribox02, and thus this construct would provide a suitable construct for X-ray structural analyses in the future.

3.3. Metal Dependence of Ribox02

Since the selection was performed in the presence of two metal ions, Zn^{2+} and Mg^{2+}, it was curious to see whether ribox02 intrinsically relies on both metal ions for its function (Figure 8a). Interestingly, the catalytic activity of ribox02 depends on both metal ions in order to exhibit the full redox activity; however, the removal of Mg^{2+} partly eliminated the activity whereas that of Zn^{2+} completely eliminated the activity (Figure 8a, lanes 7 and 8). This observation suggests that Mg^{2+} is not essential for catalytic activity, probably rather playing a structural role; however, Zn^{2+} is a catalytic ion that has a role in catalysis similar to the protein counterpart.

Despite the fact that we tried to map out the structure with Tb^{3+}-mediated cleavage, Tb^{3+} competing with Zn^{2+} or Mg^{2+} or both, we did not observe a clear competition and therefore failed to identify high-affinity metal binding sites. Since the ionic radius of Zn^{2+} is similar to that of Tb^{3+} (0.74 Å Zn^{2+} versus 0.92 Å Tb^{3+}) [60], it could be regarded as being similar to Mg^{2+}. However, one possible reason is that both Tb^{3+} and Mg^{2+} have an octahedral coordination geometry in the hydrated form, whereas Zn^{2+} tends to have a tetrahedral coordination geometry [61]. Thus, the metal ion binding site for Zn^{2+} in ribox02 may be quite different from that for Mg^{2+} in other known ribozymes. Another observation has been made, namely that the catalytic function of ribox02 is very sensitive to pH changes (unpublished data). The optimal pH conditions are in a very narrow pH range at 7.5 where the selection was conducted, while shifting the pH half a unit up or down dramatically reduces the catalytic activity with one order of magnitude. More detailed analyses are still under way, but ribox02 clearly represents a very unique ribozyme with metal ion and pH dependence. Particularly, it is curious to see if ribox uses Zn^{2+} as catalytic metal ion that coordinates to the hydroxyl group on the benzyl alcohol substrate and facilitates the hydride transfer to NAD^+ like protein enzymes (Figure 8b). In combination with the X-ray structural analysis, such a mechanism can be elucidated in the near future.

Figure 8. (**a**) Redox activity of ribox02. Lane 1, product positive control using ribox02 modified with substrate 2; lanes 2 and 3, oxidation reaction catalyzed by ribox02; lanes 4 and 5, negative controls for lane 3 where biotinylation-dependence of gel shift is shown; lanes 6–8, negative controls for lane 3 where the individual components were removed from the reaction; lane 9, a negative control using ribox02 modified with a non-oxidative toluene derivative (substrate 3), indicating the necessity of the hydroxyl group for oxidation. A, Biotinylated-substrate 2-ribox02 complexed with streptavidin; B, unreacted substrate 1-ribox02; ND, not detected. (**b**) Schematic illustration of the proposed catalytic site of combox02 with the benzyl alcohol substrate, Zn^{2+} and NAD^+. Bases in the putative catalytic core (highlighted in gray box) that interact with the substrates and cofactors are shown in bold. The black arrow indicates the hydride transfer.

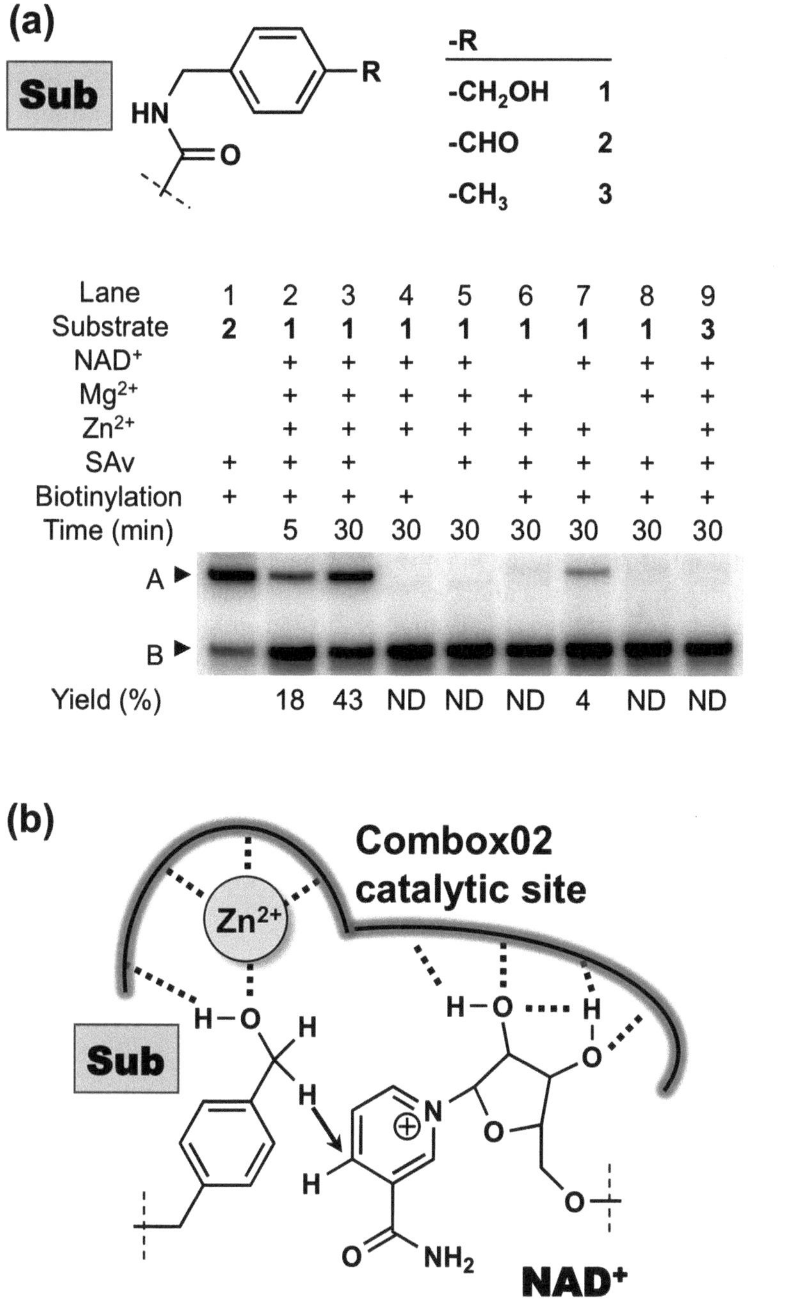
(a)
Sub
HN
O
-R
-R
-CH2OH 1
-CHO 2
-CH3 3
Lane 1 2 3 4 5 6 7 8 9
Substrate 2 1 1 1 1 1 1 1 3
NAD+
Mg2+
Zn2+
SAv
Biotinylation
Time (min) 5 30 30 30 30 30 30 30
A
B
Yield (%) 18 43 ND ND ND 4 ND ND
(b)
Combox02
catalytic site
Zn2+
Sub
H–O
H
H
N
O
H–O
H
O
O
H
O
NH2
NAD+

4. CONCLUSIONS

The metal-dependent titration and Tb^{3+}/Mg^{2+} probing studies on pre-24 have postulated that the flexizyme domains have high and low affinity metal binding sites in which the respective site can be occupied by outersphere and innersphere metal ions, respectively. To our great surprise, the X-ray structural analysis of Fx3–mini-helix RNA complexed with a hydrolytically stable substrate analog revealed that the postulation was mostly correct. Thus, the biochemical experiments pin down the possible metal binding sites and the number and coordination mode of metal ions involved in catalysis. Most importantly, flexizymes use Mg^{2+} effectively to create the cavity for the incoming substrate though Mg^{2+} does not play a direct role in interacting with the substrate. Moreover, two Mg^{2+} ions bound to the low affinity pocket at the 5′-triphosphate presumably support the constrained structure in the active form of the flexizyme and thus likely contribute to further acceleration in catalysis.

A similar approach for the investigation of a redox ribozyme did not yield fruitful information of the putative Zn^{2+} binding sites, presumably due to the difference in coordination geometry of Zn^{2+} and Tb^{3+}. In this case, we still need to wait for the visualization of Zn^{2+} in an X-ray crystal structure of the redox ribozyme, but having a compact construct of combox02 should facilitate such studies. Hopefully, these will reveal a critical role of Zn^{2+} in the ribozyme catalysis similar to the protein ADHs, and thus expand our knowledge of metal ion catalysis occurring in ribozymes.

Nevertheless, the complete studies performed on flexizymes represent a showcase for the importance of metal ion-dependent biochemical studies on artificial ribozymes coupled with X-ray analysis. We expect that accumulating knowledge generated from such studies will lead to a better understanding of the capability of metal ion catalysis in ribozymes.

ACKNOWLEDGMENTS

This work was supported by grants of the Japan Society for the Promotion of Science Grants-in-Aid for Scientific Research (S) (16101007) and Specially Promoted Research (21000005) to H.S.

ABBREVIATIONS

ADH	alcohol dehydrogenase
AMP	adenosine 5′-monophosphate

ARS	aminoacyl-tRNA synthetase
ATP	adenosine 5′-triphosphate
CBT	4-chlorobenzyl thioester
CMA	cyanomethyl ester
combox02	compact ribox02
DBE	3,5-dinitrobenzyl ester
dFx	dinitro flexizyme
EDTA	ethylenediamine-*N,N,N′,N′*-tetraacetate
eFx	enhanced flexizyme
FAD	flavin adenine dinucleotide
$FADH_2$	flavin adenine dinucleotide (reduced)
Fx3	flexizyme consisting of 45 nucleotides
NAD	nicotinamide adenine dinucleotide
NADH	nicotinamide adenine dinucleotide (reduced)
nt	nucleotide
ribox	artificial ribozyme that catalyzes a redox reaction
SELEX	systematic evolution of ligands by exponential enrichment
tRNA	transfer RNA
TSL	T stem loop

REFERENCES

1. A. M. Pyle, *J. Biol. Inorg. Chem.*, 2002, **7**, 679–690.
2. R. K. O. Sigel and A. M. Pyle, *Chem. Rev.*, 2007, **107**, 97–113.
3. X. Chen, N. Li and A. D. Ellington, *Chem. Biodivers.*, 2007, **4**, 633–655.
4. P. A. Lohse and J. W. Szostak, *Nature*, 1996, **381**, 442–444.
5. A. Jenne and M. Famulok, *Chem. Biol.*, 1998, **5**, 23–34.
6. N. Li and F. Huang, *Biochemistry*, 2005, **44**, 4582–4590.
7. H. Saito, D. Kourouklis and H. Suga, *EMBO J.*, 2001, **20**, 1797–1806.
8. G. Sengle, A. Eisenfuhr, P. S. Arora, J. S. Nowick and M. Famulok, *Chem. Biol.*, 2001, **8**, 459–473.
9. T. M. Tarasow, S. L. Tarasow and B. E. Eaton, *Nature*, 1997, **389**, 54–57.
10. B. Seelig and A. Jaschke, *Chem. Biol.*, 1999, **6**, 167–176.
11. Y. Ryu, K. J. Kim, C. A. Roessner and A. I. Scott, *Chem. Commun.*, 2006, 1439–1441.
12. S. Fusz, A. Eisenfuhr, S. G. Srivatsan, A. Heckel and M. Famulok, *Chem. Biol.*, 2005, **12**, 941–950.
13. S. Tsukiji, S. B. Pattnaik and H. Suga, *Nat. Struct. Biol.*, 2003, **10**, 713–717.
14. S. Tsukiji, S. B. Pattnaik and H. Suga, *J. Am. Chem. Soc.*, 2004, **126**, 5044–5045.
15. T. Pan and O. C. Uhlenbeck, *Biochemistry*, 1992, **31**, 3887–3895.
16. H. Murakami, H. Saito and H. Suga, *Chem. Biol.*, 2003, **10**, 655–662.
17. H. Murakami, A. Ohta, H. Ashigai and H. Suga, *Nat. Methods*, 2006, **3**, 357–359.
18. H. Saito, K. Watanabe and H. Suga, *RNA*, 2001, **7**, 1867–1878.

19. H. Saito and H. Suga, *Nucleic Acids Res.*, 2002, **30**, 5151–5159.
20. H. Xiao, H. Murakami, H. Suga and A. R. Ferré-D'Amaré, *Nature*, 2008, **454**, 358–361.
21. M. Illangasekare, G. Sanchez, T. Nickles and M. Yarus, *Science*, 1995, **267**, 643–647.
22. M. Illangasekare and M. Yarus, *J. Mol. Biol.*, 1997, **268**, 631–639.
23. M. Illangasekare, O. Kovalchuke and M. Yarus, *J. Mol. Biol.*, 1997, **274**, 519–529.
24. M. Illangasekare and M. Yarus, *Proc. Natl. Acad. Sci. USA*, 1999, **96**, 5470–5475.
25. M. Illangasekare and M. Yarus, *RNA*, 1999, **5**, 1482–1489.
26. H. Saito and H. Suga, *J. Am. Chem. Soc.*, 2001, **123**, 7178–7179.
27. P. R. Effraim, J. Wang, M. T. Englander, J. Avins, T. S. Leyh, R. L. Gonzalez, Jr. and V. W. Cornish, *Nat. Chem. Biol.*, 2009, **5**, 947–953.
28. T. Kawakami, H. Murakami and H. Suga, *Chem. Biol.*, 2008, **15**, 32–42.
29. Y. Sako, J. Morimoto, H. Murakami and H. Suga, *J. Am. Chem. Soc.*, 2008, **130**, 7232–7234.
30. Y. Goto and H. Suga, *J. Am. Chem. Soc.*, 2009, **131**, 5040–5041.
31. E. Nakajima, Y. Goto, Y. Sako, H. Murakami and H. Suga, *ChemBioChem*, 2009, **10**, 1186–1192.
32. Y. Goto, A. Ohta, Y. Sako, Y. Yamagishi, H. Murakami and H. Suga, *ACS Chem. Biol.*, 2008, **3**, 120–129.
33. Y. Goto, H. Murakami and H. Suga, *RNA*, 2008, **14**, 1390–1398.
34. Y. Yamagishi, H. Ashigai, Y. Goto, H. Murakami and H. Suga, *ChemBioChem*, 2009, **10**, 1469–1472.
35. Y. Goto, K. Iwasaki, K. Torikai, H. Murakami and H. Suga, *Chem. Commun.*, 2009, 3419–3421.
36. J. B. Murray, A. A. Seyhan, N. G. Walter, J. M. Burke and W. G. Scott, *Chem. Biol.*, 1998, **5**, 587–595.
37. A. Hampel and J. A. Cowan, *Chem. Biol.*, 1997, **4**, 513–517.
38. S. Nesbitt, L. A. Hegg and M. J. Fedor, *Chem. Biol.*, 1997, **4**, 619–630.
39. K. J. Young, F. Gill and J. A. Grasby, *Nucleic Acids Res.*, 1997, **25**, 3760–3766.
40. E. A. Curtis and D. P. Bartel, *RNA*, 2001, **7**, 546–552.
41. J. L. O'Rear, S. Wang, A. L. Feig, L. Beigelman, O. C. Uhlenbeck and D. Herschlag, *RNA*, 2001, **7**, 537–545.
42. A. Flynn-Charlebois, N. Lee and H. Suga, *Biochemistry*, 2001, **40**, 13623–13632.
43. J. M. Zhou, D. M. Zhou, Y. Takagi, Y. Kasai, A. Inoue, T. Baba and K. Taira, *Nucleic Acids Res.*, 2002, **30**, 2374–2382.
44. S. Nakano, D. J. Proctor and P. C. Bevilacqua, *Biochemistry*, 2001, **40**, 12022–12038.
45. J. H. Cate and J. A. Doudna, *Structure*, 1996, **4**, 1221–1229.
46. J. S. Kieft and I. Tinoco, Jr., *Structure*, 1997, **5**, 713–721.
47. N. Lee and H. Suga, *Biochemistry*, 2001, **40**, 13633–13643.
48. A. L. Feig, W. G. Scott and O. C. Uhlenbeck, *Science*, 1998, **279**, 81–84.
49. A. L. Feig, M. Panek, W. D. Horrocks, Jr. and O. C. Uhlenbeck, *Chem. Biol.*, 1999, **6**, 801–810.

50. A. Vaidya and H. Suga, *Biochemistry*, 2001, **40**, 7200–7210.
51. M. R. Hargittai and K. Musier-Forsyth, *RNA*, 2000, **6**, 1672–1680.
52. N. G. Walter, N. Yang and J. M. Burke, *J. Mol. Biol.*, 2000, **298**, 539–555.
53. R. K. O. Sigel, A. Vaidya and A. M. Pyle, *Nat. Struct. Biol.*, 2000, **7**, 1111–1116.
54. J. M. Berg, J. L. Tymoczko and L. Stryer, *Biochemistry*, W. H. Freeman, New York, 2002.
55. H. B. Gray and J. W. R. Ellis, in *Bioinorganic Chemistry*, eds. I. Bertini, H. B. Gray, S. J. Lippard and J. S. Valentine, University Science Books, Sausalito, California, USA, 1994, pp. 315–363.
56. H. B. White, 3rd, *J. Mol. Evol.*, 1976, **7**, 101–104.
57. S. A. Benner, A. D. Ellington and A. Tauer, *Proc. Natl. Acad. Sci. U S A*, 1989, **86**, 7054–7058.
58. R. Stoltenburg, C. Reinemann and B. Strehlitz, *Biomol. Eng.*, 2007, **24**, 381–403.
59. S. Hammes-Schiffer and S. J. Benkovic, *Annu. Rev. Biochem.*, 2006, **75**, 519–541.
60. R. D. Shannon, *Acta. Crystallogr. A*, 1976, **32**, 751–767.
61. T. Dudev and C. Lim, *J. Am. Chem. Soc.*, 2000, **122**, 11146–11153.

Met. Ions Life Sci. **2011**, *9*, 299–345

11

Metal Ion Binding and Function in Natural and Artificial Small RNA Enzymes from a Structural Perspective

Joseph E. Wedekind

Department of Biochemistry & Biophysics and Center for RNA Biology, University of Rochester School of Medicine and Dentistry, Rochester, NY 14642, USA
<joseph.wedekind@rochester.edu>

Metal Ions in Life Sciences, Volume 9 Edited by Astrid Sigel, Helmut Sigel, and Roland K. O. Sigel

Published by the Royal Society of Chemistry, www.rsc.org DOI: 10.1039/978184973251200299

ABSTRACT: Ribozymes are often perceived as part of an antiquated catalytic arsenal hearkening back to a pre-biotic RNA World that was eventually supplanted by proteins. However, recent genome-wide searches have revealed a plethora of new catalytic RNA motifs that appear to be variations on well-known themes. This suggests that ribozymes have continued to evolve in order to fulfill specific, RNA-essential biological niches. Although such ribozymes are small and catalyze one-step phosphodiester-bond scission reactions, ongoing structure and function analyses at the lab bench have demonstrated that RNA has the capacity for a diverse number of reactions such as carbon-carbon bond formation, and tRNA aminoacylation. Here we describe the fundamental structure and metal binding properties of four naturally occurring RNA enzymes: the hammerhead, hairpin, hepatitis delta virus, and *glmS* metabolite sensing ribozyme. In addition, we discuss the fold and ion coordination of three artificial ribozymes developed to probe the boundaries of RNA catalysis; these include the leadzyme, the flexizyme, and the Diels-Alder ribozyme. Our approach is to relate structure to function with the knowledge of ideal metal-ion coordination geometry that we have derived herein from surveys of high-resolution small molecule structures. An emergent theme is that natural and artificial ribozymes that catalyze single-step reactions often possess a pre-formed active site. Multivalent ions facilitate RNA active site formation, but can also provide Lewis acid functionality that is necessary for catalysis. When metal ion binding isn't possible, ribozymes make due by ionizing their bases, or by recruiting cofactors that augment their chemical functionality.

KEYWORDS: acid/base catalysis · anomalous diffraction · artificial ribozyme · cobalt hexammine · coordination geometry · Lewis acid · magnesium · manganese · pK_a shifting · phosphoryl transfer · Raman crystallography · small ribozyme

1. INTRODUCTION

A central question in the field of RNA enzymes is how a relatively simple acidic polymer with no groups that ionize at neutral pH (Figure 1A) is capable of surviving evolution, and flourishing as a catalyst of modern biological reactions [1]. The importance of this question is multifold when one considers that all life depends on catalytic RNA to decode the genetic material [2], and with one exception [3,4], to process the 5′-ends of pre-tRNAs [5]. From a global perspective, the chemistry of phosphoryl transfer is one of the most important and pervasive in nature and is ubiquitous in all kingdoms of life [6]. Despite this significance, RNA catalysis is often written off as a molecular relic that has been supplanted in modern biology by protein enzymes or protein-RNA hydrids that are remnants of a bygone world [7]. With this in mind, the approach of this chapter will be to convince the non-specialist that ribozymes are not only important to understand the origins of life, but also to expand our understanding of modern biology, human health, and pathways that lead to disease.

On the road to understanding RNA catalysis, investigators have discovered a plethora of new ribozymes in the biosphere, a richness of RNA three-dimensional structural diversity, and the invaluable contribution of divalent cations for folding and catalysis. Due to the enormous scope of this topic, the approach of this chapter is to emphasize structure-function relationships for some of the most well studied ribozymes. In all instances, we will accentuate the interplay between elegant three-dimensional folds and the binding of ions. When appropriate, we will present current kinetic and functional information to inform the reader on the mechanism of action of our representative ribozymes. Finally, we will conclude with a discussion of the commonalities of natural and artificial ribozymes, as well as thoughts on what the future may bring to this field.

1.1. The Diverse Activities of Catalytic RNA: From Bench to Genome

On the road to understanding the catalytic principles of RNA, researchers have discovered a variety of natural, as well as artificial catalysts that have helped define the boundaries of the field, as well as the origins of life. The chemical activities of such ribozymes include but are not limited to: autolytic cleaving and splicing [8], hydrolysis [9], peptide bond formation [10], tRNA aminoacylation [11,12], Diels-Alder chemistry [13], RNA-dependent RNA polymerization [14,15], redox activity [16], and a myriad of other reactions [17] that support the feasibility of a pre-biotic RNA World [18]. To propel

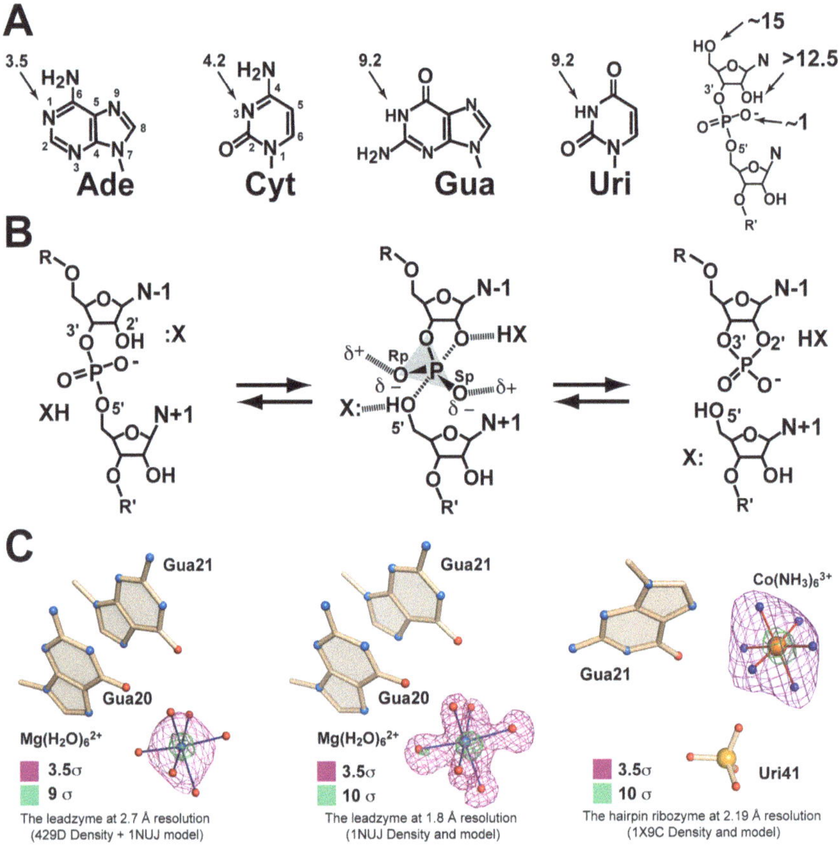

Figure 1. Schematic diagrams of the common RNA bases and their properties. (**A**) The structure and ionization points of the four RNA bases at the N1 or N3 imino nitrogens from [23]. RNA appears to have little capacity for biological activity given the base pK_a values and phosphate backbone ionization at pH ca. 1. (**B**) The chemical transesterification reaction of small ribozymes. The transition state requires 'in-line' geometry in which the 2′-nucleophile, electrophilic phosphorus and O5′-leaving group exhibit an angle of 180° (reviewed in [227]). Conservation of bond energy in the cyclic-2′,3′-phosphodiester allows the reaction to be reversible. (**C**) Representative $F_o - F_c$ omit electron density maps for ions of the leadyzme and hairpin ribozyme from the author's lab. This comparison illustrates the problem of metal ion coordination sphere assignment at various resolutions. PDB codes for each ball-and-stick model and the accompanying structure factor amplitudes, $|\mathbf{F}(hkl)|$, are indicated.

our understanding of these molecules to greater heights, many of these ribozymes have been subjected to structure-function investigations. By far the most appealing and tractable targets for analysis are those ribozymes that are relatively small and can be produced in milligram quantities by T7 polymerase *in vitro* transcription, which has facilitated kinetic, biochemical, synthetic organic, spectroscopic, and crystallographic approaches. Here, we will focus on relationships between "small" ribozyme structure and function. We will begin by looking at small ribozyme family members that catalyze a simple, reversible bond scission reaction (Figure 1B); see also Chapter 6 of this volume for a discussion of metals in relation to folding and catalysis. Subsequently, we will examine three representative artificial ribozymes that catalyze diverse biological reactions. In all cases we will explore how divalent cation coordination imparts structural stability and catalysis with the goal of identifying common traits that define principles of functionality.

Although numerous natural ribozymes have been discovered, it is useful to classify them according to size. This naming system recognizes two general classes called "large" and "small". The large ribozymes include self-splicing introns and RNase P (see also Chapter 7 of this volume). Members of the small ribozyme family include naturally occurring molecules of ~ 150 nucleotides such as the hammerhead, hairpin, VS, HDV, and *glmS* that catalyze a phosphodiester bond transesterification reaction within a cognate substrate (reviewed in [19]). Recently a variety of heretofore undetected hammerhead and HDV-like motifs were identified throughout the genomes of a diverse number of organisms in the biosphere, suggesting that catalytic RNA is much more abundant than originally suspected [20,21]. Moreover, these findings suggest that catalytic RNA motifs evolved independently in multiple contexts to fulfill particular biological niches requiring site-specific phosphoryl transfer [22], which is highly suited to ribozyme function.

1.2. The Chemical Reaction of Small Ribozymes and Use of Cofactors

The chemical reaction of small ribozymes involves activation of a 2′-OH, which serves as a nucleophile that attacks an electrophilic phosphorus (Figure 1B). The reaction proceeds via a trigonal-bipyramidal transition state in which the non-bridging oxygens of the phosphorane develop negative charge. The 5′-oxyanion leaving group must receive charge neutralization by coordination to an acid group, such as a general acid, or a Lewis base (reviewed in [23]). A major difference between small ribozymes and their large cousins is that the former use nucleobase catalysis to achieve chemical functionality [24] and can be catalytically proficient in monovalent ions alone, which are required for folding [25]. In contrast, large ribozymes require divalent ions for metal-mediated folding and catalysis. Such ions are often an integral part of the large ribozyme active site serving roles in nucleophile activation, electrostatic stabilization of the transition state, and leaving group neutralization (see Chapter 7 of this volume).

An important chemical twist in RNA biology that links the function of small ribozymes to their non-natural counterparts is the use of cofactors or small molecules to enhance chemical functionality. Whereas researchers had always envisioned that primordial ribozymes were capable of recognizing small molecules (e.g., to charge tRNA with amino acids [26]), the relationship between chemical diversity and biological functionality took a quantum leap with the discovery that the *glmS* ribozyme requires the metabolite glucosamine-6-phosphate for its biochemical activity (Section 3.4). This observation established a precedent that catalytic RNA can acquire functionality in nature by recognizing non-nucleotide, small molecule metabolites with specificity equal to or exceeding protein enzymes that utilize vitamins as catalytic cofactors. Moreover, the existence of distinct classes of riboswitches that recognize one of more than a dozen distinct small molecules (see Chapter 5 of this volume) – including *S*-adenosylmethionine, thiamine pyrophosphate (B_1), adenosyl cobalamin (B_{12}), and flavin mononucleotide (B_2) – provides a tenuous but plausible connection between modern biology, and a pre-biotic RNA World dominated by small ribozymes with an enhanced, vitamin-based chemical repertoire [27]. By comparing the structural and functional attributes of extant small ribozymes with the rich diverseness of those produced in the lab, we aspire to develop a broader perspective on current biological systems, and to envision new natural functions that have yet to be described outside the lab, but may be hidden in the plethora of non-coding RNA sequences that continue to accrue at an astonishing rate; for examples see [28–30].

2. EXPECTATIONS OF METAL BINDING AND CRYSTALLOGRAPHIC OBSERVATIONS

Before we begin, it is useful to ask, *how do site-bound divalent ions interact with RNA?* and, *what properties make divalent ions suitable for use by catalytic RNA?* Herein we will emphasize Mg^{2+} due to its prominence in living organisms, detectable at $\sim$2.5 mM in red blood cells and 0.8 mM in plasma [31]. In terms of RNA association, Draper has categorized ion binding as thermodynamically localized at specific regions or "site-bound" with kinetic lability giving rise to "diffuse" or "delocalized" properties [32]; (see also Chapter 3 of this volume). Site-bound ions can be fully hydrated such as hexaaqua magnesium, or partly to completely dehydrated. In general, RNA attracts a cloud of diffusely bound positive ions that associate with the backbone and exhibit disassociation constants characteristic of highly dynamic interactions [33,34]. Ions such as Mg^{2+} have long been known to contribute both structural stability as well as chemical functionality [35–37]. In addition to neutralizing pockets of negative charge arising from complex RNA backbone folding, metal ions have the ability to depress the acid dissociation constants of bound waters or other ligands within their primary or secondary coordination spheres. For example, Mg^{2+} can lower the pK_a of a bound water from 15.4 to 11.4 [38]. This has the effect of increasing the amount of localized hydroxide ion (a specific base). Direct alcohol-to-metal coordination can facilitate the deprotonation of hydroxyl groups for nucleophilic attack, which may entail proton transfer to a nearby base (Figure 1B, denoted as :X). Further discussion of such Lewis acid properties is presented below in the context of RNA catalysis.

2.1. The Use of Crystallography to Detect Metal Ions

Although there are many ways to detect site-bound ions in RNA (see Chapter 2 of this volume), the approach in this chapter will be to analyze metal binding sites visualized in electron density maps by X-ray diffraction (Figure 1C). Since the degree of X-ray diffraction is proportional to the number of electrons in the scattering element, metal ions can be especially prominent due to their electron richness relative to lighter atoms such as C, N, or O, which together with P, have an average scattering factor of 7.3 e^- per atom [39]. Although Mg^{2+} itself has only 10 e^- (the same as H_2O), a hydrated metal may appear especially electron-dense even at low resolution due to the contribution of innersphere ligands such as water (e.g., $Mg(H_2O)_6^{2+}$, which has 70 e^-). Ions such as Co^{3+} and Mn^{2+} are electron-dense on their own with 24 e^- and 22 e^-, respectively; these may appear quite prominent in electron density maps (Figure 1C). In addition to inferring atom identity by diffraction signal (which may not work due to partial binding occupancy), the crystallographer relies on the expected coordination geometry, and can employ the 'resonance' X-ray absorption properties of a metal. The application of X-rays to a crystal can elicit a measureable anomalous diffraction signal when the incident radiation is close to an elemental energy absorption edge; we will discuss this more in Section 3.2. The effect is a change in the scattering for the anomalously absorbing atom that can be measured from the diffraction pattern as an intensity difference, or anomalous difference (sometimes called f'' in terms of the scattering factor correction; see [39]). Tables of anomalous-scattering corrections are reported in the International Tables for Crystallography [40].

2.2. Ideal Metal Coordination Geometry from Small Molecule Structures

A shortcoming of RNA crystallography is that the spacing of the planes of atoms in the crystal (separated by a distance d_{min}) gives rise to diffraction intensities that are often insufficient to visualize the ligands within the coordination sphere of an electron density map (Figure 1C); conceptually, this is the definition of crystallographic resolution. When taken at face value, this would appear to impede the construction of an accurate three-dimensional atomic model. However, much of macromolecular crystallography relies upon assumptions of geometry and stereochemical sensibility derived from other experiments. Such restraints or constraints are imposed on the atomic model during refinement. In the case of metal ions, much knowledge can be incorporated from the known atomic-resolution structures of small molecules deposited into the CSD [41]. In this section, we conduct structural surveys to tabulate ideal geometries measured for various cation complexes (Mg^{2+}, Co^{3+}, and Mn^{2+}). In the ensuing discussions of ribozymes, we will leverage this information to infer that the reported ions should adopt similar "ideal" geometry with the caveat that inner- or outersphere ligands may be placed incorrectly in the available ribozyme models unless structures are refined with data better than 2.0 Å resolution (Figure 1C). As a further heuristic, we note that some site-bound ions and their ligands are dynamic, which means that some portion of the coordination sphere model may never be observed in electron density maps. This property is not uncommon since RNA crystals can have solvent contents as high as 70%.

2.2.1. *Mg(II) Coordination Geometry*

The Mg^{2+} ion prefers hexa-coordinate ligand geometry and $Mg(H_2O)_6^{2+}$ adopts nearly perfect octahedral interactions (Figure 1C). A survey of the CSD using ConQuest Version 1.12 produced 72 atomic-resolution crystal structures with hexaaqua Mg^{2+} (chosen from structures with $R_{factors}$ below 5%). The average Mg-to-oxygen distance for water molecules was 2.064 Å±0.025 Å. The average angle for apical to equatorial oxygen-Mg-oxygen measurements was 90.01°±2.11°. In some instances, heptacoordinate Mg^{2+} was observed, which distorts the octahedron. Crown ethers were typical in heptacoordinate cases in which the ether oxygens reside in an equatorial coordination belt of a pentagonal bipyramid. Mixed coordination spheres comprising innersphere oxygen and nitrogen ligands were searched for as well. Nitrogen ligands were contributed frequently in the context of a multidentate chelator, causing the resulting Mg^{2+} coordination sphere to be distorted from octahedral geometry. Apical-to-equatorial angles of such octahedra had average values of 90.20°±9.10° with minimum and maximum values of 72° and 126.6°, respectively. The average Mg-to-oxygen distance was 2.105 Å±0.074 Å, and the Mg-to-nitrogen ligand distances were slightly longer at 2.242 Å±0.94 Å. Such small, ligand-dependent changes in octahedral geometry are likely to be imperceptible in the context of most RNA crystal structures. These values are pertinent to the HDV, *glmS*, leadzyme, flexizyme, and Diels-Alder ribozymes (Sections 3.3 to 3.7).

2.2.2. *Mn(II) Coordination Geometry*

The Mn^{2+} ion can also adopt hexaaqua complexes, but we will restrict our consideration to the two cases embodied by the small ribozymes of this chapter. The first example is a manganese complex that comprises six innersphere oxygen ligands. In the second case we will consider penta-oxo complexes with nitrogen as the sixth ligand in an octahedral coordination sphere (e.g., N7 of a purine). In both instances the Mn^{2+} formal charge has been reduced to Mn^{+}, resulting from innersphere contact with an anionic oxygen ligand, such as the phosphate backbone of the RNA. For the all-oxygen Mn^{+} case, the average Mn-oxygen ligand distance is 2.141 Å±0.108 Å. The oxygen-Mn^{+}-oxygen apical-to-equatorial angles are 90.00°±4.763°, with minimum and maximum values of 76.0° and 104.5°, respectively. When a nitrogen atom replaces one of the oxygen ligands, the average Mn-oxygen distance is 2.074 Å±0.156 Å, whereas the average Mn-nitrogen distance is 2.110 Å±0.133 Å. Angles of the octahedron have an average value of 90.072°±6.305° with minimum and maximum values of 71.8° and 126.9°, respectively. These values are relevant for the hammerhead ribozyme discussed in Section 3.1.

2.2.3. *Cobalt(III) Hexammine Coordination Geometry*

Although $Co(NH_3)_6^{3+}$ is not an exact mimic of $Mg(H_2O)_6^{2+}$ it exhibits a comparable size, coordination geometry (Figure 1C), and hydrogen bonding capabilities [42]. Perhaps the most notable difference is that the innersphere coordination is relatively inert in terms of the exchange rates of the amine groups relative to water in hexaaqua Mg^{2+} [42]. Therefore, although $Co(NH_3)_6^{3+}$ is not a physiologically relevant ion, it is a first-choice reagent to probe the importance of innersphere metal ion coordination in ribozyme function. Indeed, the hairpin ribozyme represents an example in which the use of cobalt hexammine revealed the absence of direct ion-facilitated chemistry in the reaction (Section 3.2). This observation was corroborated subsequently by structural analysis [43]. Although undetectable by crystallography, significant RNA conformational differences have been reported in solution when $Co(NH_3)_6^{3+}$ is used for hairpin ribozyme folding in lieu of Mg^{2+} [44]. Such differences may be due to the absence of hydrogen bond donation by the cobalt amine ligands – unlike water ligands, which donate as well as accept hydrogen bonds.

A search of the CSD for amine-liganded cobalt complexes with hexacoordinate nitrogen revealed >120 structures. A clear preference for octahedral coordination geometry was observed, which is similar to oxo-Mn^{2+} and oxo-Mg^{2+} complexes. Among the hexammine cobalt complexes identified, the average Co-nitrogen distance was 1.971 Å±0.041 Å. The average apical-to-equatorial N-Co-N angle of the octahedron was 90.020°±2.159°. In light of the results in Sections 2.2.1 and 2.2.2, this observation demonstrates why $Co(NH_3)_6^{3+}$ is a suitable isosteric mimic of hexaaqua Mg^{2+} and hexaaqua Mn^{2+}. These observations are relevant to the hairpin ribozyme (Section 3.2).

2.2.4. *Mimics of Cobalt(III) Hexammine for Heavy-Atom Phasing of RNA*

In the past decade a number of crystal structures have been phased by use of $Co(NH_3)_6^{3+}$ analogs. Most notably, $Os(NH_3)_6^{3+}$ has been reported for multi-wavelength phasing or as an outright heavy-atom derivative [45–47]. Much of the original $Os(NH_3)_6^{3+}$ utilized in these experiments was produced in the laboratory of Henry Taube, who described its synthesis in 1989 [48,49]. Taube's material seeded the phasing efforts or provided proof-of-principle for a variety of exciting molecules including the P4-P6 domain of the group I intron [50], the 30S ribosome [51], and an intact group I intron [52]. More recently, Kieft et al. systematically defined the contextual importance of specific G•U wobble-paired RNA sequences to optimize localization of site-bound $Co(NH_3)_6^{3+}$, which they then used to compare the efficacy of various hexammine ions for RNA phasing [46]. $Os(NH_3)_6^{3+}$ was again distinguished as one of several isosteric mimics of $Co(NH_3)_6^{3+}$, which is illustrated by a CSD search that revealed average Os(III)-to-N distances of 2.107 Å±0.04 Å, and average axial-to-equatorial angles of 90.020°±2.753°; compare to cobalt hexammine values in Section 2.2.3. Such a small difference in metal-to-nitrogen distances between cobalt and osmium are likely to be insignificant.

Of further interest is $Ir(NH_3)_6^{3+}$, which is also isosteric with $Co(NH_3)_6^{3+}$. Average Ir-to-N distances are 2.087 Å±0.01 Å with average axial-to-equatorial angles of 90.026°±3.00°. Ir(III) and Rh(III) complexes have been employed successfully in the phasing of RNA (reviewed in [46,53]). From a practical standpoint, $Ir(NH_3)_6^{3+}$ appears better suited for heavy atom phasing because $Os(NH_3)_6^{3+}$ is not commercially available. Facile procedures have been reported for the production of $Ir(NH_3)_6(III)Cl_3$ and $Ir(NH_3)_6(III)acetate_3$ [46]. Although ruthenium hexammine is commercially available, it is sensitive to air oxidation [53]. It is also devoid of a measurable absorption edge suitable for MAD phasing. As we explore the metal binding properties of Co^{3+}, Mg^{2+}, Mn^{2+}, and other metals in the ensuing sections, we will point to specific instances of cation localization with an eye for function, but also for crystallographic phasing and interactions that may promote crystal growth.

3. METAL ION BINDING AND FUNCTION IN THE STRUCTURES OF NATURAL AND ARTIFICIAL RIBOZYMES

With an understanding of crystallographic resolution, metal-ion coordination geometry, as well as the general role of ions in RNA folding and catalysis, we are poised now to view representative small ribozyme structures and their artificial counterparts. Although it would be ideal to investigate the atomic-resolution structures of large intact supermolecular machines, such as the spliceosome and ribosome, this approach has not been routinely tractable due to the subunit complexity and plasticity of these systems; further discussion of such progress can be found in Chapters 8 and 9 of this volume. As such, much of our understanding of biological reactions must necessarily come from the investigation of model systems. In the remaining sections, we will explore the folding and metal-binding properties of natural, small ribozymes whose coordinates were retrieved from the PDB (www.pdb.org). We will then consider artificial ribozymes that exhibit broader chemical activities that are considered essential for the evolution of life. In each case, a central challenge is to understand how these catalysts gain functionality since they do not possess chemical groups that ionize near neutrality (Figure 1A), which is a prerequisite for efficient general acid/base functionality in biology [24,54], as well as electrostatic stabilization of an anionic phosphorane transition states [8], which is chemically challenging for an acidic polymer.

3.1. The Hammerhead Ribozyme: An Elusive Mechanism Falls 'In Line'

The hammerhead ribozyme was the first small ribozyme to be discovered and was isolated from small satellite RNAs derived from subviral plant pathogens [55–57]. The function of the hammerhead is self-cleavage in the context of concatenated, rolling-circle-replication intermediates, which must be reduced to unit length for viroid replication (reviewed in [58,59]). Although phylogenetic comparisons of the hammerhead motif reveal a conserved core comprising eleven nucleobases [60–62], subsequent analysis revealed that the natural ribozymes possess additionally important, non-conserved internal-loop and stem-loop elements in helices I and II. Such regions confer structural stability by formation of tertiary contacts that direct the folding landscape away from a two-step pathway, and toward a one-step process suited for the requisite in-line cleavage geometry [63]. The effect of this tertiary structure is comparable to the enhanced folding achieved by inclusion of the four-way helical junction in the hairpin ribozyme relative to minimal constructs [64]. Such peripheral elements are non-conserved, but ameliorate the need for high concentrations of divalent metal ions for folding and function [65,66]. This observation and subsequent work on these extended or 'full-length' hammerhead constructs redressed a lingering conundrum in the field. Namely, the earliest "ground-state" crystal structures of minimal hammerhead ribozymes [67] did not account for the importance of key functional groups identified as important for in-line attack in biochemical investigations (reviewed in [68,69]). Another question was, *why minimal ribozymes required non-physiological concentrations (>10 mM) of ions for in vitro activity*? In this section, we will discuss the most recent structures of full-length hammerhead ribozymes, whose robust tertiary structure between stems I and II assuages many of these earlier issues.

The hammerhead ribozyme was aptly named due to its three-stem secondary structure of approximately 55 nucleotides that adopts the shape of a hammer's head [56,70,71]. Since the original discovery of this catalytic motif, several other hammerhead ribozymes have been documented in the satellite DNA of newt [72,73], in the satellite DNA from various species of the human parasite *Schistosoma* [74], in satellite DNA from cave cricket [75], in the genome of *Arabidopsis thaliana* [76], and in the genomes of some rodents [77]. Therefore, the hammerhead motif appears relatively ubiquitous in the biosphere [21], and may have evolved independently from multiple origins [76], which is supported by the facility of its emergence by *in vitro* selection for a self-cleaving RNA motif [22].

Recently, Martick and Scott solved the crystal structure of a tertiary-stabilized variant of the hammerhead ribozyme from *Schistosoma mansoni* (Figure 2A) [78]. This produced a much clearer picture of the ribozyme that was 'in line' with decades of biochemical data [79–81]. Unlike the minimal hammerhead ribozyme crystal structures in which core residues packed at the interface of stems II and III but were physically separated from core residues in stem I [67,82], the extended hammerhead structure revealed that

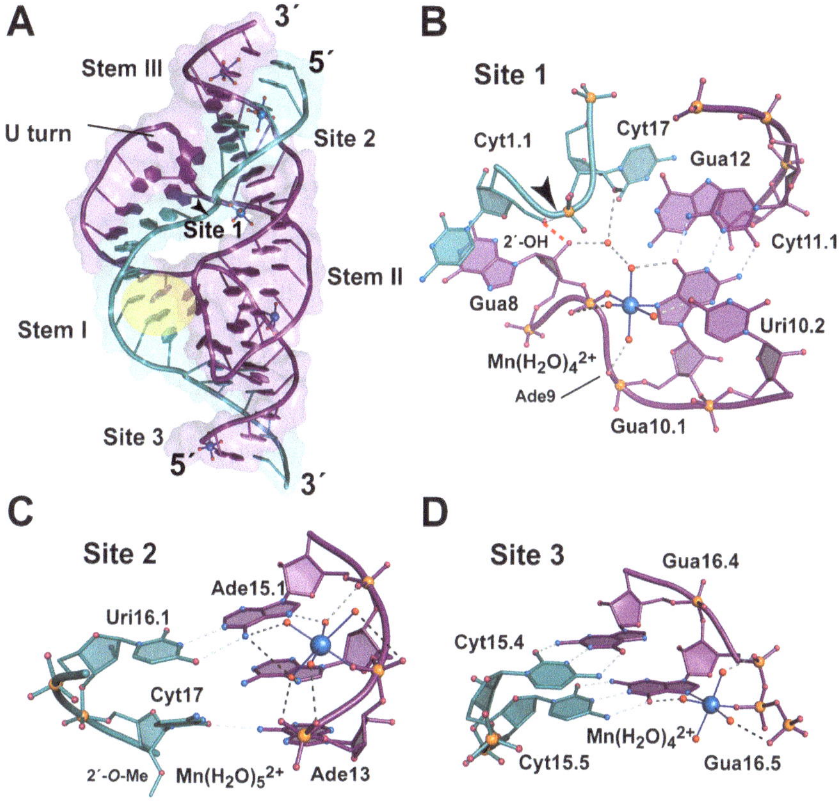

Figure 2. Cartoon diagrams of the "full-length" *Schistosoma mansoni* hammerhead ribozyme from [95] with numbering as described in [69]. (**A**) A backbone ribbon diagram and transparent surface with filled bases for the tertiary-stabilized hammerhead ribozyme (PDB entry 2OEU). The area of tertiary stabilization is highlighted as a yellow oval between stems I and II. The invariant U turn is labeled, as are the sites of Mn^{2+} binding, depicted as "sky-blue" spheres. All cartoon representations use the convention that the substrate is green and the ribozyme is purple. (**B**) Metal Site 1 is near the site of cleavage. The Mn^{2+} ion makes two innersphere contacts (solid blue lines) to N7 of Gua10.1 and the non-bridging oxygen of Ade9. Additional metal ligands are water molecules. In this structure the 2′-nucleophile at Cyt17 is capped with an inert *O*-methyl modification. Hydrogen bonds are depicted as broken gray lines; atoms for nucleotides are colored red/pink for oxygen, blue for nitrogen, and orange for phosphorus. The scissile bond is depicted here and elsewhere as a black arrowhead. (**C**) The Site 2 metal ion is located at the junction of stems I, II, and III at tandem adenines within the conserved core. The capped nucleophile is shown and its pyrimidine base is held in place by a hydrogen bond to Ade13 whose phosphate coordinates the hydration shell of the Site 2 Mn^{2+}. (**D**) The Site 3 metal ion is an artifact of *in vitro* transcription. This site illustrates the tight binding capacity of GTP for Mn^{2+}. Other ions are present in (**A**), but their immediate significance is unknown. All cartoon diagrams were rendered with the graphics program PyMol [228].

the entire core domain folds as an integrated unit [78]. The tertiary structure formed between stem II/I serves as a major source of structural stabilization. Contacts between the loop of stem II and the bulge of stem I entail several hydrogen bonds including Hoogsteen-to-Watson Crick pairings and Watson-Crick to sugar-edge interactions. The interface also features aromatic-ring stacking interactions as well as complementary-surface interfaces. Full-length hammerhead ribozymes can be 10^3-fold faster in cleavage than minimal variants assayed under comparable conditions [63,65,66,83–88]. From a structural perspective, the resulting tertiary interface has the effect of positioning Gua12 as a putative general base catalyst (Figure 2B). Cyt3 of the conserved core (not shown) pairs with Gua8, which provides its 2′-OH as a possible proton donor for the 5′-leaving group at Cyt 1.1. The importance of the Gua8 Watson-Crick face is supported by functional analysis that suggested this base was necessary for proper folding [78] rather than activity [89]. Overall, the active-site geometry of the full-length hammerhead ribozyme is nearly in-line for phosphoryl transfer ($\sim 163°$), which is close to the requisite 180° (Figure 1B).

Like other small ribozymes, the hammerhead has been show to be catalytically proficient in monovalent ions alone [25]. However, divalent metal ions are important for both the minimal and full-length hammerhead ribozymes to achieve optimal catalytic activity [85,90–92]. The full-length hammerhead ribozyme shows distinct metal binding properties for tertiary docking and cleavage [85,93,94]. A second crystal structure of the full-length hammerhead solved in the presence of Mn^{2+} revealed multiple coordination sites (Figure 2A) including one near the scissile bond [95] (Figure 2B), as well as other sites throughout the fold (Figures 2C and 2D). MD simulations by the York lab suggested that a bridging water molecule plays an important role in proton transfer in cleavage [95]. However, this work was not entirely consistent with the transition-state structure observed in biochemical studies. Phosphorothioate probing of the hammerhead revealed sensitivity at the pro-R_p non-bridging oxygens of Cyt1.1 and Ade9 [86,96–98]. Such sulfur substitution reduces innersphere coordination to oxophilic metals such as Mg^{2+}, but has little effect on softer, thiotolerant (or thiophilic) metals such as Mn^{2+} or Cd^{2+} [99] presuming there are no conformational changes [100]. Subsequent QM/MM calculations that pass through the transition state are consistent with the 2′-OH of Gua8 donating a proton to the O5′-leaving group of Cyt1.1 while maintaining its innersphere contact with Mg^{2+} [101]. A follow-up investigation also revealed how the hammerhead fold can produce an electronegative recruiting pocket that accrues highly localized concentrations of positive charge [102]. This refined mechanistic understanding better supports the phosphorothioate and biochemical data (reviewed in [19]), and properly positions the positive charge at the leaving group. Bevilacqua et al. have pointed out [23] that a 2′-OH attack produces a

$\beta_{leaving\ group}$ of nearly –1 for a 5′-oxygen equivalent [103]. This implies considerable negative charge character on the leaving group whose formation would benefit from protonation or the presence of a cation. Non-enzymatic experiments and calculations for the hammerhead and HDV ribozymes suggest that leaving group protonation is more difficult than nucleophile deprotonation [91,104]. Thus, the presence of Gua8-2′-OH/Mg^{2+} near the 5′-leaving group in the hammerhead ribozyme could play a critical role in a proton transfer, comparable to Ade38(H+) or Cyt75(H+) of the respective hairpin [105] and HDV ribozymes [106] (Sections 3.2. and 3.3). Although much progress has been made for the hammerhead ribozyme in the past 25 years, further work will be needed to fully elucidate the mechanism of action of this common catalytic motif including full rectification of mechanistic models with functional data.

3.2. The Hairpin Ribozyme: A Metal-Independent Ribozyme with an Ionized Adenine

The hairpin ribozyme is one of the best-studied members of the small ribozyme family. This minute catalyst comprises ~60 nucleotides and was discovered in the negative-polarity strands of the satellite tobacco ringspot virus RNA genome [71,107]. A great deal of functional work has been conducted on this ribozyme (reviewed in [19]), which has revealed the divalent metal independence of the reaction chemistry [108–110], as well as key nucleobases and functional groups essential for catalysis [111–117]. The natural form of the hairpin ribozyme is a four-way helical junction, which confers significant stability to the overall fold [64,118–121]. Minimal variants of the ribozyme have been prepared in the absence of the four-way junction. These comprise two essential helix-loop-helix domains that dock together to form the enzyme active site [122] (Figure 3A). Although less efficient than the four-way junction, minimal ribozymes are competent for bond cleavage and ligation, but are prone to misfolding [123–126]. Nonetheless, from a structural perspective the small size of the minimal variant provides an excellent opportunity to explore structural changes resulting from the many non-canonical functional group changes or base substitutions that can be introduced only via chemical synthesis (reviewed in [127]). Importantly, the minimal hairpin ribozyme three-dimensional structure is nearly identical to that of the four-way helical junction form, which has been attributed to the crystal contacts that appear to closely mimic the tertiary structure restraints attributed to the natural four-way junction [43,128,129].

The first high-resolution crystal structure of the hairpin ribozyme was determined in the lab of Ferré-D'Amaré [130]. The structure comprised the four-way helical junction and revealed an overall X-shaped fold with the active site situated at the confluence of the loop A and loop B domains (Figure 3A). Bases Gua8 and Ade38 from these respective internal loops were observed to flank the scissile bond located between Ade-1 and Gua + 1. Structures in the presence of the in-line attack mimic vanadium oxide suggested that Gua8 and Ade38 contribute electrostatic stabilization to the reaction and possibly serve as general acid/base catalysts [131,132]. Base rescue experiments supported an electrostatic stabilization role for these bases, which at a minimum facilitate proton transfer in the reaction [133–135]; the importance of base-mediated electrostatic stabilization has been corroborated by computational analysis as well [136]. Interpretation of the acid/base mechanism appeared consisted with earlier work on the pH-rate dependence of the reaction, which revealed a flat dependence with apparent titration points at 5.4 and 9.6 [110]. Although this observation appeared to favor acid/base catalysis suggesting two proton transfer events [137] the data were recorded from a minimal hairpin ribozyme.

Subsequently it was recognized that this profile is complicated due to ionization events associated with folding and stability rather than catalysis. In contrast, the pH-rate profile for cleavage by the natural four-way-helical junction hairpin ribozyme indicates a single titration point with an apparent pK_a of ~6.5 in a range from pH 4.5 to 10.0 [134]. Efforts to relate this apparent pK_a to microscopic pK_a values showed that Gua8 titrates at 9.6 [138], which diminishes its general base efficacy since its imino group would have little ionization at physiological pH. In contrast, Raman crystallography was used to measure directly the microscopic pK_a of Ade38 in crystals of the minimal hairpin ribozyme in its 'pre-catalytic' state.

Figure 3. Cartoon diagrams of the minimal hairpin ribozyme from [128]. (**A**) Ribbon diagram and transparent surface of the hairpin ribozyme with filled bases (PDB entry 2P7E). The minimal fold comprises two helix-loop-helix domains dubbed loop A and B. Two hexammine cobalt(III) sites (orange sphere with blue amine ligands) from the crystallization medium are shown. (**B**) Site 1 is the major site of metal binding in the S-turn. Every amine group of $Co(NH_3)_6^{3+}$ is within hydrogen bonding contact of at least one RNA atom. This site resides within the core of the loop B domain where key residue Ade38 (not shown) projects toward the scissile bond. The Co(III) site was identified in part by use of anomalous difference Fourier maps [43] that are depicted here at the 8 σ contour level as golden wire mesh. (**C**) The hexammine cobalt(III) located at site 2 is located in helix H1. Strong binding in the major groove arises from coordination of the cobalt(III) amine ligands at the O6 keto and N7 imino groups of tandem guanines Gua12 and Gua13. A sulfate from the crystallization medium interacts with the N4 exocyclic amine of Cyt-4 and promotes cross-strand interactions with the cobalt amine ligands. This site is at a crystal contact that relates two H1 helices by dyad symmetry. No other significant cobalt(III) hexammine sites were observed.

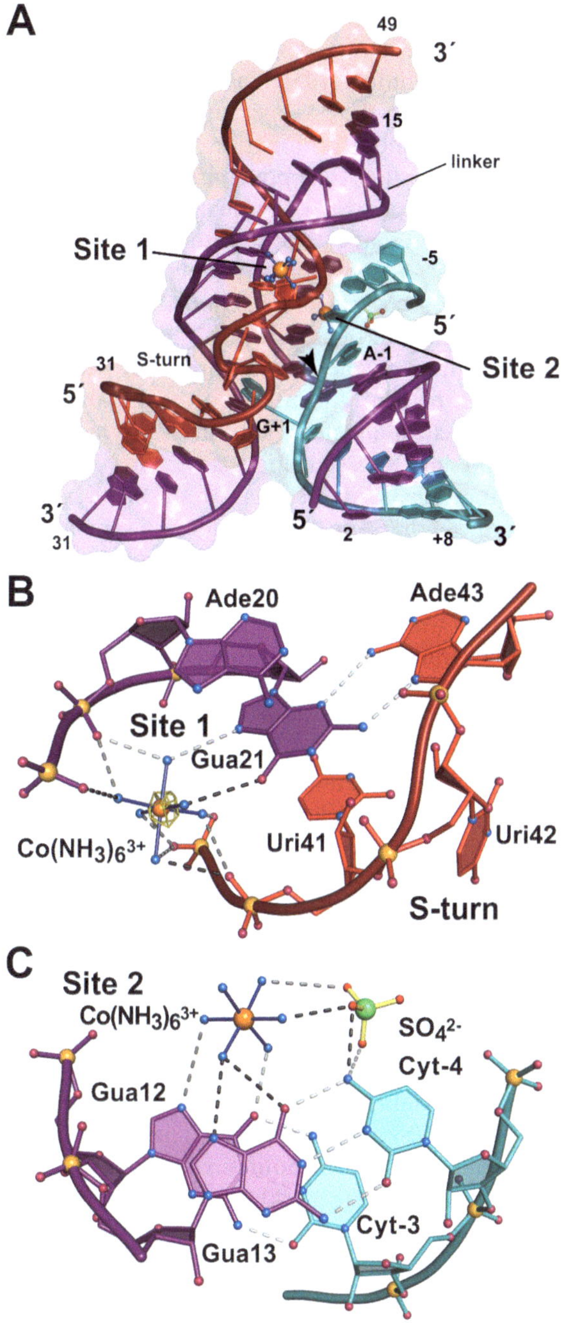
A
49
3′
15
linker
Site 1
-5
5′
S-turn
31
5′
A-1
Site 2
G+1
3′
31
5′
2
+8
3′
B
Ade20
Ade43
Site 1
Gua21
Co(NH3)6 3+
Uri41
Uri42
S-turn
C
Site 2
Co(NH3)6 3+
SO4 2-
Cyt-4
Gua12
Cyt-3
Gua13

The results revealed that the imino proton of Ade38 has a pK_a value of 5.46±0.05, which is shifted toward neutrality from a value of 3.68±0.06 measured for AMP in solution [105]. These data along with related work [139,140] are consistent with the pK_a measured in solution for the four-way junction variant, and support protonation of Ade38 for both ribozyme forms in the *pre-catalytic* state. The possibility that the apparent pK_a from the pH-rate profile corresponds to the microscopic pK_a of Ade38 is predicated on the observation that crystal structures of the minimal hairpin are nearly identical to those of the four-way junction in the presence of precatalytic substrate analogs as well as transition-state mimics [19,43,128,129]. If these pK_a values are one in the same, then the microscopic pK_a of Ade38 N1 in minimal hairpin ribozyme crystals prepared with a transition state analog is predicted to be nearly identical to the apparent pK_a of 6.5 measured for four-way helical junction hairpin ribozymes in solution since the latter value represents the transition state. Additional analysis of the microscopic pK_a for Ade38 in the context of the ligation reaction will be helpful to discern whether Ade38 N1 is indeed a general base for the back reaction, which is predicted by microscopic reversibility. Characterization of the reaction by thiol substitution of the O5′-leaving group [141] and kinetic isotope measurements would also be beneficial [142].

Unlike larger ribozymes such as the group I intron (see Chapter 7 of this volume), the hairpin ribozyme does not require multivalent ions for catalysis. Instead metals are required for proper folding [119,120,143]. A most revealing series of experiments in this regard included the demonstration that $Co(NH_3)_6^{3+}$ supported activity comparable to Mg^{2+} [108–110]. As stated in Section 2.2.3, the amine ligands of cobalt hexammine are relatively inert to exchange, which implies that no innersphere metal ion coordination to the ribozyme is necessary for chemical functionality. These observations follow from the work of Chowrira and Burke who showed that substitution of individual non-bridging oxygens in the hairpin ribozyme phosphate backbone did not elicit a significant metal-dependent effect characteristic of innersphere metal ion coordination [144,145].

To explore the mode of multivalent ion binding in the hairpin ribozyme, crystals of minimal hairpin ribozyme were prepared in the presence of cobalt hexammine (Figure 1C, right). Structures revealed one major site of binding (Site 1), and a minor site (Site 2) (Figure 3A). The major site is located in the loop B domain where the S-turn motif – which harbors the crucial base Ade38 – closely approaches the backbone of the opposing strand in the E-loop motif (Figure 3B). Every amine ligand in the Co^{3+} coordination sphere contacts the RNA, including five phosphate oxygens as well as the O6 and N7 atoms of Gua21. The identity of the Co^{3+} ion was confirmed in part by its strong anomalous scattering signal ($\Delta f''$) of 3.6 electrons collected using CuKα radiation. The anomalous difference Fourier map is contoured at the 8 σ level (i.e. seven standard deviations above the noise) in Figure 3B. Coordination of the four-way junction to Ca^{2+} ions was observed also at this site [130], and has been contrasted to cobalt hexammine in [43].

The Site 2 cobalt hexammine is located in the major groove of Helix 1 (Figure 3C). The $Co(NH_3)_6^{3+}$ ion coordinates at tandem Gua12 and Gua13 bases via N7 and O6 atoms, in a manner comparable to $Mg(H_2O)_6^{2+}$ as seen in Section 3.5 below. The presence of a sulfate ion coordinated to the N4 amine group of Cyt-4 is noteworthy because it was part of the crystallization mother liquor. This anion-binding location and its proximity to the two-fold crystallographic symmetry axis suggest that this site is the result of intermolecular crystal packing, and is not biologically relevant. Nonetheless, it provides an uncommon example of ion-pair formation in RNA, which appears to be somewhat sequence-specific.

3.3. The Hepatitis Delta Virus Ribozyme: A Ubiquitous Motif with an Ionized Cytosine and Metal-Dependent Activity

The hepatitis delta virus (HDV) is the smallest known virus to infect animals. It possesses a negative-polarity, circular-RNA genome with ~70% sequence self-complementary that causes extensive double-strandedness leading to a rod-like secondary structure [146]. The HDV genome undergoes rolling-circle replication in the host, much like the viroids that harbor the hammerhead and hairpin ribozymes, and the HDV ribozyme likewise cleaves concatenated transcripts to unit length [147]. Consequently, the HDV-infected cell contains both genomic and antigenomic RNA species that can be isolated as single-stranded circular, and linear polymeric transcripts [148]. The delta virus is not a free-living organism, but exists as a subvirus dependent on HBV. HDV can exist as a co-infection with HBV, or through infection of a previously infected HBV host cell [149]. Co-existence of the HDV and HBV in the host can exacerbate illness leading to cirrhosis [150] and hepatic carcinoma in chronic cases [151].

The HDV ribozyme consists of both genomic and antigenomic variants. Like the hammerhead, HDV-like ribozymes have been reported recently in several different genomes. Using a genome-wide search and *in vitro* selection, the Szostak lab identified four self-cleaving sequences from a human genome library, including a conserved mammalian 81-nt sequence within the intron for the gene *CPEB3*, which belongs to a family of genes responsible for polyadenylation [152]. The *CPEB3* ribozyme efficiently self-cleaves in the presence of Mg^{2+}, Mn^{2+}, and Ca^{2+}, but is refractory to cleavage in the presence of $Co(NH_3)_6^{3+}$ [152,153], suggesting a requirement for innersphere metal coordination chemistry. This feature is reminiscent of the catalytic metal requirements of the HDV ribozyme, which is the only known small ribozyme to utilize an innersphere metal ion in cleavage [25]. Notably, the HDV has surprising promiscuity in its ion preferences [154,155] and both metal-dependent and independent cleavage tracks have been described [156]. Because the *CPEB3* and HDV ribozymes share only six conserved core bases, sequence-based searches failed to identify related ribozymes outside of mammalian genomes. To overcome this barrier, the conserved ($\sim$60 nt) nested double-pseudoknot structure of the HDV was used as a restraint in secondary-structure searches involving multiple genomes. By this approach, self-cleaving RNA sequences were identified in lamprey, lancelet, mosquito, nematode, sea urchin, bacteria, and an insect virus [20]. At present the role of such ribozymes is uncertain, but their locations in mRNA transcripts suggest roles in 5′-end uncapping, IRES-mimicry and ncRNA processing [20]. These results indicate that HDV-like ribozymes are present throughout the animal kingdom, and are likely to have diverse molecular origins that evolved to fulfill specific biological functions suited exclusively to catalytic RNA.

In this chapter, we will focus on the latest structure and function developments for the HDV ribozyme, which has been investigated extensively and is reported upon in Chapter 6 of this volume. Numerous structures have been reported in states representative of the pre-catalytic [157] and product-bound states [158]. These investigations have helped to locate key catalytic residues, but the results are in conflict with respect to two proposed mechanisms of action that differ in the assignment of the general acid and general base groups (reviewed in [19]). The first HDV structure was solved by the Doudna lab and revealed a product-bound state in which the 5′-OH leaving group was within hydrogen bonding distance of essential base Cyt75, which supported a general acid role [158]. Subsequent structures of the pre-cleavage state from the same lab had a different conformation that suggested the Cyt75 imine serves as a general base in cleavage due to its close proximity to the 2′-nucleophile [157]; QM/MM and MD simulations supported the feasibility of this proposal [159,160]. However, in order to obtain crystals of the pre-catalytic complex, a Cyt75Uri mutation was made or Mg^{2+} was eliminated from crystallization [157]. These changes are potentially complicating since Cyt75 hydrogen bonds via its exocyclic amine to phosphate oxygen, which may shift its base pK_a [153,158] – an interaction that is missing in the Uri75 mutant. (Note, a comparable phosphate-to-amine interaction is present at Ade38(+) N6 of the hairpin ribozyme, and mutation of the base amine to purine causes a substantial loss in activity [134,161], possibly due to the pK_a shift this interaction normally induces at the N1 imine [105,139]). In addition, although Mg^{2+} ions are known to serve a dispensable structural role in HDV catalysis [157], they are preferred over other mono- or divalent cations [25,162]. Prior structural analyses of small ribozymes exploited both upstream and downstream substrate binding determinants (e.g., Watson-Crick pairing) to trap transition state analogs, such as vanadium oxide or nonreactive 3′-deoxy 2′,5′-phosphodiester linkages [128,131,163]. However, the HDV ribozyme has no 5′-substrate guide sequence [164,165], which has complicated efforts aimed at preparing transition-state mimics. These aspects of the HDV ribozyme have made structural analysis of Cyt75 in the context of Mg^{2+}, or other catalytic metals, a major challenge.

Recent Raman crystallography experiments by Bevilacqua, Golden, Carey et al. [106] have demonstrated that the key nucleobase Cyt75 has a pK_a shifted from a solution value of 4.2 to nearly 6.5 in HDV crystals. This observation strongly supports assignment of the Cyt75 imine as a general acid functionality in cleavage as proposed [166,167]. Along with other evidence (reviewed in [19]), the agreement of the Cyt75 microscopic pK_a in crystals with the apparent pK_a measured under comparable solution conditions provides strong evidence that Cyt75 is responsible for rate-limiting proton transfer in the reaction [106]. Moreover, the pK_a measurements from crystals demonstrated the same cationic dependence exhibited in solution, whereby the Mg^{2+} concentration is coupled anti-cooperatively with the Cyt75 pK_a value. These data suggest that the base and metal are significantly close to be electrostatically linked through space [106]. To explore the location of Mg^{2+} binding, additional Raman spectroscopy measurements were recorded on HDV ribozyme crystals. Using difference spectra, octahedral $Mg(H_2O)_n^{2+}$ (where n = <5, or a mixture thereof) was observed to exhibit innersphere coordination at ~5 of the non-bridging phosphate oxygens in the RNA backbone [168]. This information was consistent with a prior crystal structure that suggested N7 and O6 coordination at Gua [158]. In a follow-up study, kinetic analysis, pH-dependence, and N7-deazaguanine probing were each used in combination with Raman crystallography to provide insight into the binding site of Mg^{2+} at the scissile bond. The results revealed that a heretofore-unidentified hydrated Mg^{2+} binds the Gua · Uri wobble in the active site, supporting a model in which $Mg(OH^-)^+$ functions as a Brønsted base in cleavage [169]. The importance of Mg^{2+} was probed further in HDV ribozyme crystals by competition with $Co(NH_3)_6^{3+}$ binding. The outcome suggested that cobalt hexammine was capable of displacing an innersphere Mg^{2+} contact, most likely at Gua + 1 of the cleavage site, and that cobalt hexammine suppresses Cyt75 ionization (i.e., it is also anti-cooperative), suggesting it has coordination site overlap with an active site Mg^{2+} [170].

New clarity on the function of the HDV ribozyme came to light recently when Golden, Bevilacqua and coworkers determined the structure for a new crystal form of the HDV ribozyme [171]. The structure was solved by molecular replacement and refined to the relatively high resolution of 1.9 Å, providing a detailed snapshot of the pre-catalytic state. To prevent cleavage in crystals, the nucleophile and downstream bases were replaced by a 2′-deoxy modification. The structure was prepared in the presence of Mg^{2+}, which binds as a pentahydrate to the Hoogsteen edge of Gua25 and the pro-R_p non-bridging phosphate oxygen of Uri23 (Figure 4). The presence of a reverse wobble base pair between Gua25 and Uri20 is one of the novel observations of the new structure that has implications for the proposed mechanism. Importantly, the active site adopts a significantly different conformation compared to the inactive Cyt75Uri variant [157], but has fewer differences when compared to the original, cleaved HDV ribozyme structure described in [158]. One caveat to the new structure is that the Uri-1 nucleoside cannot be resolved in electron density maps (although there is evidence for the scissile phosphate), which leaves the exact in-line geometry an open question. However, if the Cyt75 base and $Mg(H_2O)_5^{2+}$ are representative of their locations in the transition state, modeling of the reaction based on these constraints would support a mechanism whereby the metal

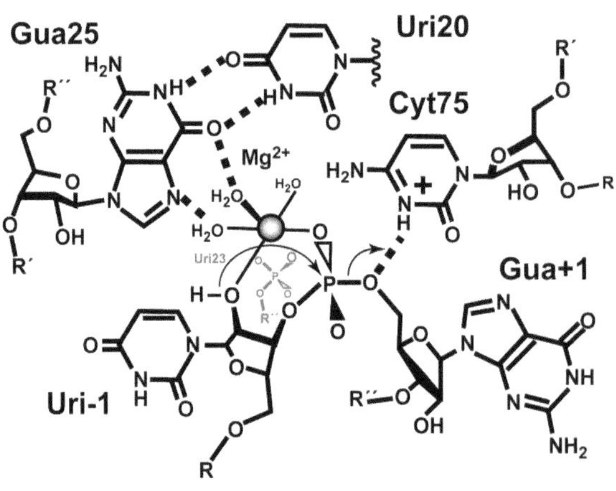

Figure 4. Schematic diagram of metal-assisted general-acid catalysis by the hepatitis delta virus ribozyme. The positions of nucleotides and the catalytic Mg^{2+} are based on a crystallographic structure and modeling adapted from [171]. Here the Mg^{2+} is proposed to coordinate directly to the 2′-nucleophile of Uri-1, which would activate the 2′-OH group by lowering its pK_a by more than a log unit from >12.5 (Figure 1A). The coordination of a non-bridging oxygen of the scissile bond to Mg^{2+} is also beneficial because it polarizes the phosphorus, shields the nucleophile from negative charge, and relieves negative charge buildup in the transition state. The restraints used to construct this model include the known in-line geometry (Figure 1B), and the observed location of the hydrated Mg^{2+} at Gua25 (as shown).

ion serves as a Lewis acid in cleavage, while Cyt75 functions as a general acid (Figure 4), as proposed previously [141,153,158,167,172].

Although understanding the exact mechanism of action of the HDV ribozyme will require additional investigation, a take home message is that small ribozymes can possess hybrid properties. Whereas large self-splicing introns usually employ multiple Mg^{2+} ion binding sites for activity, small ribozymes usually exploit bases to achieve chemical functionality [8,24,54]. In contrast, the HDV ribozyme, and presumably its newly discovered cousins such as *CPEB3*, appear to employ a nucleobase as a general acid, but a metal ion as a Lewis acid. A second take home message is that chemical groups in the active site of an RNA catalyst can be coupled through space, like the Mg^{2+} ion and the ionized Cyt75 of the HDV ribozyme. This relationship can alter the properties of the active site, and alter the pK_a of key functional groups. In the next section, we will consider how the ionization state of a metabolite cofactor is used to augment ribozyme functionality leading to gene regulation.

3.4. The *glmS* Ribozyme: A Metabolite-Sensing Gene-Regulatory Riboswitch

The *glmS* ribozyme is the most recently discovered member of the small ribozyme family [173], and it was the first self-cleaving mRNA to be identified in a free-living organism. This small catalytic motif comprises ~150 nucleotides and functions as a gene-regulatory element for numerous gram-positive bacteria [174]. Although the exact control mechanism is unknown, the ribozyme resides within the 5′-UTR of mRNA transcripts encoding the *glmS* gene product, L-glutamine: D-fructose-6-phosphate amidotransferase, which produces the metabolite glucosamine-N-6-phosphate (glcN6P) [175]. When levels of glcN6P are elevated in the cell, the *glmS* ribozyme functions as a riboswitch by binding the metabolite and inducing strand scission of the *glmS* mRNA. This cleavage reaction downregulates expression of the amidotransferase gene product causing levels of cellular glcN6P to drop, thus deactivating the ribozyme [173]. Unlike other family members, the *glmS* ribozyme is distinguished for having combined activities of RNA cleavage coupled with metabolite-sensing gene control, which distinguishes it as a novel riboswitch.

A remarkable aspect of the *glmS* ribozyme is its ability to bind selectively and avidly to glcN6P with a K_D of ~200 μM [173]. Although other analogs of this metabolite bind, such as glucosamine-6-sulfate, these require higher concentrations to elicit a full catalytic response [173,176,177]. One metabolite analog that provided significant insight into the ribozyme mechanism of action was glc6P in which a hydroxyl group replaces the amine at position C2. This analog fails to elicit ribozyme cleavage suggesting an essential role for the amino group in reaction chemistry [173,176]. Modifications of the C2 amine moiety, such as trimethylation or acetylation, ablate catalytic activity but monomethylation does not, suggesting that the amine functions to donate or accept protons in the reaction [177]. Although the C6 phosphate and C4-hydroxyl moieties of glcN6P are important for metabolite recognition by the riboswitch, neither group appears to be as essential for chemistry as the C2 amine [177,178].

Binding to the glcN6P metabolite by the *glmS* ribozyme does not appear to induce conformational changes in solution, suggesting that the enzyme active site is pre-formed or 'locked and loaded' for cleavage [173,179–182]. GlcN6P analogs with increasingly higher pK_a values at the amine group are successively poorer in catalysis or require a higher pH to reach half-maximal activity [176]. As such, it was proposed that the (Ade-1) 2′-nucleophile of the ribozyme achieves activation through proximity to the amine group (e.g., serving as X: in Figure 1A), possibly by deprotonation [183]. Like other small ribozymes, the *glmS* exhibits promiscuous metal ion requirements characteristic of a structural role [173] rather than a Lewis acid-based activity. Like the hairpin ribozyme, the *glmS* ribozyme is fully competent to cleave in the presence of cobalt hexammine [183], which is consistent with a solely structural role for multivalent ions.

Secondary structure predictions, phylogenetic sequence conservation, and biochemical probing experiments support a pseudoknotted fold for the *glmS* ribozyme [183]. Two independent groups confirmed this fold by determining crystal structures for the respective enzymes from a hotspring bacterium, *Thermoanaerobacter tengcongensis* [180], and the causative agent of anthrax, *Bacillus anthracis* [181]. Both structures revealed a compact core with a double pseudoknot structure wherein the glcN6P binding site resides between two closely packed helices in a solvent-exposed pocket [180,181,184] (Figure 5A). Sets of structures from both organisms also confirmed that the active site is pre-folded and does not exhibit significant conformational changes [181], even upon substrate cleavage in crystals [180]. The metabolite glcN6P and analogs thereof were observed to bind adjacent to the scissile bond in both structures (Figure 5B). Both structures also show a variety of metal ions coordinated throughout the core fold (Figure 5A). In the *B. anthracis* structure, metabolite recognition entails coordination of glcN6P phosphate by two hydrated Mg^{2+} ions, which are clearly observable in electron density maps [181]; these sites are denoted as Site 1 and Site 2 in Figure 5. Both are $Mg(H_2O)_6^{2+}$ ions that coordinate glcN6P via the metabolite's phosphate group. The hydrated metals bridge the RNA major groove by coordinating to the Hoogsteen edges of tandem Gua57 & Gua56, and Gua54 & Gua53. A purine at Cyt55 disrupts this run of guanines, which

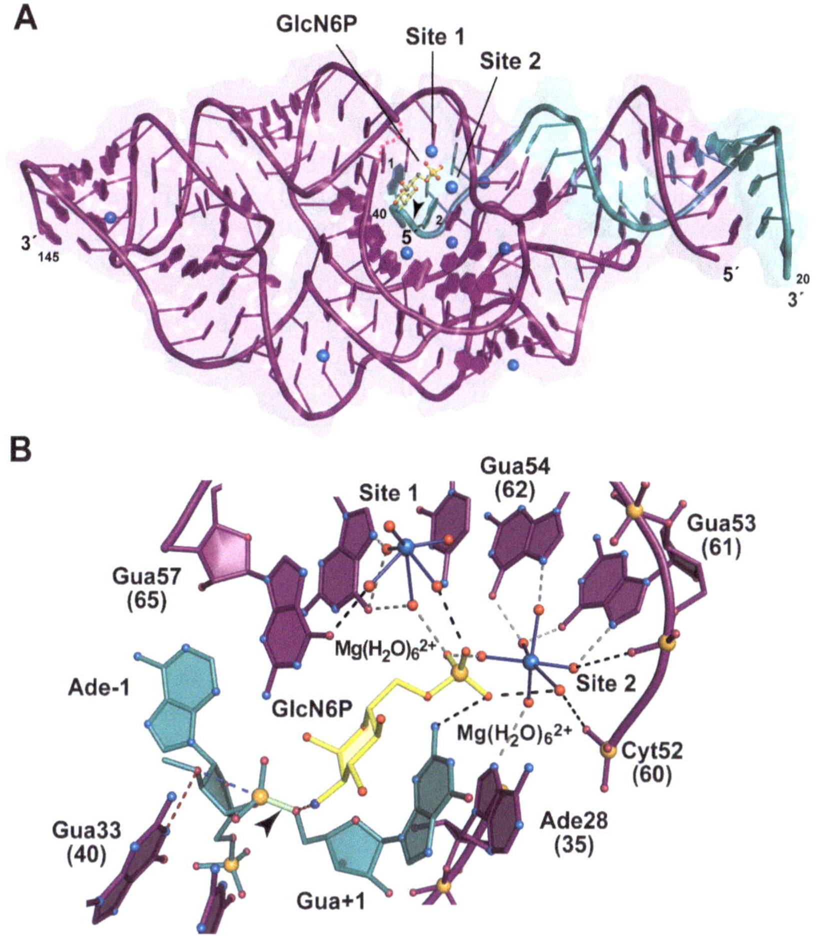

Figure 5. Cartoon diagrams of the naturally occurring *glmS* ribozymes from [181,184]. (**A**) Ribbon diagram and transparent surface of the *T. tengcongensis glmS* ribozyme with filled bases (PDB entry 2Z74). The glcN6P metabolite is depicted as a yellow ball-and-stick model in an open active site. Mg^{2+} ions are shown as "sky blue" spheres. (**B**) Close-up cartoon view of the *B. anthracis glmS* active site showing Site 1 and 2 Mg^{2+} ions that assist in glcN6P binding (PDB entry 2NZ4). The ions were modeled as hexaaqua Mg^{2+} with distorted octahedral geometry that make water-mediated contacts to the metabolite phosphate moiety. Site 1 coordinates at tandem guanines 56 and 57. The Site 2 Mg^{2+} is coordinated by tandem guanines 53 and 54. The 2′-nucleophile at Ade-1 is capped with a 2′-*O*-Me group to prevent cleavage. Nearby base Gua33 is positioned as a putative general-base catalyst. Primary numbering corresponds to *B. anthracis* with parenthetical values designating equivalent positions in *T. tengcongensis*.

encapsulates the sugar-phosphate binding pocket. Additional recognition and positioning of the scissile bond is achieved through an apparent hydrogen bond interaction between the phosphate moiety of glcN6P and the N2 exocylic amine of Gua + 1 (Figure 5B). Interaction between the Site 1 and Site 2 magnesium ions are similar in the *T. tengcongensis* structure, as is the mode of Gua + 1 recognition by the metabolite or analogs thereof.

The structure of the *B. anthracis* ribozyme demonstrates in-line geometry, despite the presence of a 2′-*O*-Me group to prevent cleavage in crystals (Figure 5B). When considering the mode of bond cleavage, crystal structures from both species revealed that the conserved base Gua33 (Gua40 in the *T. tengcongensis* numbering system) [183] is oriented close to the 2′-nucleophile whereas the C2 amine of glcN6P is poised as general acid near the O5′ leaving group of Gua + 1 [180,181,184] (Figure 5B). It is notable that the pK_a of the primary amine at C2 of glcN6P is expected to be 8.2, which means it is primarily protonated at neutral pH [181,184]. This observation and the proximity of the C2 amine to the O5′-leaving group supports a role for the amine as a general acid in substrate cleavage.

To interrogate the role of the putative general base, Gua33 (Gua40), a Gua-to-Ade mutation was introduced and analyzed for activity [185]. The results revealed a nearly 10^5-fold loss in activity suggesting a critical role for Gua33 in catalysis. When wild-type and mutant structures of the *T. tengcongensis* ribozyme were solved in the presence of a 3′-deoxy, 2′,5′ transition state analog, both structures were identical, and no change in the positions of the non-bridging phosphate oxygens were observed relative to the pre-catalytic structure [185] unlike the hairpin ribozyme [128]. Therefore, tighter binding of the transition state does not appear to be as pronounced a strategy by which the *glmS* ribozyme achieves bond cleavage, as compared to the hairpin ribozyme [131]. Importantly, the Gua33Ade mutation did not disrupt the apparent in-line geometry of the reaction (reviewed in Figure 1B and see Figure 5B), suggesting that the N1 position at base 33 plays an important role in chemistry (i.e., an important functional group is deprived from the active site when Gua is mutated to Ade, and loss of activity is not due to local misfolding). Crystal structures of the respective Ade33 and Gua33 variants demonstrate that Ade33 N1 is farther away from the 2′-nucleophile than Gua33, whose equivalent imine is within hydrogen bonding distance (3.2 Å). As such, placement of the Gua33 N1 imino moiety near the nucleophile serves an equally important role in catalysis as compared to the amine group of glcN6P [185]. This interdependence of functional groups is more pronounced than in the hairpin ribozyme in which the N1 imine of Ade38(+) appears critical for cleavage activity [105,140], whereas ablation of Gua8 has an important, but more subtle effect on activity [134,135]. Additional investigations of the *glmS* ribozyme will be necessary to account for the role of Gua33, and how glcN6P favors catalysis as a possible general acid. Measurements of the microscopic pK_a values of these groups within crystals could shed some light on this issue [186] or increasing the leaving-group ability by replacement of oxygen with sulfur, as reported for the HDV ribozyme [141].

3.5. The Leadzyme: An Artificial Lead-Dependent RNA-Cleaving Motif

The leadzyme is a small *in vitro* selected catalytic RNA derived from tRNA that catalyzes a two-step reaction in the presence in Pb^{2+} and Mg^{2+} resulting in phosphodiester bond cleavage [9,187]. Due to its small size the leadzyme motif is prevalent in many RNA sequences, which led to the proposition that it serves as a novel source of Pb^{2+} toxicity [188]. Although the first step of the leadzyme reaction is comparable to the bond-scission reaction characteristic of natural small ribozymes (Figure 1B), the ensuing step involves hydrolysis of the cyclic 2′,3′-phosphodiester in a manner comparable to ribonuclease A [189]. The premise of developing the leadzyme as a model system was predicated in part on the observation that site-specific cleavage occurred in crystal structures of yeast $tRNA^{Phe}$ derivatized with the electron-dense heavy atom Pb^{2+}. When the crystal structure of tRNA was solved, Pb^{2+} was observed to bind at three distinct locations [190,191] and elicited strong scission between residues Uri17 and Gua18 of the D loop [191,192]. It was hypothesized that Pb^{2+} binding increases the localized concentration of hydroxide ions due to its exceptional Lewis acid properties, which can lower the pK_a of bound water from 15.4 to 7.9 [38]. As such, a well-positioned Pb^{2+} ion can increase the reactivity of RNA by activating a 2′-OH group for nucleophilic in-line attack.

The leadzyme ribozyme core comprises an internal bulged loop of ten nucleotides making it one of the smallest known catalytic RNA motifs [9,187]. The fold of the enzyme has been investigated extensively by NMR [193], computational modeling [194], conformational probing [195,196], time-resolved spectroscopy [197], and X-ray crystallography [198,199]. In a highly revealing analysis of conformation and activity, Bevilacqua and colleagues performed kinetic assays on three respective 8BrGua variants of the leadzyme that were shown to adopt different *syn* conformations in models derived by MC-Sym computation, NMR spectroscopy, and crystallography techniques [196]. The results demonstrated that 8BrGua24 is hyperactive, whereas 8BrGua7 and 8BrGua9 had reduced activity. Surprisingly, only the MC-Sym computational model reported a syn Gua24 conformation making it most relevant to the phosphoryl transfer reaction. At present, there are no experimentally determined leadzyme structures in relevant pre-catalytic or other states close to the required in-line geometry.

A major take-home lesson from the leadzyme analysis is that energy landscape of RNA folding is shallow and extremely diverse [200], making ribozymes best described by an ensemble of conformations in solution at any given time [197]. With this in mind, we now turn our attention to metal binding by the leadzyme with the expectation that the structures discussed are likely to be different than the activated state, but still representative of solution conformations available in the presence of divalent ions (Figure 6A). An NMR analysis of the leadzyme revealed no significant structural shifts resulting from Pb^{2+} or Mg^{2+} titrations, suggesting the leadzyme is prefolded prior to metal ion binding [201]. These results appear consistent with X-ray crystallographic analyses in which various metals were either co-crystallized (Mg^{2+}) or soaked (Ba^{2+}, Sr^{2+}, Pb^{2+}) into crystals to locate divalent binding sites relevant to stability and catalysis [198,199]. The observation of two distinct leadzyme ribozyme conformations in the asymmetric volume supports the solution flexibility of the leadzyme. A pairwise superposition of these crystallographically-independent molecules revealed an overall deviation of 2.0 Å for all atoms, with a maximum rmsd in the internal bulged loop of 7.6 Å at Gua7. The large rmsd difference between structures revealed that only a single conformation was competent to bind Ba^{2+} or Sr^{2+} ions (conformation 2) in a location tantalizingly close to the 2′-nucleophile of Cyt6 (Figure 6A). Although an imperfect electronic match for Pb^{2+} [202], Ba^{2+}, and Sr^{2+} were employed because they have comparable ionic radii to lead, and are inert in leadzyme cleavage. Their electron richness also makes these metals easy to identify in electron-density maps as well.

The overall mode of Sr^{2+} binding to the leadzyme was defined in electron density maps calculated to 1.8 Å resolution [198]. The Sr^{2+} ion was partially hydrated but made at least two innersphere contacts via the N1 imine of Ade23 and the non-bridging oxygen of Gua7 (Figure 6B). A search of the CSD revealed Sr^{2+} prefers eight ligands that are arranged in a distorted pentagonal bipyramid in which there are one or two axial ligands. Coordinating atoms may include oxygen and nitrogen at average distances of 2.621 Å±0.093 Å and 2.789 Å±0.114 Å, respectively. In contrast to Sr^{2+}, Pb^{2+} adopts a variety of coordination geometries involving 6 to 10 ligands. Average Pb^{2+}-to-oxygen and Pb^{2+}-to-nitrogen coordination distances from small molecule structures were 2.727 Å±0.156 Å and 2.681 Å±0.139 Å, respectively. This observation suggests that the 2′-OH group of Cyt6 is not a true innersphere ligand of Sr^{2+} at Site 1 (Figure 6B) due to a long coordination distance of 3.8 Å. Moreover, the observed mode of Sr^{2+} binding may have identified atoms relevant to Pb^{2+} binding but the catalytic lead must coordinate to a significantly different structure to fulfill the requisite in-line attack geometry. In support of this notion, leadzyme crystals soaked in high concentrations of Pb^{2+} lost their diffraction properties and demonstrated partial cleavage of substrate upon dissolution, consistent with an ion-induced conformational

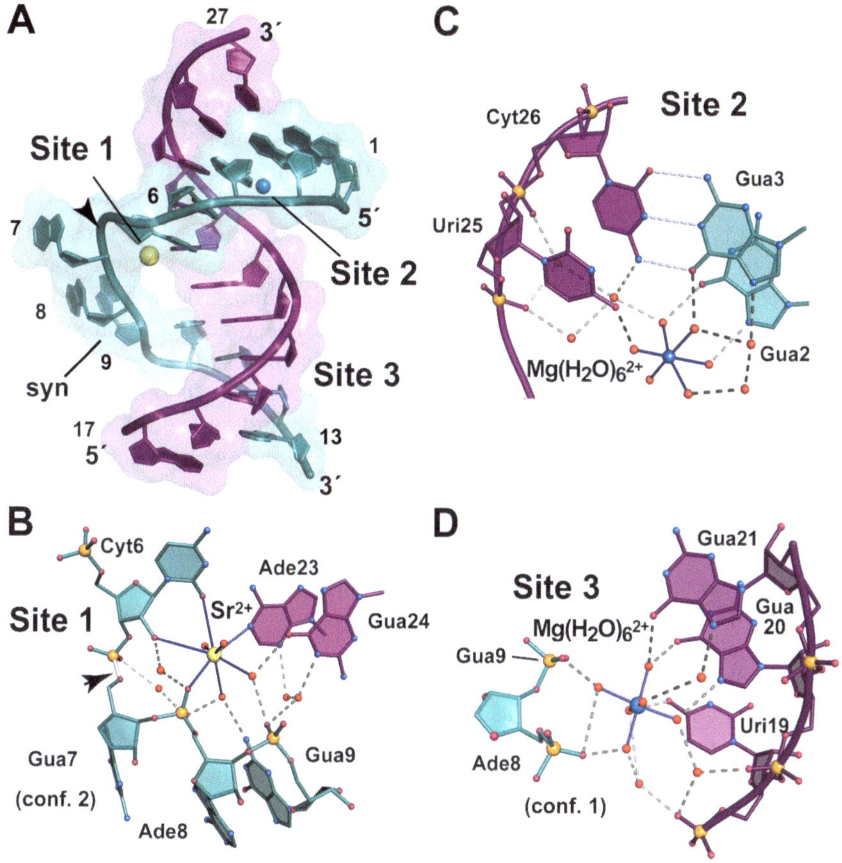

Figure 6. Cartoon diagrams of an *in vitro* selected lead-dependent ribozyme variant from [198] that catalyzes a two-step phosphodiester bond cleavage and hydrolysis reaction. (**A**) Ribbon diagram and transparent surface of the leadzyme with filled bases (PDB entry 1NUV). A Sr^{2+} ion is depicted as a yellow sphere at the site of cleavage; Mg^{2+} ions are shown as "sky blue" spheres. Numbering is based on the nomenclature used in [193]. (**B**) Metal binding Site 1 near the cleavage site is occupied by Sr^{2+}. This site also accommodates Ba^{2+} but neither Pb^{2+} nor Mg^{2+} has been observed in this location [198,199]. The 2′-nucleophile of Cyt6 would benefit from contact to Pb^{2+}, which would lower its pK_a from ~13 to 7.9. However, this structure cannot represent the transition state conformation due to several factors, including poor in-line geometry and the absence of a *syn* conformation at Gua24. The requisite transition state might also be expected to place the Pb^{2+} in direct coordination with the non-bridging oxygens of the scissile bond (like the HDV ribozyme) or the O5′-leaving group. (**C**) Mg^{2+} coordinates at tandem guanine bases in a hexaaqua form at Site 2. This site is observed in both conformations of the ribozyme in the asymmetric volume. (**D**) The Site 3 Mg^{2+} also binds at tandem guanine as a hexaaqua species, but is observed in only one of the two conformations in the asymmetric unit. The role of this ion is to neutralize negative charge from the bulged loop harboring the cleavage site, which is in close contact with the major groove.

change [199]. Further structure and function analysis of the leadzyme will be required to identify the true mode of Pb^{2+}-mediated activation of this small catalytic motif, which is likely to influence the local structure.

In addition to Pb^{2+} binding, the leadzyme reaction requires Mg^{2+} ions for RNA folding. Too much Mg^{2+} (~10 mM) results in inhibition of cleavage [203], consistent with stabilization of one or more inactive RNA conformations [198]. The refined structure of the leadzyme provides an opportunity to examine both stabilizing Mg^{2+} ions, as well as those that appear to compete with Pb^{2+} at a site distant from the scissile bond. Ions in the former category were observed at Site 2 in the major groove, coordinated to tandem Gua at positions 2 and 3 (Figure 6C). The ion observed at Site 2 is distinctly hexaaqua Mg^{2+} with waters that bridge bases at the junction of a helix and the bulged-loop active site. This ion-binding site is present in both conformations of the leadzyme in the asymmetric unit and seems to fulfill a role in RNA stabilization. In contrast, the hexaaqua Mg^{2+} at Site 3 (Figure 1C and 6D) is present in only one of the conformations in the asymmetric unit. The other conformation binds Sr^{2+}, as well as Pb^{2+} [198,199], which coincides with the observation of site-bound Ba^{2+} or Sr^{2+} at the nucleophile (Site 1). The role of the Site 3 $Mg(H_2O)_6^{2+}$ appears to be neutralization of electrostatic repulsion between the backbone of the bulged loop (Gua7-Ade8-Gua9) and the major grove of the opposing strand (Figure 6D). Specifically, the $Mg(H_2O)_6^{2+}$ ion binds at the Hoogsteen edge of tandem Gua bases 20 and 21 as well as the O4 keto oxygen of Uri19. Importantly, Mg^{2+} was not observed to bind at Site 1, suggesting that competitive inhibition of Pb^{2+} in high Mg^{2+} [203] was not occurring at this location [198]. As such, it was hypothesized that binding of Mg^{2+} at Site 3 may favor a specific active site conformation (e.g., conformation 1), which precludes binding of a structural Pb^{2+} in a nearby pocket close to Site 3. If true, this ion selectivity would provide an example of a cation-specific allosteric site that predisposes the leadzyme to adopt catalytically active versus inactive conformations depending on the ratio of Pb^{2+} to Mg^{2+} in the solution [198]. The ability of ions to control RNA structure is not surprising since a Mg^{2+}-sensing gene-regulatory element was described recently [204] (see also Chapter 5 of this volume). We now turn our attention to more complex *in vitro* selected ribozymes that significantly expand the catalytic repertoire of RNA molecules, further supporting the case for a pre-biotic RNA World.

3.6. The Flexizyme: An Artificial Aminoacyl tRNA Synthetase Ribozyme

The flexizyme is a small *in vitro* selected ribozyme that aminoacylates specific tRNAs by Watson-Crick base recognition within the substrate acceptor stem [205]. Like class II aminoacyl tRNA synthetases the flexizyme preferentially targets the 3′-OH group and forms canonical base pairs to the unpaired CCA sequence, which does not entail RNA bending. The latter feature differentiates class II enzymes from class I enzymes in which a kink is generated in the single-stranded region. The flexizyme's modest 45-nucleotide size, its 10^5 rate acceleration over the uncatalyzed reaction, and its regioselectivity make this small artificial ribozyme a model system to investigate the molecular determinants necessary and sufficient for RNA-mediated tRNA charging. Thus far, the flexizyme has revealed key molecular hallmarks comparable to the evolutionary traits of extant tRNA recognition enzymes, as well as catalytic attributes hypothesized to signal RNA replication by primitive aminoacyl tRNA synthetases (i.e., replicases) in a primordial RNA World [26].

The core flexizyme fold comprises a series of co-axially-stacked, A-form helices and an irregular helix that harbors the active site (Figure 7A). The crystal structure was solved by Ferré-D'Amaré and colleagues using Se-Met MAD phasing in which the U1A spliceosomal protein was added as a crystallization and phasing scaffold [205]. A major innovation was the covalent attachment of a tRNA minihelix acceptor stem that allows visualization of the enzyme-substrate complex (Figure 7A). The walls of the active site are formed by a series of non-canonical base pairs between Ade23 · Gua48, Gua24 · Uri47 and Gua25 · Uri44; Ade45 and Uri46 contribute unpaired bases as well. Difference Fourier analysis provided additional insight into the potential location of L-Phe (ethyl ester) binding (Figure 7B), although no model could be placed definitively into the electron density [205].

A hydrated Mg^{2+} stabilizes the core flexizyme fold by binding in the major groove at Site I (Figure 7B). Although the 2.8 Å resolution of the structures is insufficient to resolve the entire Mg^{2+} coordination sphere, two innersphere waters could be built into electron density maps. Here the hydrated ion mediates contacts between the Hoogsteen edge of Gua25 and the non-bridging phosphate oxygens of Ade23 and Gua24, which flank the site of 3′-OH aminoacylation (Figure 7B). In this case, the metal ion is not believed to participate in catalysis, but knits together disparate parts of the active site. No change in metal ion coordination was observed in the presence or absence of the L-Phe analogue.

Two additional, strongly occupied divalent metal ion coordination sites were located within the flexizyme helical core. The strongest sites are located at the 5′-end of the RNA transcript, which comprises the triphosphate group arising from the start of transcription. Well-coordinated Mg^{2+} at a terminal GTP moiety has been described for the hammerhead ribozyme [95], which presumably facilitates crystallographic stabilization; this is a recurring theme (Figure 2D).

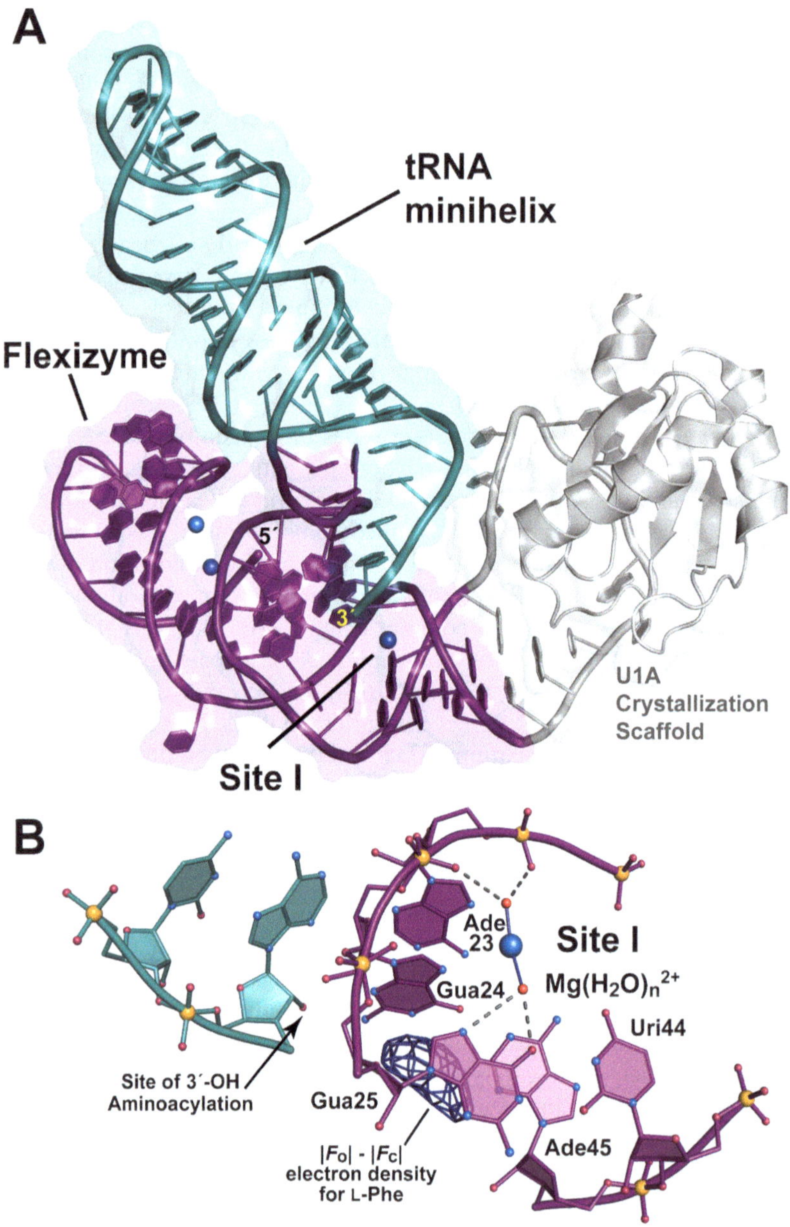
A
tRNA
minihelix
Flexizyme
5´
3´
U1A
Crystallization
Scaffold
Site I
B
Ade
23
Site I
$Mg(H_2O)_n^{2+}$
Gua24
Uri44
Site of 3´-OH
Aminoacylation
Gua25
$|F_O| - |F_C|$
electron density
for L-Phe
Ade45

Overall the results provide insight into the minimal core required for regioselective tRNA aminoacylation. A key question is whether the active site of the flexizyme is subject to induced-fit conformational changes along the reaction coordinate, which would be similar to its class II protein counterparts [206,207]. Alternatively, the active site may be preformed and rigid when binding small molecules like the preceding small ribozymes (Sections 3.1 to 3.4) and the Diels-Alder ribozyme [208] (Section 3.7, below). Additional exploration of these questions would help to define the flexizyme mechanism of action. At a minimum, we can hypothesize that the metal ions bolster the stability of the active site structure, and may accommodate conformational changes in the catalytic core through subtle variations in the metal ion coordination sphere, as postulated for the hammerhead and HDV ribozymes (Sections 3.1 and 3.3).

3.7. The Diels-Alder Ribozyme: An Artificial Carbon-Carbon Bond-Making Catalyst

Formation of carbon-carbon bonds is critical for life. Diels-Alder ribozymes catalyze such a reaction, which is seen largely as evidence that an RNA World could exist with a sufficiently rich chemical diversity that a complex metabolism could emerge [209]. The Diels-Alder reaction is a standard method in organic chemistry to form a six-membered ring. The reaction entails [4 + 2] cycloaddition between a conjugated 1,3-diene and a dienophile (substituted alkene) leading to partially hydrogenated benzene (cyclohexene). The reaction is noteworthy because it occurs in nature and can produce exquisite stereospecificity [210,211]. Two independent groups have generated RNA-based Diels-Alder-ases by use of selection methods [212,213]. However, only the enzyme of Seelig and Jäschke is a true ribozyme because it comprises normal, unmodified nucleotides. The best variant of their

Figure 7. Cartoon diagrams of the artificial flexizyme ribozyme that has aminoacyl tRNA synthetase activity [205]. (**A**) Ribbon diagram and transparent surface of the flexizyme (purple) fused to a substrate minihelix (green) recognized by Watson-Crick base pairing upstream of the CCA target site (PDB entry 3CUL); Mg^{2+} ions are shown as "sky blue" spheres. The U1A crystallization scaffold is shown for perspective. (**B**) Close-up view of the flexizyme active site showing the Site 1 Mg^{2+} ion, which was modeled with a partially filled coordination sphere. The ion is distant from the 3′-OH of the substrate that becomes charged with L-Phe, suggesting the metal fulfills a structural role. An omit electron density map (blue mesh) contoured at 2.5 σ indicates the putative position of L-Phe soaked into crystals (PDB entry 3CUN). The electron density map was retrieved from the electron density server [229].

enzyme produced a 10^4-fold rate acceleration over the uncatalyzed reaction [213], which occurred in the context of a modestly-sized 49-mer. A 38-mer enzyme was also produced that catalyzed true multiple turnover chemistry on short anthracene-labeled oligonucleotide conjugates [213,214]. Only eleven nucleotides of the core fold are conserved, which is typical of the natural small ribozymes we have seen in Sections 3.1 to 3.4.

The core fold of the Diels-Alder ribozyme comprises three helices joined in a nested double pseudoknot (Figure 8A). Serganov and coworkers in the Patel lab solved the structure by 2′-Se-CH_3 MAD-phasing (using techniques described in [215]) in the presence and absence of product [208]. This work revealed that the ribozyme possesses a pre-organized active site, reminiscent of natural small ribozymes (Sections 3.2 to 3.4) and corroborative of solution analyses [216,217]. In the crystal structure, the 5′-end of the substrate strand (Figure 8A, green) is covalently linked to the cyclohexene product adduct by a flexible (but disordered) linker. The product is bound in a mostly hydrophobic pocket stabilized by aromatic stacking interactions with flanking nucleobases. Three hydrogen bonds are also observed between the product and the ribozyme. The high degree of complementarity between the ribozyme and the product accounts for the high stereospecificity of the reaction. This degree of fit is expected to improve for the enzyme transition state complex [208].

The Diels-Alder ribozyme requires metal ions for function [213]. NMR titration data indicated that a majority of the RNA is folded when two equivalents of Mg^{2+} are present [208]. Inspection of the crystal structure reveals six octahedrally coordinated Mg^{2+} ions in the tertiary fold, albeit the 3.0 Å resolution of the data preclude definitive assignment of the ligand sphere. Of the six divalent metals, Site 2 was deemed most important because of a structural contribution in which it bridges a sharply kinked backbone region where six bases flank the product-binding pocket (Figures 8A and 8B). Site 1 also contributes in this role, and both metals make direct as well as water-mediated contacts to the RNA. Site 3 binds in the major groove at tandem Gua 27 and 28 (Figure 8C). This site appears to stabilize a transition point between substrate recognition helix 1 and the floor of the product-binding site. Site 5 stabilizes unpaired base Gua24 by binding its Hoogsteen edge (Figure 8D), establishing a complementary crevice for the hydrophobic pentanyl moiety of the product formed in part by the base's sugar edge. The Site 4 (Figure 8E) and Site 6 (not shown) Mg^{2+} ions reside in the major groove of helix 1, which knits together the substrate RNA strand (Figure 8A, green) and the 5′-substrate recognition sequence of the ribozyme. The Site 4 metal binds at tandem Gua bases via water-mediated contacts to the Hoogsteen edge of these bases.

The observation that the Site 5 Mg^{2+} is situated very close to the product, but does not actually make direct or water-mediated contacts (Figure 8D),

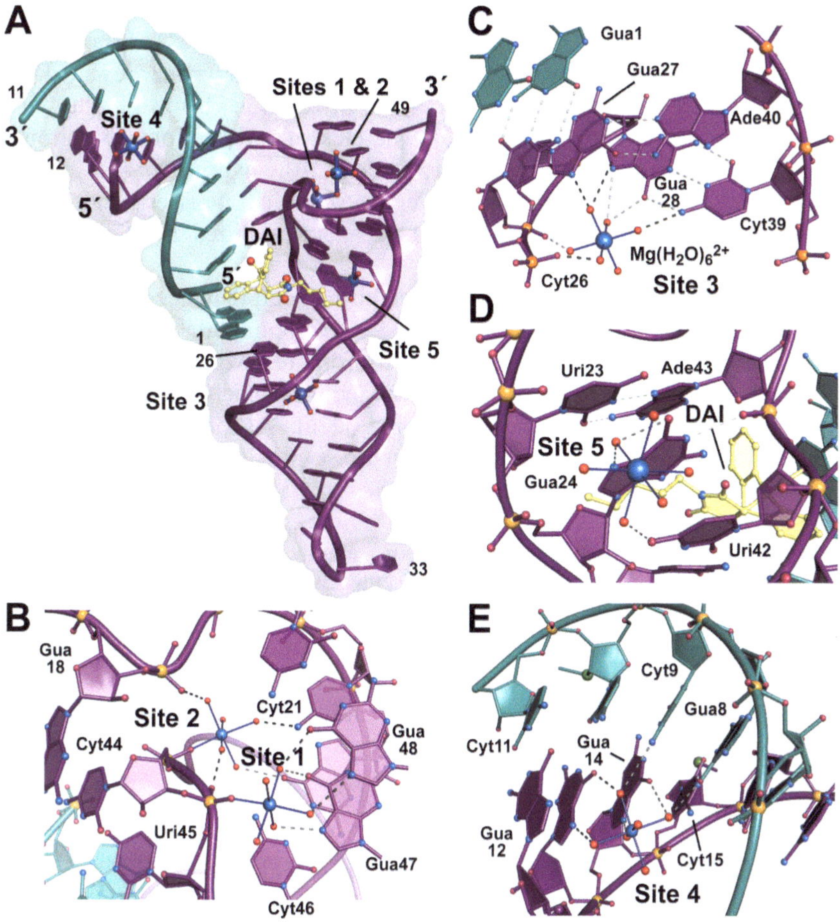

Figure 8. Cartoon diagrams of the non-natural, carbon-carbon bond forming Diels-Alder ribozyme from [208]. (**A**) Ribbon diagram and transparent surface of the Diels-Alder ribozyme (PDB entry 1YLS). The overall fold is a nested pseudoknot with a preformed hydrophobic active site that complements the shape to the hydrophobic product (yellow ball-and-stick model DAI). Multiple Mg^{2+} ions ("sky blue") were observed in the structure and were modeled as ideal hexaaqua octahedra. (**B**) The Site 2 metal ion was deemed most important for fulfilling a structural role that neutralizes the backbone charge at a sharp kink that approaches the major groove edge of Cyt21 and Gua48, which form the top of the active site pocket; Site 1 also helps in this role. Here direct contacts between the Mg^{2+} ions and the backbone were modeled for lack of resolution. (**C**) The Site 3 Mg^{2+} binds at tandem guanines 27 and 28 that form the bottom of the hydrophobic active site. (**D**) The Site 5 Mg^{2+} ion binds to the major groove edges of unpaired Gua24 and Uri42, which form a cross-strand stack that flanks the active site pocket. (**E**) The Site 4 Mg^{2+} ion binds at tandem guanines 13 and 14 of the ribozyme strand (purple). This interaction stabilizes the substrate recognition helix (green) that is covalently attached to the Diels-Alder reactant.

has implications for the proposed mechanism of action for Diels-Alder ribozyme catalysis. Although activity analyses showed a distinct Mg^{2+} dependence [213], structures of the Diels-Alder ribozyme revealed no ordered Mg^{2+} ions situated to participate in reaction chemistry, thus suggesting a solely structural role [217]. The lack of suitable ionizable groups typical of general base catalysts in RNA led to the hypothesis that a metal ion would be essential for Diels-Alder ribozyme catalysis especially since there are numerous precedents for Lewis-acid-catalyzed reactions in the organic literature [211].

Significantly, in the biological realm both metal-dependent and metal-independent Diels-Alder reactions exist. Macrophomate synthetase is a putative Diels-Alderase enzyme that coordinates Mg^{2+} to activate the dienophile [210,218,219]. In contrast, the artificial Diels-Alderase catalytic antibody 1E9 does not employ an ion [220]. At present, the current mechanism of action for the Diels-Alder ribozyme does not suggest a need for a Lewis acid. Rather, function is believed to arise from a variety of effects imparted by the RNA upon substrate binding; these include proximity, orientation, shape complementarity, and electrostatic stabilization. Some parallels have been drawn between the Diels-Alder ribozyme and the naturally occurring hairpin ribozyme, which employs many comparable catalytic strategies in its metal-free active site [221]. Future work on this ribozyme may be wise to consider crystallization of the ribozyme with a transition state mimic, or the use of computational analyses and alternative substrates to tease out plausible mechanistic pathways.

4. CONCLUSIONS AND FUTURE PROSPECTS

In this chapter we briefly reviewed the structural and functional properties for several naturally occurring small ribozymes including the hammerhead, hairpin, HDV, and *glmS* varieties. We also touched upon the fold and activity of three artificial ribozymes – the leadzyme, the flexizyme, and the Diels-Alder ribozyme – that serve as model systems to probe the boundaries of RNA chemistry and the emergence of catalysis in a pre-biotic RNA World. In all instances, we observed that metals such as Mg^{2+} are essential to establish a diverse number of three-dimensional folds. Significantly, most natural small ribozymes appear to utilize tertiary contacts to produce a stable fold that allows them to function at physiological, free-Mg^{2+} ion concentrations of 0.5 to 1.8 mM. This observation is especially relevant for the hammerhead and hairpin ribozymes whose early analyses were encumbered by the use of minimal constructs that were overly sensitive to nucleobase changes, thus confounding the ability to deconvolute folding from catalytic effects (reviewed in [19]). Often such minimal ribozymes required non-physiologically high levels of ions for stability as well. It is clear that numerous other hammerhead-like and HDV-like (*CPEB3*) ribozymes exist in the biosphere, and that the lessons reviewed here will be beneficial in the characterization of new family members.

In terms of the location and mode of multivalent ion binding, the preponderance of metals were observed to coordinate waters as $Mg(H_2O)_n^{2+}$ (where n = 4 to 6). Most frequently these ions were site-bound at tandem purines, especially guanine, which has been noted as a preferred cation binding site elsewhere [222]. Analogs of $Mg(H_2O)_6^{2+}$ such as $Co(NH_3)_6^{3+}$ and $Mn(H_2O)_6^{2+}$ were observed to bind similarly, which is not surprising due to their preferred octahedral coordination geometry and comparable metal-to-ligand interaction distances. Batey, Kieft et al. have exploited these observations to derive a series of RNA sequences selective for hexammine ion binding in order to facilitate crystallographic phasing [46].

With respect to catalytic function, we observed a series of metal-free and metal-occupied active sites. In the case of the hammerhead the role of the Mg^{2+} appears to be twofold. First it is proposed to act as a Lewis acid by depressing the pK_a of the Gua8 2′-OH, which serves as a general acid. Simultaneously, it coordinates a non-bridging oxygen of the scissile bond to provide electrostatic stabilization to the transition-state phosphorane [101]. A nearby Mg^{2+} fortifies the local fold near the 2′-nucleophile. Likewise, the HDV ribozyme, and possibly its *CPEB3* cousins, possess similar qualities in which a base, Cyt75(H+), serves as a general acid, but Mg^{2+} activates the 2′-nucleophile at Uri-1 by simultaneously functioning as a Lewis acid and conferring electrostatic stabilization to the phosphorane [171]. In contrast, the *glmS* ribozyme appears to use Gua33 as a general base and the N2-amino group of the metabolite, glcN6P, as a general acid. It is noteworthy that Gua is a central player in at least three of the small ribozymes reviewed in this chapter [223]. Two Mg^{2+} ions come into play in the glmS ribozyme by coordinating the metabolite phosphate group, which effectively positions glcN6P in the active site. This use of metals illustrates how RNA accrues specificity, affinity, and functionality from small molecules, which provides an important connection between modern day gene-regulatory elements (riboswitches) and ancient ribozymes of the primordial RNA World [27].

Our observations from natural and artificial ribozymes also revealed how metals can influence the stability of unusual structural elements whose folds are essential to orient key base or other functional groups for catalysis (e.g., the *glmS* and hairpin ribozymes). In the case of the hairpin ribozyme, $Co(NH_3)_6^{3+}$ resides within an E-loop motif that harbors the critical Ade38(H+) residue, which has been proposed to serve as a general acid in the cleavage reaction [105,139]. Some similarities are observed in the leadzyme, flexizyme, and Diels-Alder ribozyme in which $Mg(H_2O)_n^{2+}$ ions are located at key topological switch points in the RNA backbone that are critical for active site formation, but the ions do not participate directly in catalysis. In the case of the leadzyme, the disparate conformations observed in the crystal structure suggest that Mg^{2+} binding in lieu of Pb^{2+} (Site 3) may prohibit adoption of a conformation necessary for catalysis (i.e., a negative allosteric effector site) [198]. There appears to be no restriction on the number of ions that support active site formation in artificial ribozymes, nor is there a correlation between the number of site-bound ions and a ribozyme's size. One Mg^{2+} is observed in the flexizyme, two in the leadzyme (depending on the conformation adopted) and four in the Diels-Alder ribozyme. The number of site-bound ions in crystal structures may be significantly different from the ion atmosphere model of ions present in solution (reviewed in Section 2.1).

With the exception of the leadzyme ribozyme (and possibly the flexizyme), the preponderance of ribozymes reviewed here exhibits a preformed active site (i.e., high shape complementarity to the transition state). It has been pointed out that such ribozymes are suited to pre-folding because they each bind a relatively simple, single-step transition state [221]. In contrast, although the leadyzme is one of the simplest catalysts in terms of size, its structure is complex, possibly arising from its two-part reaction mechanism whereby the scissile bond is first cleaved and the resulting cyclic phosphodiester is hydrolyzed [187]. As such this tiny artificial ribozyme must accommodate two different transition states, which may be a source of active site heterogeneity. Other more accentuated multistep reactions that require core plasticity include spliceosome-mediated intron splicing [224] and RNA-dependent RNA polymerization [15]. These observations have implications for the complexity and strategies that must be entertained in future ribozyme structure determinations in order to capture defined and relevant conformational states.

As a final take home message, we observed that small ribozymes could possess hybrid properties. Although the majority of small ribozymes exploits bases to achieve chemical functionality [8,24,54], the large self-splicing introns employ multiple Mg^{2+} ions. The HDV and the hammerhead appear however to exhibit aspects of both families. Although they each employ RNA functional groups as general acid/base catalysts, Mg^{2+} functions not only as a Lewis acid but also as a means to confer "transition state stabilization". The binding of such ions has a predictable (but not obvious) impact on the ionization states of nearby bases, as demonstrated exquisitely by the anti-cooperativity of Cyt75(H+) and Mg^{2+} in the HDV ribozyme (Section 3.3). This electrostatic coupling can alter the properties of the active site, and influence the pK_a of key functional groups [225]. For the HDV ribozyme, this charge antagonism has been framed in the context of ground-state destabilization [226] that is relieved in the transition state, thus serving as a driving influence for catalysis [167]. Such crosstalk has been proposed to provide a basis to lower the pK_a of the 2′-nucleophile in the hairpin ribozyme, which is spatially close to an ionized N1 at Ade38 [105]. Similar scenarios can be envisioned for the hammerhead and *glmS* ribozymes, which must depress N1 imine pK_a values from Gua12 and Gua33, respectively, from 9.2 toward neutrality. At present the answers to these questions, and the exact push and pull of electrons for the artificial ribozymes, remain open issues that deserve considerably more attention.

ACKNOWLEDGMENTS

I thank Philip Bevilacqua and Barbara Golden for sharing unpublished data. I thank David B. McKay for supplying the leadzyme ribozyme data. I thank Clara Kielkopf and Doug Turner for stimulating conversations about RNA structure and chemistry. I am grateful to members of the Wedekind lab for helpful discussions. I am also grateful to Nils Walter, Dave Perrin, Darrin York, Mike Harris and Dan Herschlag for sharing their expertise on RNA catalysis. This work was supported by Public Health Service NIH Grants R01 GM063162 and S10 RR026501 to J.E.W.

ABBREVIATIONS

8BrGua	8-bromo-guanine nucleotide
CSD	Cambridge Structural Database
DAI	(3AS,9AS)-2-pentyl-4-hydroxymethyl-3A,4,9,9A-tetra-hydro-4,9[1′,2′]-benzeno-1H-benz[F]isoindole-1,3(2H)-dione
e^-	electron
glcN6P	glucosamine-N-6-phosphate
glmS	the gene encoding L-glutamine: D-fructose-6-phosphate amidotransferase
GTP	guanosine 5′-triphosphate
HBV	hepatitis B virus
HDV	hepatitis delta virus
MAD	multiwavelength anomalous diffraction
MD	molecular dynamics
mRNA	messenger RNA
ncRNA	non-coding RNA
PDB	Protein Database
QM/MM	quantum mechanical molecular mechanics
rmsd	root-mean-square distance displacement

tRNA	transfer RNA
UTR	untranslated region of mRNA
VS	Varkud satellite

REFERENCES

1. P. C. Bevilacqua, *Biopolymers*, 2004, **73**, 69–70.
2. H. F. Noller, *Annu. Rev. Biochem.*, 1991, **60**, 191–227.
3. L. Randau, I. Schroder and D. Soll, *Nature*, 2008, **453**, 120–123.
4. I. U. Heinemann, D. Soll and L. Randau, *FEBS Lett.*, 2010, **584**, 303–309.
5. J. L. Chen and N. R. Pace, *RNA*, 1997, **3**, 557–560.
6. F. H. Westheimer, *Science*, 1987, **235**, 1173–1178.
7. F. H. Westheimer, *Nature*, 1986, **319**, 534–535.
8. M. J. Fedor and J. R. Williamson, *Nat. Rev. Mol. Cell Biol.*, 2005, **6**, 399–412.
9. T. Pan and O. C. Uhlenbeck, *Biochemistry*, 1992, **31**, 3887–3895.
10. J. S. Weinger, K. M. Parnell, S. Dorner, R. Green and S. A. Strobel, *Nat. Struct. Mol. Biol.*, 2004, **11**, 1101–1106.
11. H. Murakami, H. Saito and H. Suga, *Chem. Biol.*, 2003, **10**, 655–662.
12. N. Li and F. Huang, *Biochemistry*, 2005, **44**, 4582–4590.
13. J. C. Schlatterer, F. Stuhlmann and A. Jäschke, *ChemBioChem*, 2003, **4**, 1089–1092.
14. P. J. Unrau and D. P. Bartel, *Nature*, 1998, **395**, 260–263.
15. W. K. Johnston, P. J. Unrau, M. S. Lawrence, M. E. Glasner and D. P. Bartel, *Science*, 2001, **292**, 1319–1325.
16. S. Tsukiji, S. B. Pattnaik and H. Suga, *J. Am. Chem. Soc.*, 2004, **126**, 5044–5045.
17. R. Fiammengo and A. Jäschke, *Curr. Opin. Biotechnol.*, 2005, **16**, 614–621.
18. G. F. Joyce, *Nature*, 2002, **418**, 214–221.
19. M. J. Fedor, *Annu. Rev. Biophys.*, 2009, **38**, 271–299.
20. C. H. Webb, N. J. Riccitelli, D. J. Ruminski and A. Luptak, *Science*, 2009, **326**, 953.
21. S. Graf, R. Przybilski, G. Steger and C. Hammann, *Biochem. Soc. Trans.*, 2005, **33**, 477–478.
22. K. Salehi-Ashtiani and J. W. Szostak, *Nature*, 2001, **414**, 82–84.
23. P. C. Bevilacqua, T. S. Brown, S. Nakano and R. Yajima, *Biopolymers*, 2004, **73**, 90–109.
24. P. C. Bevilacqua and R. Yajima, *Curr. Opin. Chem. Biol.*, 2006, **10**, 455–464.
25. J. B. Murray, A. A. Seyhan, N. G. Walter, J. M. Burke and W. G. Scott, *Chem. Biol.*, 1998, **5**, 587–595.
26. A. M. Weiner and N. Maizels, *Proc. Natl. Acad. Sci. USA*, 1987, **84**, 7383–7387.
27. J. C. Cochrane and S. A. Strobel, *RNA*, 2008, **14**, 993–1002.
28. S. R. Eddy, *Nat. Rev. Genet.*, 2001, **2**, 919–929.
29. F. F. Costa, *Gene*, 2007, **386**, 1–10.
30. R. J. Taft, K. C. Pang, T. R. Mercer, M. Dinger and J. S. Mattick, *J. Pathol.*, 2010, **220**, 126–139.
31. M. J. Laires, C. P. Monteiro and M. Bicho, *Front. Biosci.*, 2004, **9**, 262–276.

32. V. K. Misra and D. E. Draper, *Biopolymers*, 1998, **48**, 113–135.
33. D. E. Draper, *RNA*, 2004, **10**, 335–343.
34. D. E. Draper, *Biophys. J.*, 2008, **95**, 5489–5495.
35. V. J. DeRose, *Curr. Opin. Struct. Biol.*, 2003, **13**, 317–324.
36. A. M. Pyle, *J. Biol. Inorg. Chem.*, 2002, **7**, 679–690.
37. R. K. O. Sigel and A. M. Pyle, *Chem. Rev.*, 2007, **107**, 97–113.
38. A. L. Feig and O. C. Uhlenbeck, in *The RNA World*, eds. R. F. Gesteland, T. R. Cech and J. F. Atkins, Cold Spring Harbor Press, Cold Spring Harbor, 1999, p. 709.
39. J. L. Smith, in *CCP4 Study Weekend*, eds. K. S. Wilson, G. Davies, A. W. Ashton and S. Bailey, CCLRC, Daresbury Laboratory, 1997, pp. 1–10.
40. D. C. Creagh, in *International Tables for Crystallography: mathematical, physical and chemical tables*, ed. E. Prince, International Union of Crystallography, Chester, UK, 2006, pp. 241–258.
41. F. H. Allen, *Acta Crystallogr. B*, 2002, **58**, 380–388.
42. R. W. Jou and J. A. Cowan, *J. Am. Chem. Soc.*, 1991, **113**, 6685–6686.
43. S. Alam, V. Grum-Tokars, J. Krucinska, M. L. Kundracik and J. E. Wedekind, *Biochemistry*, 2005, **44**, 14396–14408.
44. D. Lambert, J. E. Heckman and J. M. Burke, *Biochemistry*, 2006, **45**, 829–838.
45. J. A. Doudna, *Nat. Chem. Biol.*, 2005, **1**, 300–303.
46. A. Y. Keel, R. P. Rambo, R. T. Batey and J. S. Kieft, *Structure*, 2007, **15**, 761–772.
47. V. Ramakrishnan, *Biochem. Soc. Trans.*, 2008, **36**, 567–574.
48. P. A. Lay, R. H. Magnuson and H. Taube, *Inorg. Chem.*, 1989, **28**, 3001–3007.
49. P. A. Lay and H. Taube, *Inorg. Chem.*, 1989, **28**, 3561–3564.
50. J. H. Cate and J. A. Doudna, *Structure*, 1996, **4**, 1221–1229.
51. D. E. Brodersen, W. M. Clemons, Jr., A. P. Carter, B. T. Wimberly and V. Ramakrishnan, *Acta Crystallogr. D Biol. Crystallogr.*, 2003, **59**, 2044–2050.
52. B. L. Golden, A. R. Gooding, E. R. Podell and T. R. Cech, *Science*, 1998, **282**, 259–264.
53. E. Ennifar, P. Walter and P. Dumas, *Nucleic Acids Res.*, 2003, **31**, 2671–2682.
54. J. A. Doudna and J. R. Lorsch, *Nat. Struct. Mol. Biol.*, 2005, **12**, 395–402.
55. J. M. Buzayan, W. L. Gerlach and G. Bruening, *Nature*, 1986, **323**, 349–353.
56. C. J. Hutchins, P. D. Rathjen, A. C. Forster and R. H. Symons, *Nucleic Acids Res.*, 1986, **14**, 3627–3640.
57. G. A. Prody, J. T. Bakos, J. M. Buzayan, I. R. Schneider and G. Bruening, *Science*, 1986, **231**, 1577–1580.
58. R. H. Symons, *Annu. Rev. Biochem*, 1992, **61**, 641–671.
59. R. H. Symons, *Trends in Biochemi.cal Sciences*, 1989, **14**, 445–450.
60. J. M. Buzayan, W. L. Gerlach and G. Bruening, *Proc. Natl. Acad. Sci. USA*, 1986, **83**, 8859–8862.
61. C. C. Sheldon and R. H. Symons, *Nucleic Acids Res.*, 1989, **17**, 5679–5685.
62. D. E. Ruffner, G. D. Stormo and O. C. Uhlenbeck, *Biochemistry*, 1990, **29**, 10695–10702.
63. J. C. Penedo, T. J. Wilson, S. D. Jayasena, A. Khvorova and D. M. Lilley, *RNA*, 2004, **10**, 880–888.

64. E. Tan, T. J. Wilson, M. K. Nahas, R. M. Clegg, D. M. Lilley and T. Ha, *Proc. Natl. Acad. Sci. USA*, 2003, **100**, 9308–9313.
65. M. De la Pena, S. Gago and R. Flores, *EMBO J.*, 2003, **22**, 5561–5570.
66. A. Khvorova, A. Lescoute, E. Westhof and S. D. Jayasena, *Nat. Struct. Biol.*, 2003, **10**, 708–712.
67. H. W. Pley, K. M. Flaherty and D. B. McKay, *Nature*, 1994, **372**, 68–74.
68. D. B. McKay, *RNA*, 1996, **2**, 395–403.
69. J. E. Wedekind and D. B. McKay, *Annu. Rev. Biophys. Biomol. Struct.*, 1998, **27**, 475–502.
70. A. C. Forster and R. H. Symons, *Cell*, 1987, **50**, 9–16.
71. A. C. Forster and R. H. Symons, *Cell*, 1987, **49**, 211–220.
72. L. M. Epstein and J. G. Gall, *Cell*, 1987, **48**, 535–543.
73. L. M. Epstein and J. G. Gall, *Cold Spring Harbour Symp. Quant. Biol.*, 1987, **52**, 261–265.
74. G. Ferbeyre, J. M. Smith and R. Cedergren, *Mol. Cell. Biol.*, 1998, **18**, 3880–3888.
75. A. A. Rojas, A. Vazquez–Tello, G. Ferbeyre, F. Venanzetti, L. Bachmann, B. Paquin, V. Sbordoni and R. Cedergren, *Nucleic Acids Res.*, 2000, **28**, 4037–4043.
76. R. Przybilski, S. Graf, A. Lescoute, W. Nellen, E. Westhof, G. Steger and C. Hammann, *Plant Cell*, 2005, **17**, 1877–1885.
77. M. Martick, L. H. Horan, H. F. Noller and W. G. Scott, *Nature*, 2008, **454**, 899–902.
78. M. Martick and W. G. Scott, *Cell*, 2006, **126**, 309–320.
79. J. A. Nelson and O. C. Uhlenbeck, *Mol. Cell*, 2006, **23**, 447–450.
80. J. A. Nelson and O. C. Uhlenbeck, *RNA*, 2008, **14**, 605–615.
81. R. Przybilski and C. Hammann, *ChemBioChem*, 2006, **7**, 1641–1644.
82. W. G. Scott, J. T. Finch and A. Klug, *Cell*, 1995, **81**, 991–1002.
83. M. D. Canny, F. M. Jucker, E. Kellogg, A. Khvorova, S. D. Jayasena and A. Pardi, *J. Am. Chem. Soc.*, 2004, **126**, 10848–10849.
84. M. D. Canny, F. M. Jucker and A. Pardi, *Biochemistry*, 2007, **46**, 3826–3834.
85. J. A. Nelson, I. Shepotinovskaya and O. C. Uhlenbeck, *Biochemistry*, 2005, **44**, 14577–14585.
86. E. M. Osborne, J. E. Schaak and V. J. Derose, *RNA*, 2005, **11**, 187–196.
87. M. Roychowdhury-Saha and D. H. Burke, *RNA*, 2006, **12**, 1846–1852.
88. I. V. Shepotinovskaya and O. C. Uhlenbeck, *Biochemistry*, 2008, **47**, 7034–7042.
89. J. Han and J. M. Burke, *Biochemistry*, 2005, **44**, 7864–7870.
90. S. C. Dahm, W. B. Derrick and O. C. Uhlenbeck, *Biochemistry*, 1993, **32**, 13040–13045.
91. Y. Takagi, M. Warashina, W. J. Stec, K. Yoshinari and K. Taira, *Nucleic Acids Res.*, 2001, **29**, 1815–1834.
92. J. L. O'Rear, S. Wang, A. L. Feig, L. Beigelman, O. C. Uhlenbeck and D. Herschlag, *RNA*, 2001, **7**, 537–545.
93. N. K. Kim, A. Murali and V. J. DeRose, *J. Am. Chem. Soc.*, 2005, **127**, 14134–14135.
94. J. L. Boots, M. D. Canny, E. Azimi and A. Pardi, *RNA*, 2008, **14**, 2212–2222.
95. M. Martick, T. S. Lee, D. M. York and W. G. Scott, *Chem. Biol.*, 2008, **15**, 332–342.

96. G. Slim and M. J. Gait, *Nucleic Acids Res.*, 1991, **19**, 1183–1188.
97. M. Koizumi and E. Ohtsuka, *Biochemistry*, 1991, **30**, 5145–5150.
98. S. Wang, K. Karbstein, A. Peracchi, L. Beigelman and D. Herschlag, *Biochemistry*, 1999, **38**, 14363–14378.
99. V. L. Pecoraro, J. D. Hermes and W. W. Cleland, *Biochemistry*, 1984, **23**, 5262–5271.
100. K. Suzumura, Y. Takagi, M. Orita and K. Taira, *J. Am. Chem. Soc.*, 2004, **126**, 15504–15511.
101. T. S. Lee, C. Silva Lopez, G. M. Giambasu, M. Martick, W. G. Scott and D. M. York, *J. Am. Chem. Soc.*, 2008, **130**, 3053–3064.
102. T. S. Lee, G. M. Giambasu, C. P. Sosa, M. Martick, W. G. Scott and D. M. York, *J. Mol. Biol.*, 2009, **388**, 195–206.
103. D. M. Perreault and E. V. Anslyn, *Angew. Chem. Int. Ed. Engl.*, 1997, **36**, 432–450.
104. D. M. Zhou and K. Taira, *Chem. Rev.*, 1998, **98**, 991–1026.
105. M. Guo, R. C. Spitale, R. Volpini, J. Krucinska, G. Cristalli, P. R. Carey and J. E. Wedekind, *J. Am. Chem. Soc.*, 2009, **131**, 12908–12909.
106. B. Gong, J. H. Chen, E. Chase, D. M. Chadalavada, R. Yajima, B. L. Golden, P. C. Bevilacqua and P. R. Carey, *J. Am. Chem. Soc.*, 2007, **129**, 13335–13342.
107. A. Hampel, R. Tritz, M. Hicks and P. Cruz, *Nucleic Acids Res*, 1990, **18**, 299–304.
108. A. Hampel and J. A. Cowan, *Chem. Biol.*, 1997, **4**, 513–517.
109. K. J. Young, F. Gill and J. A. Grasby, *Nucleic Acids Res.*, 1997, **25**, 3760–3766.
110. S. Nesbitt, L. A. Hegg and M. J. Fedor, *Chem. Biol.*, 1997, **4**, 619–630.
111. S. P. Ryder, A. K. Oyelere, J. L. Padilla, D. Klostermeier, D. P. Millar and S. A. Strobel, *RNA*, 2001, **7**, 1454–1463.
112. S. P. Ryder and S. A. Strobel, *J. Mol. Biol.*, 1999, **291**, 295–311.
113. S. P. Ryder and S. A. Strobel, *Nucleic Acids Res.*, 2002, **30**, 1287–1291.
114. R. Pinard, K. J. Hampel, J. E. Heckman, D. Lambert, P. A. Chan, F. Major and J. M. Burke, *EMBO J.*, 2001, **20**, 6434–6442.
115. R. Pinard, D. Lambert, N. G. Walter, J. E. Heckman, F. Major and J. M. Burke, *Biochemistry*, 1999, **38**, 16035–16039.
116. B. M. Chowrira, A. Berzal-Herranz, C. F. Keller and J. M. Burke, *J. Biol. Chem.*, 1993, **268**, 19458–19462.
117. A. Berzal-Herranz, S. Joseph, B. M. Chowrira, S. E. Butcher and J. M. Burke, *EMBO J.*, 1993, **12**, 2567–2573.
118. A. I. Murchie, J. B. Thomson, F. Walter and D. M. Lilley, *Mol. Cell*, 1998, **1**, 873–881.
119. F. Walter, A. I. Murchie and D. M. Lilley, *Biochemistry*, 1998, **37**, 17629–17636.
120. F. Walter, A. I. Murchie, J. B. Thomson and D. M. Lilley, *Biochemistry*, 1998, **37**, 14195–14203.
121. T. J. Wilson, Z. Y. Zhao, K. Maxwell, L. Kontogiannis and D. M. Lilley, *Biochemistry*, 2001, **40**, 2291–2302.
122. S. E. Butcher, J. E. Heckman and J. M. Burke, *J. Biol. Chem.*, 1995, **270**, 29648–29651.
123. J. A. Esteban, N. G. Walter, G. Kotzorek, J. E. Heckman and J. M. Burke, *Proc. Natl. Acad. Sci. USA*, 1998, **95**, 6091–6096.

124. D. Klostermeier and D. P. Millar, *Biochemistry*, 2000, **39**, 12970–12978.
125. D. Klostermeier and D. P. Millar, *Biochemistry*, 2001, **40**, 11211–11218.
126. N. G. Walter, J. M. Burke and D. P. Millar, *Nat. Struct. Biol.*, 1999, **6**, 544–549.
127. R. C. Spitale and J. E. Wedekind, *Methods*, 2009, **49**, 87–100.
128. A. T. Torelli, J. Krucinska and J. E. Wedekind, *RNA*, 2007, **13**, 1052–1070.
129. C. MacElrevey, R. C. Spitale, J. Krucinska and J. E. Wedekind, *Acta Crystallogr. D Biol. Crystallogr.*, 2007, **63**, 812–825.
130. P. B. Rupert and A. R. Ferré-D'Amaré, *Nature*, 2001, **410**, 780–786.
131. P. B. Rupert, A. P. Massey, S. T. Sigurdsson and A. R. Ferré-D'Amaré, *Science*, 2002, **298**, 1421–1424.
132. A. R. Ferré-D'Amaré and P. B. Rupert, *Biochem. Soc. Trans.*, 2002, **30**, 1105–1109.
133. Y. I. Kuzmin, C. P. Da Costa, J. W. Cottrell and M. J. Fedor, *J. Mol. Biol.*, 2005, **349**, 989–1010.
134. Y. I. Kuzmin, C. P. Da Costa and M. J. Fedor, *J. Mol. Biol.*, 2004, **340**, 233–251.
135. L. L. Lebruska, Y. I. Kuzmine and M. J. Fedor, *Chem Biol.*, 2002, **9**, 465–473.
136. K. Nam, J. Gao and D. M. York, *RNA*, 2008, **14**, 1501–1507.
137. P. C. Bevilacqua, *Biochemistry*, 2003, **42**, 2259–2265.
138. L. Liu, J. W. Cottrell, L. G. Scott and M. J. Fedor, *Nat. Chem. Biol.*, 2009, **5**, 351–357.
139. R. C. Spitale, R. Volpini, M. G. Heller, J. Krucinska, G. Cristalli and J. E. Wedekind, *J. Am. Chem. Soc.*, 2009, **131**, 6093–6095.
140. R. C. Spitale, R. Volpini, M. V. Mungillo, J. Krucinska, G. Cristalli and J. E. Wedekind, *Biochemistry*, 2009, **48**, 7777–7779.
141. S. R. Das and J. A. Piccirilli, *Nat. Chem. Biol.*, 2005, **1**, 45–52.
142. Q. Dai, J. K. Frederiksen, V. E. Anderson, M. E. Harris and J. A. Piccirilli, *J. Org. Chem.*, 2008, **73**, 309–311.
143. M. J. Fedor, *J. Mol. Biol.*, 2000, **297**, 269–291.
144. B. M. Chowrira and J. M. Burke, *Nucleic Acids Res.*, 1992, **20**, 2835–2840.
145. B. M. Chowrira and J. M. Burke, *Biochemistry*, 1991, **30**, 8518–8522.
146. K. S. Wang, Q. L. Choo, A. J. Weiner, J. H. Ou, R. C. Najarian, R. M. Thayer, G. T. Mullenbach, K. J. Denniston, J. L. Gerin and M. Houghton, *Nature*, 1986, **323**, 508–514.
147. T. B. Macnaughton, S. T. Shi, L. E. Modahl and M. M. Lai, *J. Virol.*, 2002, **76**, 3920–3927.
148. P. J. Chen, G. Kalpana, J. Goldberg, W. Mason, B. Werner, J. Gerin and J. Taylor, *Proc. Natl. Acad. Sci. USA*, 1986, **83**, 8774–8778.
149. S. Makino, M. F. Chang, C. K. Shieh, T. Kamahora, D. M. Vannier, S. Govindarajan and M. M. Lai, *Nature*, 1987, **329**, 343–346.
150. M. Rizzetto, *Hepatology*, 1983, **3**, 729–737.
151. G. Fattovich, G. Giustina, E. Christensen, M. Pantalena, I. Zagni, G. Realdi and S. W. Schalm, *Gut*, 2000, **46**, 420–426.
152. K. Salehi-Ashtiani, A. Luptak, A. Litovchick and J. W. Szostak, *Science*, 2006, **313**, 1788–1792.
153. I. H. Shih and M. D. Been, *Annu. Rev. Biochem.*, 2002, **71**, 887–917.

154. Y. A. Suh, P. K. Kumar, K. Taira and S. Nishikawa, *Nucleic Acids Res.*, 1993, **21**, 3277–3280.
155. S. Nakano, A. L. Cerrone and P. C. Bevilacqua, *Biochemistry*, 2003, **42**, 2982–2994.
156. S. Nakano, D. J. Proctor and P. C. Bevilacqua, *Biochemistry*, 2001, **40**, 12022–12038.
157. A. Ke, K. Zhou, F. Ding, J. H. Cate and J. A. Doudna, *Nature*, 2004, **429**, 201–205.
158. A. R. Ferré-D'Amaré, K. Zhou and J. A. Doudna, *Nature*, 1998, **395**, 567–574.
159. P. Banas, L. Rulisek, V. Hanosova, D. Svozil, N. G. Walter, J. Sponer and M. Otyepka, *J. Phys. Chem. B.*, 2008, **112**, 11177–11187.
160. M. V. Krasovska, J. Sefcikova, N. Spackova, J. Sponer and N. G. Walter, *J. Mol. Biol.*, 2005, **351**, 731–748.
161. J. A. Grasby, K. Mersmann, M. Singh and M. J. Gait, *Biochemistry*, 1995, **34**, 4068–4076.
162. A. T. Perrotta and M. D. Been, *Biochemistry*, 2006, **45**, 11357–11365.
163. A. T. Torelli, R. C. Spitale, J. Krucinska and J. E. Wedekind, *Biochem. Biophys. Res. Commun.*, 2008, **371**, 154–158.
164. A. T. Perrotta and M. D. Been, *Nucleic Acids Res.*, 1996, **24**, 1314–1321.
165. M. D. Been and G. S. Wickham, *Eur. J. Biochem.*, 1997, **247**, 741–753.
166. A. T. Perrotta, I. Shih and M. D. Been, *Science*, 1999, **286**, 123–126.
167. S. Nakano, D. M. Chadalavada and P. C. Bevilacqua, *Science*, 2000, **287**, 1493–1497.
168. B. Gong, Y. Chen, E. L. Christian, J. H. Chen, E. Chase, D. M. Chadalavada, R. Yajima, B. L. Golden, P. C. Bevilacqua and P. R. Carey, *J. Am. Chem. Soc.*, 2008, **130**, 9670–9672.
169. J. H. Chen, B. Gong, P. C. Bevilacqua, P. R. Carey and B. L. Golden, *Biochemistry*, 2009, **48**, 1498–1507.
170. B. Gong, J. H. Chen, P. C. Bevilacqua, B. L. Golden and P. R. Carey, *Biochemistry*, 2009, **48**, 11961–11970.
171. J. H. Chen, R. Yajima, D. M. Chadalavada, E. Chase, P. C. Bevilacqua and B. L. Golden, *Biochemistry*, 2010, **49**, 6508–6518.
172. S. A. Strobel, *Nature Chem. Biol.*, 2005, **1**, 5–6.
173. W. C. Winkler, A. Nahvi, A. Roth, J. A. Collins and R. R. Breaker, *Nature*, 2004, **428**, 281–286.
174. J. E. Barrick, K. A. Corbino, W. C. Winkler, A. Nahvi, M. Mandal, J. Collins, M. Lee, A. Roth, N. Sudarsan, I. Jona, J. K. Wickiser and R. R. Breaker, *Proc. Natl. Acad. Sci. USA*, 2004, **101**, 6421–6426.
175. S. Milewski, *Biochim. Biophys. Acta*, 2002, **1597**, 173–192.
176. T. J. McCarthy, M. A. Plog, S. A. Floy, J. A. Jansen, J. K. Soukup and G. A. Soukup, *Chem. Biol.*, 2005, **12**, 1221–1226.
177. J. Lim, B. C. Grove, A. Roth and R. R. Breaker, *Angew. Chem. Int. Ed. Engl.*, 2006, **45**, 6689–6693.
178. J. C. Cochrane, S. V. Lipchock, K. D. Smith and S. A. Strobel, *Biochemistry*, 2009, **48**, 3239–3246.
179. K. J. Hampel and M. M. Tinsley, *Biochemistry*, 2006, **45**, 7861–7871.

180. D. J. Klein and A. R. Ferré-D'Amaré, *Science*, 2006, **313**, 1752–1756.
181. J. C. Cochrane, S. V. Lipchock and S. A. Strobel, *Chem. Biol.*, 2007, **14**, 97–105.
182. R. A. Tinsley, J. R. Furchak and N. G. Walter, *RNA*, 2007, **13**, 468–477.
183. A. Roth, A. Nahvi, M. Lee, I. Jona and R. R. Breaker, *RNA*, 2006, **12**, 607–619.
184. D. J. Klein, S. R. Wilkinson, M. D. Been and A. R. Ferré-D'Amaré, *J. Mol. Biol.*, 2007, **373**, 178–189.
185. D. J. Klein, M. D. Been and A. R. Ferré-D'Amaré, *J. Am. Chem. Soc.*, 2007, **129**, 14858–14859.
186. B. Gong, J. H. Chen, R. Yajima, Y. Chen, E. Chase, D. M. Chadalavada, B. L. Golden, P. R. Carey and P. C. Bevilacqua, *Methods*, 2009, **49**, 101–111.
187. T. Pan and O. C. Uhlenbeck, *Nature*, 1992, **358**, 560–563.
188. M. Z. Barciszewska, M. Szymanski, E. Wyszko, J. Pas, L. Rychlewski and J. Barciszewski, *Mutat. Res.*, 2005, **589**, 103–110.
189. R. T. Raines, *Chem. Rev.*, 1998, **98**, 1045–1066.
190. J. R. Rubin and M. Sundaralingam, *J. Biomol. Struct. Dyn.*, 1983, **1**, 639–646.
191. R. S. Brown, B. E. Hingerty, J. C. Dewan and A. Klug, *Nature*, 1983, **303**, 543–546.
192. R. S. Brown, J. C. Dewan and A. Klug, *Biochemistry*, 1985, **24**, 4785–4801.
193. C. G. Hoogstraten, P. Legault and A. Pardi, *J. Mol. Biol.*, 1998, **284**, 337–350.
194. S. Lemieux, P. Chartrand, R. Cedergren and F. Major, *RNA*, 1998, **4**, 739–749.
195. K. R. Julien, M. Sumita, P. H. Chen, I. A. Laird-Offringa and C. G. Hoogstraten, *RNA*, 2008, **14**, 1632–1643.
196. R. Yajima, D. J. Proctor, R. Kierzek, E. Kierzek and P. C. Bevilacqua, *Chem. Biol.*, 2007, **14**, 23–30.
197. B. M. Kadakkuzha, L. Zhao and T. Xia, *Biochemistry*, 2009, **48**, 3807–3809.
198. J. E. Wedekind and D. B. McKay, *Biochemistry*, 2003, **42**, 9554–9563.
199. J. E. Wedekind and D. B. McKay, *Nat. Struct. Biol.*, 1999, **6**, 261–268.
200. R. Russell, X. Zhuang, H. P. Babcock, I. S. Millett, S. Doniach, S. Chu and D. Herschlag, *Proc. Natl. Acad. Sci. USA*, 2002, **99**, 155–160.
201. P. Legault, C. G. Hoogstraten, E. Metlitzky and A. Pardi, *J. Mol. Biol.*, 1998, **284**, 325–335.
202. E. Westhof and T. Hermann, *Nat. Struct. Biol.*, 1999, **6**, 208–209.
203. T. Pan, B. Dichtl and O. C. Uhlenbeck, *Biochemistry*, 1994, **33**, 9561–9565.
204. C. E. Dann, 3rd, C. A. Wakeman, C. L. Sieling, S. C. Baker, I. Irnov and W. C. Winkler, *Cell*, 2007, **130**, 878–892.
205. H. Xiao, H. Murakami, H. Suga and A. R. Ferré-D'Amaré, *Nature*, 2008, **454**, 358–361.
206. S. Cusack, A. Yaremchuk and M. Tukalo, *EMBO J.*, 1996, **15**, 2834–2842.
207. L. Moulinier, S. Eiler, G. Eriani, J. Gangloff, J. C. Thierry, K. Gabriel, W. H. McClain and D. Moras, *EMBO J.*, 2001, **20**, 5290–5301.
208. A. Serganov, S. Keiper, L. Malinina, V. Tereshko, E. Skripkin, C. Hobartner, A. Polonskaia, A. T. Phan, R. Wombacher, R. Micura, Z. Dauter, A. Jäschke and D. J. Patel, *Nat. Struct. Mol. Biol.*, 2005, **12**, 218–224.
209. G. F. Joyce and L. E. Orgel, in *The RNA World*, eds. R. F. Gesteland and J. F. Atkins, Cold Spring Harbor Laboratory Press, Cold Spring Harbor, 1993, pp. 1–25.

210. T. Ose, K. Watanabe, T. Mie, M. Honma, H. Watanabe, M. Yao, H. Oikawa and I. Tanaka, *Nature*, 2003, **422**, 185–189.
211. E. M. Stocking and R. M. Williams, *Angew. Chem. Int. Ed. Engl.*, 2003, **42**, 3078–3115.
212. T. M. Tarasow, S. L. Tarasow and B. E. Eaton, *Nature*, 1997, **389**, 54–57.
213. B. Seelig and A. Jäschke, *Chem. Biol.*, 1999, **6**, 167–176.
214. B. Seelig, S. Keiper, F. Stuhlmann and A. Jäschke, *Angew. Chem. Int. Ed. Engl.*, 2000, **39**, 4576–4579.
215. Q. Du, N. Carrasco, M. Teplova, C. J. Wilds, M. Egli and Z. Huang, *J. Am. Chem. Soc.*, 2002, **124**, 24–25.
216. F. Stuhlmann and A. Jäschke, *J. Am. Chem. Soc.*, 2002, **124**, 3238–3244.
217. S. Keiper, D. Bebenroth, B. Seelig, E. Westhof and A. Jäschke, *Chem. Biol.*, 2004, **11**, 1217–1227.
218. T. Ose, K. Watanabe, M. Yao, M. Honma, H. Oikawa and I. Tanaka, *Acta Crystallogr. D Biol. Crystallogr.*, 2004, **60**, 1187–1197.
219. J. M. Serafimov, D. Gillingham, S. Kuster and D. Hilvert, *J. Am. Chem. Soc.*, 2008, **130**, 7798–7799.
220. J. Xu, Q. Deng, J. Chen, K. N. Houk, J. Bartek, D. Hilvert and I. A. Wilson, *Science*, 1999, **286**, 2345–2348.
221. J. N. Pitt and A. R. Ferré-D'Amaré, *Nat. Struct. Mol. Biol.*, 2005, **12**, 206–208.
222. B. E. Hingerty, R. S. Brown and A. Klug, *Biochim. Biophys. Acta*, 1982, **697**, 78–82.
223. J. C. Cochrane and S. A. Strobel, *Acc. Chem. Res.*, 2008, **41**, 1027–1035.
224. E. J. Sontheimer, *Nat. Struct. Biol.*, 2001, **8**, 11–13.
225. K. Chin, K. A. Sharp, B. Honig and A. M. Pyle, *Nat. Struct. Biol.*, 1999, **6**, 1055–1061.
226. W. P. Jencks, *Catalysis and Enzymology*, Dover, New York, 1987.
227. G. A. Soukup and R. R. Breaker, *RNA*, 1999, **5**, 1308–1325.
228. W. L. DeLano, *DeLano Scientific*, San Carlos, California, 2002.
229. G. J. Kleywegt, M. R. Harris, J. Y. Zou, T. C. Taylor, A. Wahlby and T. A. Jones, *Acta Crystallogr. D Biol. Crystallogr.*, 2004, **60**, 2240–2249.

Met. Ions Life Sci. **2011**, *9*, 347–377

12

Binding of Kinetically Inert Metal Ions to RNA: The Case of Platinum(II)

Erich G. Chapman, Alethia A. Hostetter, Maire F. Osborn, Amanda L. Miller, and Victoria J. DeRose

Department of Chemistry, University of Oregon, Eugene OR 97403, USA

< derose@uoregon.edu >

Metal Ions in Life Sciences, Volume 9 Edited by Astrid Sigel, Helmut Sigel, and Roland K. O. Sigel

Published by the Royal Society of Chemistry, www.rsc.org DOI: 10.1039/978184973251200347

ABSTRACT: In this chapter several aspects of Pt(II) are highlighted that focus on the properties of Pt(II)-RNA adducts and the possibility that they influence RNA-based processes in cells. Cellular distribution of Pt(II) complexes results in significant platination of RNA, and localization studies find Pt(II) in the nucleus, nucleolus, and a distribution of other sites in cells. Treatment with Pt(II) compounds disrupts RNA-based processes including enzymatic processing, splicing, and translation, and this disruption may be indicative of structural changes to RNA or RNA-protein complexes. Several RNA-Pt(II) adducts have been characterized *in vitro* by biochemical and other methods. Evidence for Pt(II) binding in non-helical regions and for Pt(II) cross-linking of internal loops has been found. Although platinated sites have been identified, there currently exists very little in the way of detailed structural characterization of RNA-Pt(II) adducts. Some insight into the details of Pt(II) coordination to RNA, especially RNA helices, can be gained from DNA model systems. Many RNA structures, however, contain complex tertiary folds and common, purine-rich structural elements that present suitable Pt(II) nucleophiles in unique arrangements which may hold the potential for novel types of platinum-RNA adducts. Future research aimed at structural characterization of platinum-RNA adducts may provide further insights into platinum-nucleic acid binding motifs, and perhaps provide a rationale for the observed inhibition by Pt(II) complexes of splicing, translation, and enzymatic processing.

KEYWORDS: chemotherapeutic · cisplatin · metal coordination · metals · platinum · ribonucleic acids · ribosome · ribozymes · RNA · translation · tRNA

1. INTRODUCTION

Nucleic acids are targets of metal-based therapeutic agents, the most extensively studied being the Pt(II) anticancer compounds [1–6]. *Cis*-diamminedichloro Pt(II) (cisplatin) and others of the class of square planar Pt(II) compounds form stable complexes with DNA. In the most common adducts, the *cis*-diammine Pt(II) moiety is coordinated to N7 imino nitrogen ligands of neighboring purine nucleobases to create a bidentate Pt(II)-oligonucleotide adduct. Recognition of cellular Pt(II)-DNA adducts ultimately leads to apoptosis, one basis of the antitumor properties of these compounds.

Much less studied are the interactions of Pt(II) and related metal compounds with RNA. As reviewed in this article, a growing list of evidence suggests that understanding the effects of metal coordination to cellular RNAs may be an important part of a comprehensive description of the molecular mechanisms involved in drug activity. Early experiments found that incubation of cells with Pt(II) compounds resulted in significant levels of Pt(II) bound to extracted RNA [7]. Correspondingly, treatment of cell extracts with cisplatin has been shown to inhibit important RNA-dependent processes such as translation [8–11] and splicing [12,13], activities that are both mediated by RNA structure and RNA-protein interactions. Additionally, cisplatin treatment has been shown to inhibit the activity of the Group I intron, a complexly folded ribozyme [13]. This observation is similar to recent findings describing the ability of Pt(II) compounds to coordinate to tRNA [14,15] and across RNA internal loops [16] *in vitro*. These studies and others highlighting how Pt(II) may affect RNA biology are reviewed in more detail in the following sections. Although these reports suggest novel interactions between Pt(II) compounds and structured RNAs, very few molecular-level investigations into the mechanisms underlying Pt(II) coordination to RNA have been reported, even as the field of RNA biology has grown considerably. Such interactions are of great interest, given growing recognition of the enormous influence of RNA-controlled cellular processes [17].

There have been several recent reviews of the interactions of RNA with other metal ions [18–22]. In general, these involve RNA metal sites that are under thermodynamic equilibrium, with relatively fast ligand exchange kinetics between hexahydrated metal ions and RNA ligands. The square-planar Pt(II) compounds represent a different class of metals that have very defined ligand preferences for both type and geometry, very slow to inert ligand exchange, and binding that is therefore under kinetic rather than thermodynamic control in most settings. In this review we will first provide an overview of the cellular distribution of Pt(II) drugs and evidence for Pt-RNA interactions based on *in vivo* and cell extract studies. We will then summarize what is currently known about specific interactions between RNA and Pt(II) compounds. Because DNA has historically been considered the target of Pt(II) compounds there is an extensive literature on details of DNA and Pt(II) anticancer compounds [1–6]. In some instances we will draw on these findings to accentuate similarities and differences in the metallobiochemistry of RNA and DNA.

2. Pt(II) COMPOUNDS: PROPERTIES AND BIOLOGICAL DISTRIBUTION

2.1. General Properties of Pt(II) Compounds

Pt(II) demonstrates a preference for 'soft' ligands, such as nitrogen and sulfur σ-donors, which are typically arranged in a square-planar coordination geometry. When coordinated to these types of ligands Pt(II) complexes exhibit very slow ligand dissociation kinetics. As is the case for the majority of $16e^-$ complexes, ligand exchange reactions typically occur through an associative mechanism proceeding through a trigonal bipyramidal transition state [23].

Cisplatin (**1**, Figure 1), the foremost member of biologically active Pt(II) complexes, has two kinetically inert ammine ligands and two more readily exchangeable chloride ligands. In therapeutic contexts, cisplatin is delivered intravenously where an approximately 100 mM concentration of Cl^- ions in the bloodstream inhibits ligand exchange until cisplatin has entered a cell. Once inside the cell, a lower chloride concentration of approximately 4–12 mM facilitates the exchange of chloride ligands, producing the aquated species seen in Figure 2 with a half life of approximately 2 hours [2]. For Pt(II) complexes in general, many factors including the identity and geometry of the Pt(II) ligands, pH, and the surrounding ionic environment influence the equilibria, mechanisms and rates of these reactions [24]. Once

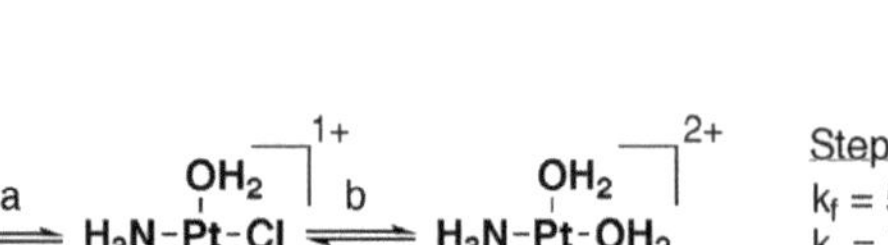

Figure 1. The three FDA-approved Pt(II) therapeutics.

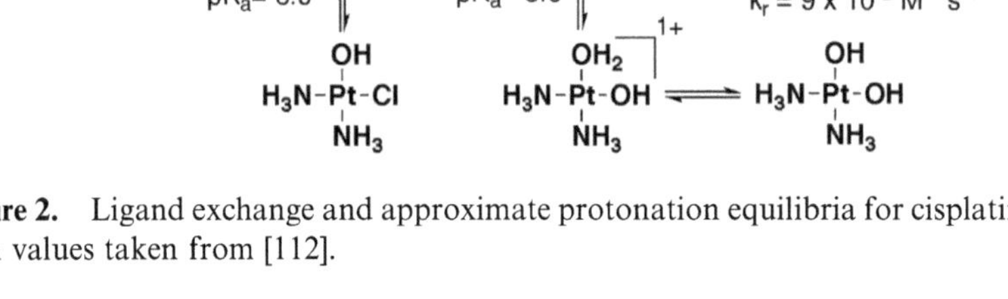

Figure 2. Ligand exchange and approximate protonation equilibria for cisplatin (**1**), with values taken from [112].

positively charged, Pt(II) complexes undergo further ligand substitution reactions and are ultimately bound to a variety of N- and S-containing molecules, such as glutathione, histidine and cysteine residues of proteins, and imino nitrogens on nucleic acid nucleobases.

Despite the synthesis and screening of many platinum-centered molecules [3,25], in addition to cisplatin, only two other Pt(II) complexes have received FDA approval: carboplatin (*cis*-diammine (cyclobutanedicarboxylato) platinum(II)) (**2**), and oxaliplatin (*cis*-oxalato-(trans-l)-1,2-(diaminocyclohexane)platinum(II)), (**3**) (Figure 1). Nedaplatin (*cis*-diammine(glycolato)-platinum(II)) is currently administered in Japan [3]. These derivatives exhibit different pharmaceutical properties than cisplatin. It is hypothesized that the dicarboxylate ligand on carboplatin functions to slow hydration of the platinum center and that the chiral diammine of oxaliplatin tunes the lipophilicity and steric parameters of the drug. These compounds emphasize the two major subcategories of platinum derivatization work, namely the tuning of pharmacokinetics and molecular recognition, and depict the importance of both kinetic and structural aspects of platinum coordination chemistry in biological systems [3,25].

2.2. Classes of Platinum Binding Targets

Pt(II) complexes have the potential to bind a wide range of molecular targets. In biological systems these targets include small molecules like glutathione, membrane phospholipids, RNA, DNA, and proteins [5]. An early study by Pascoe and Roberts [8] sought to address which classes of biomolecules are targeted by Pt complexes in living cells by employing atomic absorption spectroscopy (AAS) to assay the amount of Pt bound to the RNA, DNA, and protein components of HeLa cells following treatment with cisplatin and its non-pharmacologically active counterpart, transplatin (*trans*-diamminedichloro Pt(II)). When considered on a Pt(II) per gram of biomolecule basis, the results of this study show that significantly more Pt is bound to RNA than to either DNA or protein for both Pt complexes. Interestingly, a noticeable difference in cellular uptake between the *cis*- and *trans*-isomers was also observed. At low micromolar concentrations, where only cisplatin was observed to be cytotoxic, close to twice as much Pt from transplatin was found bound to RNA, DNA, and protein fractions, although at higher concentrations this difference was less pronounced. In addition to differential uptake, the isomeric complexes also displayed different extents of DNA interstrand cross-linking; when analyzed by density gradient, cisplatin was shown to form 10-fold more interstrand cross-links than transplatin [8].

A similar and more recent study by Miyahara and coworkers [7] also assayed cisplatin binding to biomolecules in HeLa cells. By measuring the incorporation of ^{195m}Pt from labeled cisplatin into the protein, RNA, and DNA fractions of HeLa cells the researchers determined that the majority of ^{195m}Pt was bound to trichloroacetic acid insoluble protein fractions. Both nucleic acids displayed a similar, however lower, extent of drug binding [7]. When the experiment was repeated using transplatin, it was again observed that higher amounts of ^{195m}Pt(II) were bound to all three classes of macromolecules, with the largest increase seen for RNA [26]. In addition to these studies, significant differences in tissue accumulation [27], cellular accumulation [28,29] and DNA binding [30–34] for different Pt complexes has been observed by AAS and inductively coupled plasma mass spectrometry (ICP-MS). The differences observed in these studies indicate that, as is observed for cisplatin and transplatin, there could be important and pronounced variance in the way Pt drugs bind to cellular biomolecules.

2.3. Drug Localization in the Cell

Characterizing the spatial distribution of Pt(II) binding within a cell provides additional information regarding which types of the cellular machines and architectures Pt(II) complexes may target. Major cellular targets, including those important for RNA processes, are highlighted in Figure 3 and the accompanying studies are portrayed in Table 1. AAS and ICP-MS have been used as tools to measure Pt drug accumulation in several different types of organelles. For cisplatin, Pt accumulation in intact mitochondria [35], and drug binding to mitochondrial DNA have been quantified using AAS and by immunodetection techniques [36–38]. Similarly, the accumulation of cisplatin and several other Pt(II) complexes in the nuclei of drug-treated cells has also been measured [39]. More recently, Pt accumulation has been detected in vesicles by ICP-MS following the treatment of cells with cisplatin, carboplatin, and oxaliplatin [40,41]. The importance of Pt(II) accumulation in these types of cellular compartments is currently unknown, however, understanding where in the cell the drug binds may provide further information regarding which types of RNA may be targeted by Pt(II) complexes as well as insight into biological processing of drug-damaged biomolecules.

Direct imaging techniques have also provided a powerful means to study platinum distribution in treated cells. These techniques divide into two main categories: elemental imaging, which directly measures the location of the Pt atoms in the cell, and fluorescent tagging, which identifies drug binding locations using the fluorescent properties of a covalent Pt(II) conjugated fluorophore.

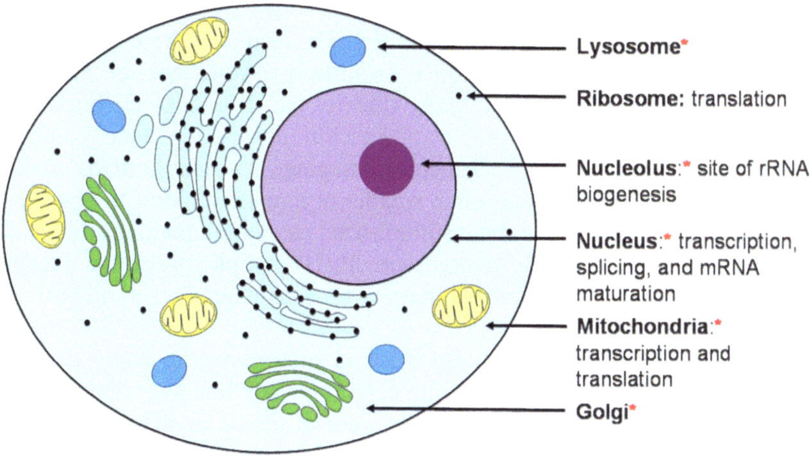

Figure 3. The organelles that accumulate Pt drugs in cancerous cells (indicated with a red star) and the locations of important RNA processes within the cell. References are given in Table 1.

Table 1. Organelles in which Pt(II) drug accumulation has been identified.

Organelle	Pt accumulation observed by elemental imaging	Pt accumulation observed by fluorescent label
Nucleus	[43–49,52]	[54,56–58]
Nucleolus	[44–46,50]	[58]
Mitochondria	[45,46]	[57]
Lysosome	[53]	[57]
Golgi	Not observed	[54,57,58]

2.3.1. Elemental Imaging Techniques

Elemental imaging techniques are capable of directly detecting Pt nuclei while the drug is in the cell and are therefore broadly applicable in the study of Pt(II) localization [42]. The majority of the studies summarized below use characteristic X-ray fluorescence bands to specifically identify Pt(II) in the presence of other physiological metals. Excitation is typically achieved using either an electron beam, as in electron microprobe analysis and X-ray microanalysis, or by using an X-ray beam, as in X-ray fluorescence and synchrotron radiation-induced X-ray emission (SRIXE) studies. A similar technique, electron microscopy, locates Pt via its electron-dense nature.

This range of techniques has been applied to a variety of cancerous and non-cancerous cell lines and tissue samples. The results of these studies are in many cases conflicting; however, it is important to note that the significant variations observed are most likely due to the different cell lines, drug concentrations, and sample preparation techniques used in these studies. In addition, the resolution of elemental imaging is often limited, making identification of Pt accumulation in smaller organelles difficult to observe. It is important to note that in these summarized elemental imaging studies, the cell lines were continuously treated with Pt(II) complexes for the duration of the experiment, and thus incubation time can be used as a basis for comparison.

Perego and coworkers [43] used electron microscopy to study the early localization of cisplatin in an ovarian carcinoma cell line over times ranging from 5–30 minutes. Platinum deposits were observed at the plasma membrane, nuclear envelope, and in deposits scattered throughout the cytoplasm and nuclear matrices. Interestingly, the authors also observed Pt deposits spanning through the membranes themselves [43].

After 4–5 hr of drug treatment Pt is typically observed to accumulate in cell nuclei where in addition to DNA replication, transcription and critical RNA-processing events also take place. Following 4 hour treatment with cisplatin, Khan and Sadler have observed Pt binding in the nucleolus and on the inner edge of nuclear membrane of HeLa cells using a combination of electron microscopy and X-ray probe microanalysis [44]. Similarly, after 4 hours of drug treatment Kiyozuka et al. [45] also identified Pt binding to the nucleolus and at the periphery of the nucleus in two ovarian carcinoma cell lines. In this study the authors note Pt(II) accumulation in mitochondria [45] which is supported by similar findings by Meijer et al. [46] who observed Pt-DNA binding in mitochondrial DNA and in dense heterochromatin and granules surrounding the nucleoli. Interestingly, Pt-DNA binding is observed to take place in a cell cycle-dependent manner [46]. In a contrasting study, Ortega et al. report uniform Pt distribution throughout human ovarian cancer cells following treatment for 5 hr with cisplatin [47]. At longer timepoints, Hambley and coworkers observe that cisplatin, several Pt(IV) prodrugs, and a Br-tagged cisplatin analogue accumulate exclusively in the nucleus of ovarian carcinoma cells [48,49].

Platinum accumulation in non-cancerous tissues has been studied in order to understand the dose-limiting side effects of Pt(II) complexes. In these tissues, different platinum accumulation patterns have been observed, which may be relevant in assessing which RNA-dependent processes are likely affected in different tissues. In human fibroblasts treated with cisplatin for 2 hr, Pt preferentially localized to the nucleolus [50], as is observed in many cancerous cell lines. However, rabbit bone marrow treated with cisplatin for 10 and 20 hr showed Pt accumulation in the cytoplasm, but not the nucleus [51]. In animal models, Pt distribution has been shown to be tissue-specific. In rat models Pt accumulation has been observed in the vesicles and microbodies of liver cells and within the microbodies, lysosomes, and nuclear matrix of kidney cells [52,53].

2.3.2. *Fluorescently Labeled Platinum Compounds*

Fluorescently tagged platinum compounds have been used for visualizing the cellular localization of platinum drugs in real time. These drug conjugates typically utilize the chelating ligand ethylenediamine (en) as an anchor for attaching labels such as fluorescein [42]. The effects of attaching a large, non-polar fluorophore on the biological distribution and processing of platinum drugs must be taken into account in interpreting these studies.

In one of the first studies of this type, Reedjik and coworkers [54] used a carboxyfluorescein diacetate-tagged [Pt(en)Cl_2] complex to monitor localization of the compound within human osteosarcoma cells. In these experiments, cells were treated with the complex for 30 minutes, washed, and subsequently imaged. Initially observed throughout the cell, the Pt(II) complex accumulated in the nucleus after 1–2 hr and after 6–8 hr the compound appeared to migrate out of the nucleus and into Golgi bodies. These organelles seem to be the ultimate destination for this compound at extended time points. It is interesting to note that very little difference was observed in how this compound and similar fluorescently-labeled dinuclear Pt(II) compounds localized in an ovarian carcinoma cell line [54,55].

Howell and coworkers [56] have also used fluorescently-labeled Pt(II) compounds to study platination in a human ovarian carcinoma cell line. Following treatment with low micromolar concentrations of the complex, the Pt(II)-fluorophore is observed at the periphery of the cellular membrane, in the nucleus, and in small vesicular structures scattered throughout the cytoplasm. Supporting biological assays show that while the Pt(II)-fluorophore conjugate is about 4-fold less potent than cisplatin, Pt(II)-resistant cell lines are similarly insensitive to the two complexes, suggesting that the complexes may be similarly processed *in vivo* [56]. Further work by Howell and coworkers has used fluorescent Pt(II) complexes in concert with specific small molecule inhibitors to show that these compounds were first sequestered by lysosomes, subsequently transferred to Golgi apparatus and finally into secretory vesicles [57]. The accumulation of Pt(II) complexes in Golgi bodies has similarly been observed by Gottesman and coworkers using a different Pt(II) fluorophore-cisplatin conjugate in studies that also identify platination occurring at nucleosomes and within the nucleolus [58]. In this

Table 2. Inhibition of cell growth by Pt(II) compounds.

Drug	Common name	Chemical structure	Organism	IC_{50} (μM)	Ref.
Cis-diamminedichloroplatinum(II)	cisplatin	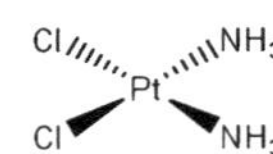	HCC	12 (1 hr)	[59]
			S. cerevisiae	600 (9 hr)	[60]
			S. pombe	260 (48 hr)	[61]
			MEF	~20 (72 hr)	[62]
			HOC	0.21 (96 hr)	[63]
Trans-diamminedichloroplatinum(II)	transplatin	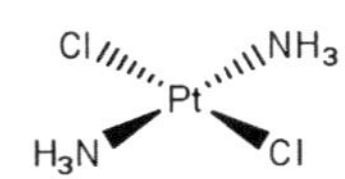	HCC	265 (1 hr)	[59]
Tetraammineplatinum(II) dichloride		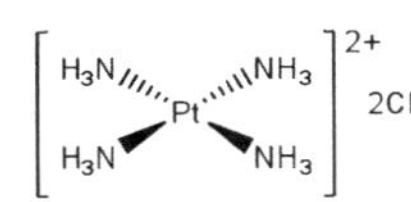		Inactive	[64]

Cis-diammine(cyclobutane-1,1-dicarboxylate-*O,O′*)platinum(II)	carboplatin	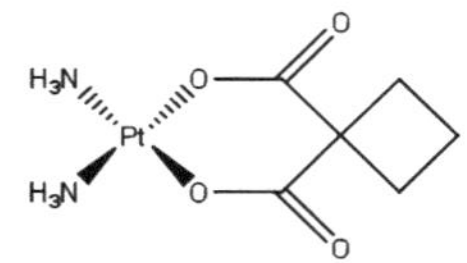	HCC MEF HOC	98 (1 hr) ~75 (72hr) 0.35 (96 hr)	[59] [62] [63]
[(1R,2R)-cyclohexane-1,2-diamine](ethanedioato-O,O′)platinum(II)	oxaliplatin	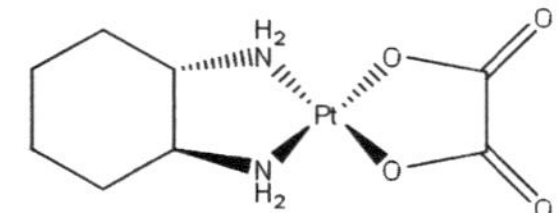	HCC *S. pombe* MEF HOC	17 (1 hr) >1000 (48 hr) ~90 (72 hr) 0.12 (96 hr)	[59] [61] [62] [63]
Tetrachloro[d,l-trans]1,2-diaminocyclohexane platinum(IV)	tetraplatin	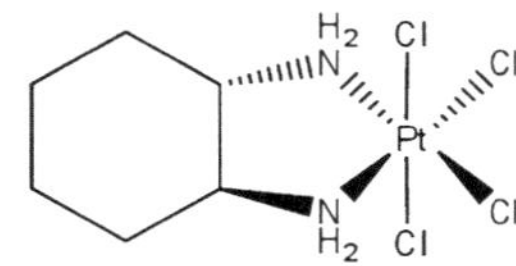	HCC *S. pombe* HOC	34 (1 hr) 7 (48 hr) 0.15 (96 hr)	[59] [61] [63]

MEF = murine embryonic fibroblasts, HOC = human ovarian carcinoma, HCC = human colon carcinoma.

case, in a 2 hr treatment the Pt(II)-fluorophore seemingly accumulates more in the cytosol than within the nucleus.

The approach of fluorescently tagging Pt drugs has produced, over several studies, a more uniform picture of Pt(II)-conjugate localization than has been observed from the direct Pt(II) imaging-based techniques, although the influence of the attached fluorophore may affect the outcome of these studies. Nonetheless, these findings combined with those of the elemental imaging techniques are beginning to form a picture of the cellular components involved in platinum binding and processing, particularly for cancerous cells. Initially, cisplatin and other Pt(II) drugs enter the cell and accumulate to varying degrees in the vesicles and organelles of the cytoplasm, including lysosomes, Golgi, and mitochondria. At later times Pt accumulates in the nucleus, often accumulating along the periphery of the nucleus and in nucleoli. Depending on the treatment conditions and cell type this nuclear accumulation may become greater than cytoplasmic accumulation at 1–4 hr. Finally, export from the cell may involve the Golgi and vesicles of the secretory export pathway.

Given these sites of Pt accumulation (summarized in Figure 3 and Table 1), it is likely that particular RNA targets for Pt(II) drugs are located in the nucleus, nucleolus, and mitochondria. The nucleolus is the site of rRNA biogenesis, while transcription, splicing, and mRNA maturation all occur in the nucleus. In addition, mitochondria transcribe and translate their own genes. Pt complexation may disrupt any or all of these important RNA-based processes.

3. Pt(II) COMPOUNDS AND RNA PROCESSES

3.1. Influence of Pt(II) Compounds on the Rate of Cellular DNA, RNA, and Protein Synthesis

Over the past 30 years, researchers have worked towards a comprehensive picture of the inhibitory effects of platinum compounds on important metabolic processes. The effects of a range of platinum drugs on overall cell growth have been established for several eukaryotic cell lines, with comparative studies summarized in Table 2 [59–64]. As in both whole cell and DNA Pt accumulation studies (Section 2.2), the IC_{50} concentrations of these drugs vary significantly. Cisplatin and oxaliplatin, two of the three platinum compounds which are currently FDA approved, exhibit the lowest IC_{50} values among the complexes listed in Table 2. The effects of factors governing drug activity such as ligand geometry and identity are also portrayed in Table 2, where simple changes give rise to very different biological activities.

In studies in *Saccharomyces cerevisiae* and a human placental cell line, the rates of synthesis of DNA, RNA, and proteins in platinum-treated cells were measured through incorporation of isotopically labeled nucleotides or amino acids [60,64]. Dose-dependent inhibition of DNA synthesis was observed at lower Pt(II) concentrations than those required to reduce the rate of either RNA or protein synthesis ($<25\,\mu M$ in the human placenta cell line). However, at slightly higher concentrations, RNA-dependent processes also become significantly affected by cisplatin [64].

3.2. Influence of Pt(II) Compounds on RNA Transcription, Splicing, and Translation

RNA-based processes in cells are influenced by treatment with Pt(II) compounds (Figure 4). Each of these processes depends heavily on complex RNA structures as well as on RNA-protein and protein-protein interactions. With the exception of carefully designed RNA polymerase experiments in which a single Pt(II) adduct is specifically placed within a DNA template

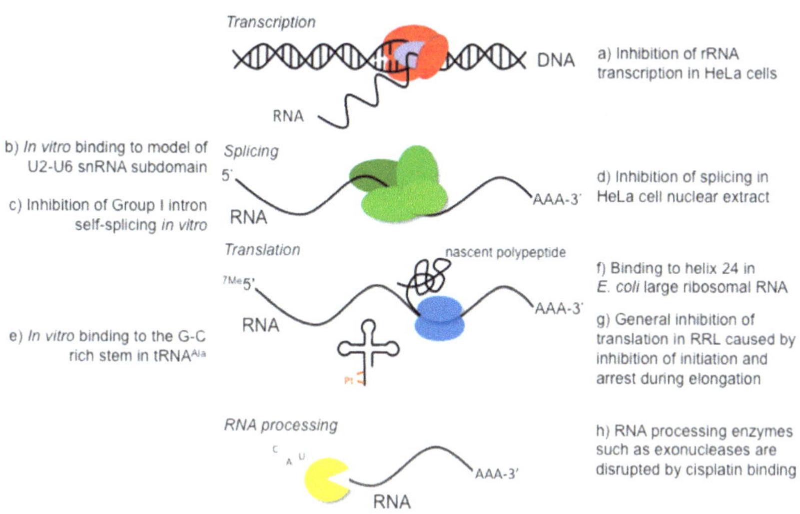

Figure 4. RNA processes inhibited by Pt(II) compounds as determined from studies in cells, cell extracts, and *in vitro* studies. References: (a) [67–69], (b) [16], (c) [13], (d) [12], (e) [15], (f) [70], (g) [9–11], (h) [71].

[65,66], the studies in Figure 4 have been conducted in chemically complex contexts where platination may interfere in many possible ways. Pt(II) adducts formed with RNA or DNA templates, RNA products, or ribonucleoprotein complexes all have the potential to disrupt these RNA-dependent processes.

Platinum interference has been observed at almost every step of the RNA lifecycle, from transcription of pre-mRNA to translation of mature mRNAs into functional proteins. Cisplatin has been observed to preferentially block transcription of ribosomal RNA in HeLa cells and drug treatment results in a redistribution of the RNA polymerase I transcription machinery [66]. In mammalian cells, arrest of RNA polymerase II at a platinum-DNA lesion has been demonstrated *in vitro* [67], in cellular extracts [65] and recently, *in vivo* [68]. The structural basis for this inhibition has also been reported, in which it was found that transcriptional inhibition was a result of the inability of the DNA lesion to enter the active site of the enzyme. Interestingly, this mechanism of polymerase stalling is unique from that of similar DNA lesions that result from UV irradiation [69].

Splicing is a critical step in the RNA lifecycle and is responsible for the successful maturation of pre-mRNA transcripts. In HeLa cell nuclear extract, a dose-dependent inhibition of pre-mRNA splicing was observed after treatment with tetraplatin (also known as ormaplatin (tetrachloro[d,l-*trans*]1,2-diaminocyclohexane platinum(IV))) and cisplatin. The inhibition of splicing observed in these experiments can be correlated to a disruption in formation of spliceosomal complexes on pre-mRNA [12]. An inhibition of splicing has also been observed following cisplatin treatment of the protein-independent self-splicing *Tetrahymena* rRNA Group I intron, possibly through the formation of an interstrand cisplatin-RNA cross-link [13].

Efficient translation is highly dependent on intact and functional RNA molecules such as ribosomes, messenger RNA, and tRNA. Platinum disruption of translation has been described to a great extent in rabbit reticulocyte lysate. Early studies by Rosenberg and Sato demonstrated that Pt(II)-bound mRNA added to platinum-free cellular lysate inhibits protein synthesis by 85%, while Pt(II)-incubated lysate treated with platinum-free mRNA showed a 19% decrease in translation [9]. In a later study, it was found that the rate of translational inhibition by cisplatin matched that of NaF, an inhibitor of translational initiation. In addition, the polysome profiles following both drug treatments showed an apparent decrease in polysome intensity [10]. This was interpreted to mean that cisplatin disrupted translation by preventing initiation. Heminger and coworkers later revisited this study and noted a slightly different result, in which the steady accumulation of higher order polysomes was observed on mRNA templates following Pt(II) treatment [11]. They concluded that their data showed Pt(II) inhibition of elongation, not initiation. However, despite the appearance of higher-order polysomes, the apparent intensity of the polysome fractions decreases upon Pt(II) treatment in both studies in conjunction with an increase in the population of free ribosomes. Therefore, it is likely that platinum in cell extracts interferes in both initiation *and* elongation in ribosomes to cause an overall decrease in protein synthesis.

Recent work has documented specific platinum binding events within the translation machinery. Stable cisplatin adducts have been mapped to various locations within *E. coli* and *S. cerevisiae* ribosomal RNA through primer extension of RNA extracted from treated cells. Cisplatin forms stable adducts on helix 24 of the 16S subunit [70] and within the sarcin-ricin loop of the 25S subunit [113]. The effects of platinum binding in these locations on overall protein synthesis have not been investigated. Cisplatin also binds to purine-rich regions of transfer RNA *in vitro* [15] which, if it occurs in cells, could influence translation and other tRNA-dependent processes.

3.3. Influence of Pt(II) Compounds on RNA Processing Enzymes

Regulation of the transcriptome includes RNA surveillance and regulated degradation, processes that may also be influenced by RNA-Pt(II) interactions. *In vitro* RNA-platinum adduct formation has been found to inhibit the function of several RNA processing enzymes [71]. The activities of 5'-to-3' and 3'-to-5' exonucleases and a purine-specific endoribonuclease are inhibited by specific Pt(II) adducts installed on short RNA oligonucleotides (Figure 5). Primer extension by reverse transcriptase has also been shown to halt at sites of cisplatin-RNA adducts. These data underline the importance of characterizing interactions between platinum and RNA to further understand how the platinum-based drugs exert cytotoxicity *in vivo*.

4. *IN VITRO* STUDIES OF RNA-Pt(II) ADDUCTS

4.1. Mechanistic Studies

In vitro studies have been used to investigate the details of Pt(II)-RNA binding and may provide insight into the observed inhibition of RNA processes. On the mechanistic level, once cationic Pt(II) species are

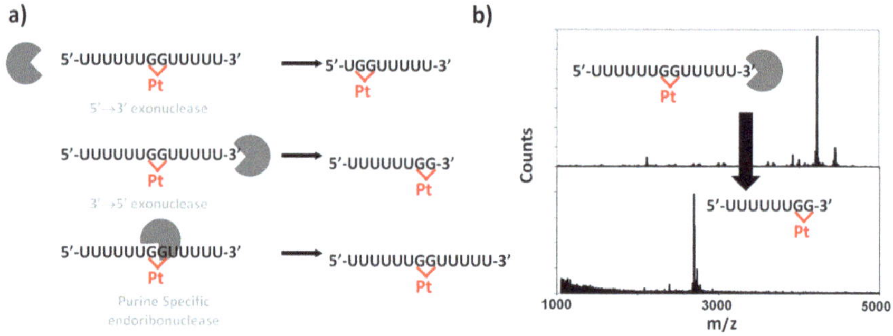

Figure 5. The enzymatic processing of platinated RNAs. (**a**) Schematic of the products resulting from 5'→3' (top), 3'→5' (center) exonuclease or the purine specific endoribonuclease U2 (bottom) digestion of platinated oligonucleotides. (**b**) MALDI-MS spectra of a platinated 5'-U_6-GG-U_5-3' RNA (top) and the platinated 5'-GG-U_5-3' fragment left following nuclease digestion. Adapted with permission from [71]; copyright 2010, American Chemical Society.

formed (Section 2.1), aquated metal complexes enter the condensed cationic atmosphere surrounding a negatively charged nucleic acid [16,72,73]. Here, hydrated cations diffuse along the polyanionic biopolymer, transiently coordinating to RNA and DNA. In this cation atmosphere Pt(II) complexes encounter a variety of potential coordination environments, including positions along the negatively charged phosphodiester backbone, before forming kinetically inert adducts with DNA and RNA nucleobases [74–77]. Here may be the first level at which chemical and structural differences between RNA and DNA influence the Pt(II) coordination properties of each nucleic acid. Two recent studies characterizing the rate of reaction between *cis*-$[Pt(NH_3)_2Cl(OH_2)]^+$ and relatively short RNA and DNA hairpins (including the constructs shown in Figure 6), using dPAGE [16] or HPLC [73] methods to monitor reaction kinetics, have revealed that RNA oligonucleotides react 2- to 6-fold faster than DNAs of analogous size and sequence. While additional considerations regarding oligonucleotide structure and flexibility should be made, differences in the electrostatic surfaces projected by the more compact A-form helical structure adopted by RNA, and that of the more extended B-form helical structure of DNA [78,79] may contribute to the observed differences in rate. Because cellular RNAs have lifetimes ranging from tens of hours to several weeks (as summarized in [16]), the platination rates of $\sim$2–6 $M^{-1}s^{-1}$ measured in these studies imply that Pt-damaged RNAs may accumulate in a cell and disrupt the function of important RNA processes and by extension that RNA targeting may contribute to the effects of Pt(II) anti-tumor drugs and other metallopharmaceuticals.

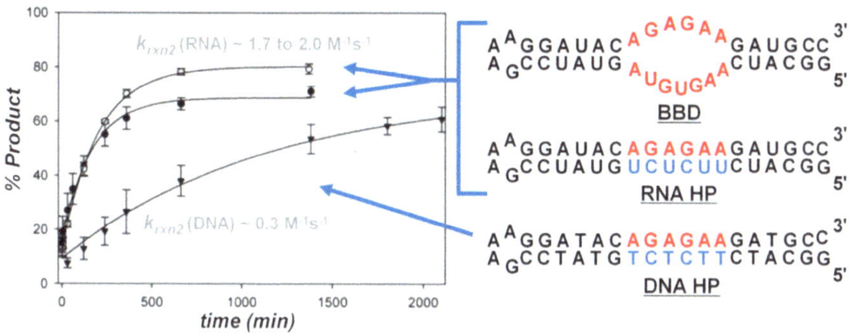

Figure 6. Comparison of the reaction rates for RNA and DNA oligonucleotides showing faster platination of RNA sequences. Adapted with permission from [16]; copyright 2009, American Chemical Society.

4.2. Pt(II) Adducts Formed with RNA

Pt(II) complexes demonstrate a strong preference for forming coordinate-covalent bonds with "soft" nucleophilic positions on DNA and RNA nucleobases. Accordingly, Pt(II) coordination is most commonly observed at the N7 position of guanine and adenine. This feature holds especially true for studies using duplex nucleic acids where other potential Pt(II) ligands such as the N1 of adenine and N3 of cytosine [18] are precluded from platinum binding by their participation in Watson-Crick base pairs. Generally, Pt(II) complexes with two open coordination sites form macrochelate complexes between proximal purine nucleotides, which results in a variety of intramolecular adducts. Perhaps the best characterized examples of this type of Pt(II) coordination are the adducts formed by cisplatin on DNA, which primarily take place at 1,2 d(R*pR*) (R = purine, * denotes platination) and 1,3 d(G*pNpG*) sequences [5]. While several of the structural features of these types of adducts as well as Pt(II) coordination in non-canonical forms of DNA are discussed in the next section, it is interesting to note that the repetitive structure of the DNA double helix seems to offer a limited number of potential Pt(II) coordination geometries. In contrast RNA exhibits a diverse array of secondary and tertiary architectures (Figure 7) [21] which include specific and pre-organized metal-ion binding sites [19–20,22] as well as solvent-excluded folds where nucleobase pK_a's can be significantly offset from those observed in free solution [80,81]. Consequently, these structures greatly expand the chemical space in which physiological metal complexes find appropriate coordination geometries.

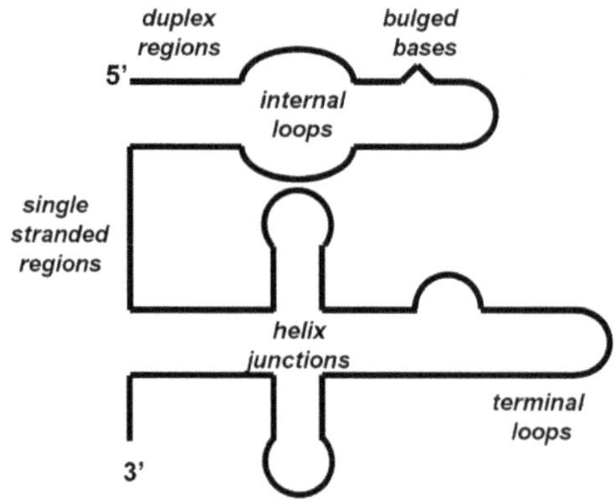

Figure 7. Schematic representation of RNA secondary structures.

4.3. Examples of Pt(II) Binding to RNA Structures

Seeking to understand how cisplatin may target RNA structures, a number of recent studies have described the formation of Pt(II) adducts with isolated RNAs (Figure 8). These studies examine cases ranging from the platination of relatively short single-stranded RNAs [71] to studies involving coordination of cisplatin to ribosomes [20]. In accordance with the observations above, an emerging feature of these studies seems to be ability of Pt(II) complexes to bind in non-Watson-Crick base paired regions of RNA.

Several preliminary studies, conducted shortly after the discovery of cisplatin's antitumor activity, sought to identify Pt(II) binding sites with $tRNA^{Phe}$ by reacting the RNA with Pt(II) complexes in solutions mimicking crystallization conditions or by directly soaking into pre-formed RNA crystals [82–84]. In the earliest of these studies, Clark and coworkers reacted both cis- and transplatin with $tRNA^{Phe}$ at 4°C for 3 days in solutions containing high concentrations of Mg^{2+}, spermine, and 1,6-*n*-hexane-diol [82]. Using thin-layer chromatography to identify RNA fragments produced by nuclease digestion of the platinated $tRNA^{Phe}$, the authors identified a major transplatin binding site within the purine rich anticodon loop of the RNA and a secondary, minor Pt(II) binding site within loop III. Surprisingly, under the conditions employed in this study, cisplatin was not observed to coordinate to tRNA. A second investigation conducted by Sundaralingam and coworkers using pre-formed $tRNA^{Phe}$ crystals produced somewhat different results [83]. While only relatively low (5.5 Å) resolution structures were obtained, the authors were able to identify both transplatin adducts noted previously, along with additional Pt(II) binding sites at G18, located in the dihydrouridine loop and at A73 located in the molecule's acceptor

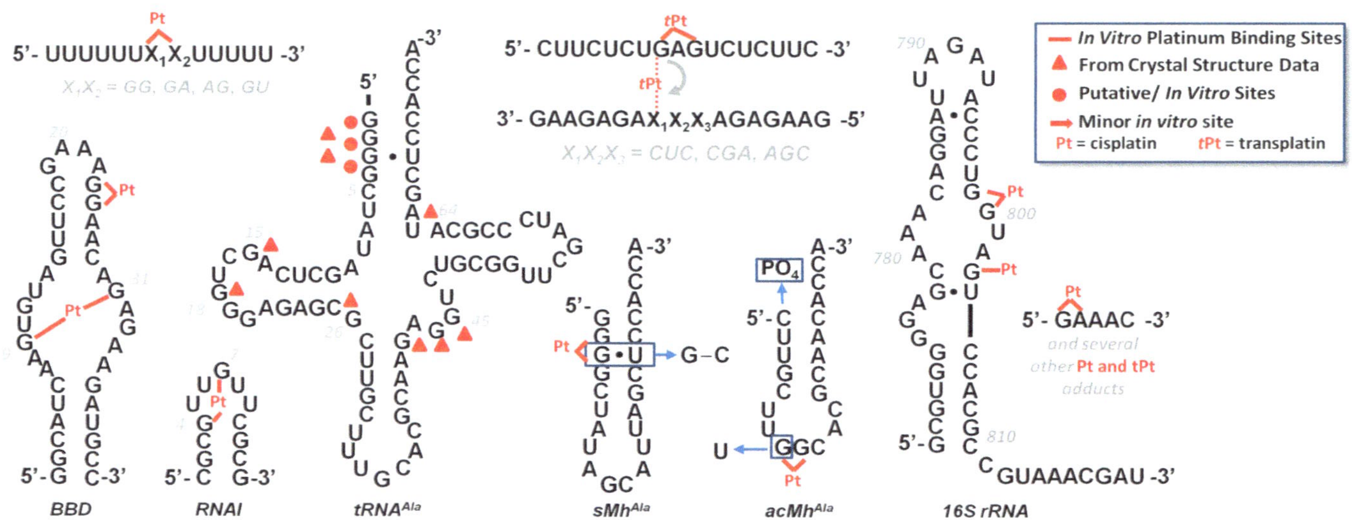

Figure 8. Sequences and secondary structures of RNAs in which regions or sites of Pt(II) coordination have been identified. RNA structures that have been studied include the model 5'$U_5XXU_5$3' (X = A or G), [71]), a purine-rich internal loop derived from the spliceosome (BBD, [16]), an RNA hairpin (RNAI, [73]), tRNA, [14,82–84], hairpin models for the tRNA acceptor stem (sMh^{Ala}, [14,15]) and anticodon loop ($acMh^{Ala}$, [15]), intrastrand-to-interstrand cross-linking [88–90], a site on the ribosomal small subunit [70], and a small purine-rich strand [97]. Pt(II) site identification has relied on mapping techniques with the exception of low-resolution crystallography of tRNA. While $tRNA^{Ala}$ is shown, several of the crystallographic studies mentioned in the text were performed using $tRNA^{Phe}$. The binding sites identified in these studies are indicated using triangles and are overlayed on the $tRNA^{Ala}$ sequences.

stem. In contrast to the earlier report, however, this study also identifies cisplatin-derived adducts at both G15 and G18 in the D arm of $tRNA^{Phe}$. In an independent, but very similar, study Dewan later reported a cisplatin-soaked $tRNA^{Phe}$ 6 Å structure with four Pt binding sites [84]. While the low resolution of these crystals, reported to be unavoidable due to the structural distortions caused by drug binding, prevented a detailed analysis, electron density indicative of Pt(II) binding is apparent at G3-G4, C25-mG26, G42-G43-A44-G45, and at A64-G65. The combined results of these three studies studies are overlayed on the secondary structure of $tRNA^{Ala}$ in Figure 8.

More recent biochemical studies conducted by Elmroth and coworkers have employed full length $tRNA^{Ala}$ as well as hairpin models of the molecule's anticodon loop and acceptor stem to make several interesting conclusions regarding the sequence and structural specificity of platinum binding to these RNAs [14,15]. By monitoring changes in the dPAGE mobility of platinated RNAs, it was found that in the hairpin RNA modeling the acceptor stem (sMH^{Ala}; Figure 8), replacement of a native G · U wobble pair with a standard G-C base pair inhibits platinum binding to the G-rich stem of the RNA. Additionally, attachment of 5'-terminal phosphate in the same region alters the distribution of products observed upon platination. In the hairpin model of the $tRNA^{Ala}$ anticodon stem, substitution of a natural UG sequence in the terminal loop for a GG sequence increases the rate of platination observed for this RNA and suggests that, similar to what is observed in DNA, Pt(II) complexes target neighboring purine sequences in RNA. Combined, these results highlight the influence of sequence and local structure in determining *in vitro* Pt(II) interactions with RNA.

Because non-Watson-Crick base-paired regions of RNA form the core of many functional RNAs, understanding how Pt(II) complexes coordinate to these structures is important in determining how Pt(II) antitumor drugs may affect RNA processes. The spliceosome is an incredibly complex RNA-protein machine responsible for removing intronic sequences from pre-mRNAs. As described in Section 3.2, Pt(II) compounds inhibit spliceosome activity in cell extracts, but specific sites of Pt(II) interactions were not determined. Highlighted earlier in Chapter 8 of this volume, the catalytic core of the spliceosome is proposed to be comprised of the U2-U6 snRNA complex. In efforts to understand how cisplatin coordinates to structured regions of RNA, our lab has recently characterized how the drug binds to a 41-nt RNA subdomain of the U2-U6 complex, termed 'BBD' for 'branch-bulge domain' [16]. This BBD RNA consists of a hairpin structure in which base paired regions flank a purine-rich, asymmetric internal loop (Figure 8). Reaction of this RNA with an aquated form of cisplatin results in a novel Pt-induced intramolecular cross-link. Alkali hydrolysis mapping was used to identify two G's located in opposite sides of the molecule's purine-rich internal loop as the nucleotides involved in forming this previously unidentified type of Pt(II) adduct. While the structure of this RNA is currently unknown, the nucleotides involved in this cross-link have been proposed as metal ion binding sites in the full U2-U6 snRNA complex [85,86]. In light of this information, it is thought-provoking to speculate that cisplatin may compete with native metal ion binding sites in RNA. This type of mimicry would be particularly interesting as metal ions often mediate important tertiary contacts between interacting RNA domains [19,87] where cross-linking could potentially disrupt the dynamic function of these structures.

In another recent example of Pt(II) coordination to a complex RNA structure, Rijal and Chow have reported how Pt(II)-metal complexes may be used as chemical probes for identifying solvent-exposed purine nucleotides in intact bacterial ribosomes [70]. Stable platinum adducts formed within helix 24 of *E. coli* ribosomes were identified by primer extension as stop sites caused by platination. Using this method, it was found that G nucleotides in several non-Watson-Crick base-paired regions of the RNA were targeted. This targeting provides another example of platinum coordination within regions of complex RNA structure, notably even in the presence of competing GG sequences in duplex regions of the same RNA. As mentioned in previous sections, aside from general inhibitory effects it is currently unknown how Pt(II) binding affects the function of these cellular RNAs.

4.4. Transplatin Cross-Linking and Pt(II) Drug Conjugates

The nucleic acid coordination properties of Pt(II) complexes have also inspired research seeking to use these types of compounds as sequence-specific RNA cross-linking reagents and as scaffolds upon which to build new types of RNA-targeted drug conjugates.

In studies aimed at understanding how transplatin distorts nucleic acid structure differently than its *cis* counterpart, it was discovered that pre-formed 1,3 G*pNpG* transplatin-derived adducts on DNA readily rearrange into intermolecular cross-links when annealed to complementary oligonucleotides [88–90]. Subsequent mechanistic studies have shown that isomerization reactions between these 1,3 G*pNpG* transplatin adducts and complementary oligonucleotides are sequence-specific [91], can result in unusual G-N7* to C-N3* cross-links [89], and are also observed using 2'-OMe [92] and chemically modified RNAs [93]. Further work has shown that transplatin-modified 2'-OMe RNAs [94] and cisplatin-modified RNAs [95,96] can be used as antisense oligos for attenuating gene expression in cell lysates and in living cells.

Similarily, cisplatin, transplatin, and $[Pt(NH_3)_3Cl]^+$ have been used as mechanistic probes for studying the Mn^{2+}-catalyzed cleavage reaction of GAAACp, which takes place in the presence of a poly(U) sequence, to produce guanosine 2'-3' cyclic phosphate and AAACp [97]. Nuclease digestion was used to identify platinum coordination to nucleotides within the GAAACp sequence, and a variety of platinated products were found. Evidence for platination of every nucleobase, including to the terminal C, is observed in this small RNA motif.

Finally, work in advancing RNA as a drug target [98] has prompted synthesis of a pair of Pt-drug conjugates. By tethering neomycin B and guanidinoneomycin B to a square planar Pt(II) complex, Tor and coworkers [99] have shown that Pt(II)-drug conjugates demonstrate substantial selectivity in the targeting of RNA versus DNA in direct competition assays employing calf-thymus DNA. Hydrolysis mapping of the drug adducts formed with the structured Rev Response Element (RRE) RNA revealed Pt(II) cross-linking between sequentially distant but spatially proximal G nucleotides, similar to the cross-linking that is observed in work mentioned above showing Pt(II) cross-linking across the internal loop of the BBD motif [16]. In addition to this work, a recent report describes biophysical characterization of the interactions of a Pt(II)-estradiol conjugate with tRNA [100].

5. STRUCTURAL FEATURES OF Pt(II)-NUCLEIC ACID ADDUCTS

In an effort to understand molecular mechanisms for the therapeutic effects of Pt(II) drugs, significant work has been directed towards characterizing structural aspects of Pt(II)-nucleic acid adducts. While the majority of these investigations have been conducted in the context of DNA, relatively simple models have been employed which may also provide insight into Pt(II) coordination to RNA. The model systems used in these investigations range from simple Pt(II)-nucleobase complexes to duplexed oligonucleotides (Figure 8). In general, Pt(II) coordination seems capable of disrupting the conformation of chelated nucleic acids, in many instances causing severe distortion from native nucleic acid structure. In DNA duplexes, structural distortions, such as the widening of the minor groove, are recognized by cellular proteins and start a cascade of events leading to apoptosis. Due to a lack of information regarding Pt(II) coordination to RNA, here we focus on the fundamental aspects of Pt(II) binding to nucleic acid structures that have been discovered by studying DNA and simple models. Of additional interest would be studies involving Pt(II) binding to non-canonical DNA structures, as these might provide relevant examples for the diverse conformations of oligonucleotides that are anticipated to form Pt-RNA adducts [120].

5.1. Pt-Nucleobase, -Nucleoside, and -Nucleotide Complexes

A variety of *cis*-platinum(II) square planar complexes with nucleobases, nucleotides, nucleosides, and short oligomers have been synthesized and characterized by NMR and X-ray crystallography. Interestingly, the conformations of nucleobase ligands in these complexes varies with the degree of conformational flexibility, which decreases in moving from models of nucleobases to oligonucleotides. In simple *cis*-Pt(II)-nucleobase complexes, Pt(II) is primarily observed coordinated at the N7 atom of purine bases. Under these circumstances the two nucleobases are capable of adopting two distinct conformations, either a "head-to-tail" (HT) conformation or a "head-to-head" conformation (HH). A HT conformation is the scenario in which a five-membered imadazole ring of one purine nucleobase is oriented towards the 6-membered pyrimidine ring of the adjacent purine. In contrast, in the HH orientation the imadazole rings point towards each other. The *cis*-$[(NH_3)_2PtG_2]^{2+}$ (G = guanine) complex can adopt both a HH conformation and a HT conformation. The HH conformation is favored when the G residues are tethered (i.e., via a phosphodiester bond), causing the rotation around the Pt-G bond to be restricted. This is the case in double-stranded DNA models studied to date [101,102]. However, in the case of single stranded Pt(II)-DNA and mononucleotides, the orientation is generally HT, most likely due to greater thermodynamic stability of this conformation which causes it to be favored when the rotation about the Pt-G bond is less restricted. Since non-duplex forms of RNA are sometimes less conformationally restrained than duplexed RNAs and DNAs, it is possible that platination of non-duplex RNA oligonucleotides may have very different structural effects. It seems likely that both HH and HT conformations may be observed in Pt-RNA adducts, depending on the secondary and tertiary structures of the RNA under investigation.

An important structural feature of these nucleotide systems is the angle at which the nucleobase ligands are oriented relative to each other, and the potential for hydrogen bonding between the nucleobases and the other Pt(II) ligands. HT conformers have significant canting (observed by NMR) that moves the C8 of one ligand closer to its neighboring purine. Canting is thought to occur due to stabilization by dipole-dipole interactions between the electron-rich O6 of one guanine and the electron-poor H8 of the other guanine [103]. From the resulting nucleobase orientations, additional hydrogen bonding interactions with other Pt(II) ligands can occur. It has been observed that *cis* nucleobases are capable of hydrogen bonding with ammine ligands on platinum. In particular, in Pt(II) complexes, guanine O6, uracil O4, cytosine O2, and thymine O2 provide stabilizing hydrogen bonds with *cis* NH_3 ligands [104]. Studies of nucleotides (mono-, di-, and triphosphate) have shown an additional stabilizing hydrogen bonding interaction between one ammine ligand of *cis*-diammine Pt(II) and an oxygen of the phosphodiester 5' to the platinated base. In $Pt(NH_3)_2(pGpG)$, hydrogen bonding between the phosphodiester and ammine ligands causes distortion of the phosphate backbone [105]. The tertiary structures of non-duplexed RNA can often depend on a relatively small number of stabilizing interactions, and such directed hydrogen bonding interactions from Pt(II) ligands may have potential to disrupt the energetic balance between competing RNA conformers.

5.2. Platinum Adducts of Canonical and Non-Canonical DNA Motifs

The interaction of cisplatin with canonical double helix DNA has been well studied because it is thought to be an initiator of antitumor activity [5], and it is useful to summarize the findings of these studies in order to provide a basis for comparison with potential RNA structures. In the case of canonical B-form DNA, Pt(II) drugs bind in the major groove, resulting in pinching of the major groove and concurrent widening of the minor groove. These perturbations create a hybrid A/B-form around the platination site, where hallmark characteristics of A-form duplexes include slightly closer spacing between adjacent purine N7 sites (Figure 9), a narrower major groove, a wider and shallower minor groove, and dominant C3'-endo sugar pucker conformation. Interestingly, for the vast majority of structurally

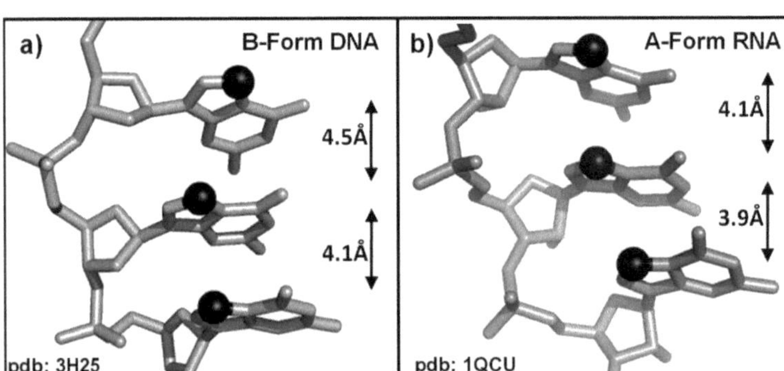

Figure 9. Structures of representative 5'-GGG-3' sequences taken from crystallographically characterized duplex regions representing (**a**) B-form DNA, PDB 3H25 [110] and (**b**) A-form RNA, PDB 1QCU [111]. The N7 atoms of each G are represented as black spheres and the distance between neighboring N7's is noted.

characterized Pt-DNA adducts, the nucleotide 5' to the platination site adopts a C3'-endo sugar pucker, which is atypical of DNA helical regions but may favor Pt(II)-RNA interactions. In the case of the NMR solution structure of a DNA dodecamer, perturbations by Pt(II) coordination in the major groove result in helix bending by approximately 78° and significantly more A-form character in the sugar pucker conformations near the platination site [106]. It is not yet known how platination may affect the dimensions of RNA A-form helices or subsequent changes in RNA-protein recognition.

While a number of studies have biochemically characterized the non-canonical DNA forms (i.e., A-forms, Z-forms, quadruplex, etc.), currently, there are not many structures published of these motifs binding with platinum complexes. One interesting structure observed by X-ray crystallography is of a Z-DNA-$Pt(NH_3)_3$ interaction. This monoadduct structure shows very little distortion caused by platinum binding; however, the presence of guanine O6-NH_3 hydrogen bonding in the crystal structure is similar to observations for canonical platinum-DNA adducts [107].

6. CONCLUDING REMARKS

The profound effects of Pt(II) complexes on RNA structure and activity derive from unique properties of the Pt(II) ion, including very slow ligand exchange and specific thermodynamic preferences for 'soft' ligands and square-planar geometries. These properties result in formation of extremely stable, kinetically trapped complexes that predominantly (though not exclusively) involve purine nucleobases in close proximity, a situation that is frequently encountered in complex RNA structures. Although this review has focused on Pt(II) complexes because of their antitumor properties, it should be noted that other metal ions with kinetically 'inert' (very slow) ligand exchange properties are of interest for studies of RNA structure and function [18–22]. By far the most commonly used 'exchange-inert' ion for such studies has been Co(III) hexammine, $(Co(NH_3)_6)^{3+}$, which has very slow ligand exchange kinetics by virtue of its low-spin d^6 electronic structure [114]. Cobalt hexammine has been used as a pseudo-mimic of hydrated $(Mg(OH_2)_6)^{2+}$ because the complex has advantages for detection by X-ray and NMR spectroscopies [115]. Cobalt hexammine also has found use as a probe in the metal-activation of ribozymes, because it is generally thought that reactivity supported by cobalt hexammine indicates that there is no requirement for an innersphere metal-RNA coordination event [22,114,116]. As has been noted, however, this is a trivalent cation, and in general may substitute for hexaaqua Mg^{2+} preferentially in areas of more negative electrostatic potential [115]. Because they bind tightly to RNA, the hexammine complexes of Co(III), Ru(III), and Ir(III) are used in approaches to solving the phase problem in X-ray crystallographic studies of RNA, and a $(Co(NH_3)_6)^{3+}$ binding RNA 'chip' has even been developed for this purpose [117]. It is of interest to note that X-ray crystallographic studies of cobalt hexammine-soaked oligonuclotides have found evidence for inner-sphere coordination of the cobalt ion to RNA purine N7 and phosphate sites [118], suggesting surprising exchange of the ammine ligands. Whether this ligand exchange might be expected in solution conditions, or perhaps is an outcome of X-ray exposure and potential reduction to the more labile Co(II), is not known [118]. Unlike the coordinatively saturated cobalt hexammine, cobaltic pentammine $(Co(NH_3)_5(OH_2))^{3+}$ binds readily to DNA [119] but has not been explored for RNA studies.

In this chapter several aspects of Pt(II) have been highlighted that focus on the properties of Pt(II)-RNA adducts and the possibility that they influence RNA-based processes in cells. Cellular distribution of Pt(II) complexes results in significant platination of RNA, and localization studies find Pt(II) in the nucleus, nucleolus, and a distribution of other sites in cells. Treatment with Pt(II) complexes disrupts RNA-based processes including enzymatic processing, splicing, and translation, and this disruption may be indicative of structural changes to the RNA or RNA-protein complexes. Several RNA-Pt(II) adducts have been characterized *in vitro* by biochemical and other methods. Although adduct sites have been identified, there currently exists very little in the way of detailed structural characterization of RNA-Pt(II) adducts. In addition to helical regions and complex tertiary folds, common, purine-rich structural elements such as cross-strand purine stacks [108] or GNRA tetraloops [109], which present suitable Pt(II) nucleophiles in unique arrangements, seem to hold the potential to form novel types of platinum-RNA adducts. Future research aimed at structural characterization of platinum-RNA adducts may provide further insights into platinum-nucleic acid binding, and perhaps provide a rationale for the observed inhibition by Pt(II) complexes of splicing, translation, and enzymatic processing, all of which are RNA processes that are central to cell activities.

ACKNOWLEDGMENTS

The authors are grateful for support from the National Institutes of Health (GM058096 to VJD and 5T32GM007759 to MFO) and from the University of Oregon.

ABBREVIATIONS

AAS	atomic absorption spectroscopy
BBD	branch-bulge domain
dPAGE	denaturing polyacrylamide gel electrophoresis
en	ethylenediamine
GNRA	the 4-nucleotide loop capping and stabilizing an oligonucleotide hairpin, with sequence 5'-G (guanine), N (any nucleobase), R (any purine), A (adenine)
HH	head-to-head
HPLC	high-performance liquid chromatography
HT	head-to-tail
IC_{50}	half-maximal inhibitory concentration
ICP-MS	inductively coupled plasma mass spectrometry
mRNA	messenger RNA
NMR	nuclear magnetic resonance
FDA	Food and Drug Administration
RRE	rev response element in HIV-1 RNA
rRNA	ribosomal RNA
SRIXE	synchrotron radiation-induced X-ray emission
tRNA	transfer RNA

REFERENCES

1. B. Rosenberg, L. Van Camp, J. E. Trosko and V. H. Mansour, *Nature*, 1969, **222**, 385–386.
2. Y. Jung and S. J. Lippard, *Chem. Rev.*, 2007, **107**, 1387–1407.
3. J. Reedijk, *Proc. Natl. Acad. Sci.*, 2003, **100**, 3611–3616.
4. J. Reedijk, *Chem. Rev.*, 1999, **99**, 2499–2510.
5. E. R. Jamieson and S. J. Lippard, *Chem. Rev.*, 1999, **99**, 2467–2498.
6. G. A. Natile and L. G. Marzilli, *Coord. Chem. Rev.*, 2006, **250**, 1315–1331.
7. M. Akoboshi, K. Kawai, H. Maki, K. Akuta, Y. Ujeno and T. Miyahara, *Jpn. J. Cancer Res.*, 1992, **83**, 522–526.
8. J. M. Pascoe and J. J. Roberts, *Biochem. Pharmacol.*, 1974, **23**, 1345–1357.
9. J. Rosenberg and P. Sato, *Mol. Pharmacol.*, 1987, **33**, 611–616.
10. J. M. Rosenberg and P. H. Sato, *Mol. Pharmacol.*, 1993, **43**, 491–497.
11. K. A. Heminger, S. D. Hartson, J. Rogers and R. L. Matts, *Arch. Biochem. Biophys.*, 1997, **344**, 200–207.
12. T. D. Schmittgen, J.-F. Ju, K. D. Danenberg and P. V. Danenberg, *Int. J. Oncol.*, 2003, **23**, 785–789.
13. P. V. Danenberg, L. C. Shea, K. D. Danenberg and T. Horikoshi, *Nucl. Acids Res.*, 1991, **19**, 3123–3128.

14. P. Papsai, T. Persson, J. Aldag and S. K. C. Elmroth, *Dalton Trans.*, 2006, 3515–3517.
15. P. Papsai, A. S. Snygg, J. Aldag and S. K. C. Elmroth, *Dalton Trans.*, 2008, 5225–5234.
16. A. A. Hostetter, E. G. Chapman and V. J. DeRose, *J. Am. Chem. Soc.*, 2009, **131**, 9250–9257.
17. J. S. Mattick, *Nat. Rev. Genet.*, 2004, **5**, 316–323.
18. B. Lippert, *Coord. Chem. Rev.*, 1999, **182**, 263–295.
19. V. J. DeRose, S. Burns, N.-K. Kim and M. Vogt, in *Comprehensive Coordination Chemistry II*, Elsevier, St. Louis, 2003, pp. 787–813.
20. E. Freisinger and R. K. O. Sigel, *Coord. Chem. Rev.*, 2007, **251**, 1834–1851.
21. P. B. Moore, *Ann. Rev. Biochem.*, 2003, **68**, 287–300.
22. A. M. Pyle, *J. Biol. Inorg. Chem.*, 2002, **7**, 679–690.
23. R. H. Crabtree, *The Organometallic Chemistry of the Transition Metals*, 4th ed., John Wiley and Sons, New York, 2005.
24. S. Ahmad, A. A. Isab and S. Ali, *Trans. Met. Chem.*, 2006, **31**, 1003–1016.
25. E. Wong and C. M. Giandomenico, *Chem. Rev.*, 1999, **99**, 2451–2466.
26. M. Akaboshi, K. Kawai, H. Maki, K. Akuta, Y. Ujeno, K. Ono and T. Miyahara, *Nuc. Med. Biol.*, 1994, **20**, 389–393.
27. D. Esteban-Fernández, J. M. Verdaguer, R. Ramírez-Camacho, M. A. Palacios and M. M. Gómez-Gómez, *J. Anal. Toxicol.*, 2008, **32**, 140–146.
28. P. Kabolizadeh, J. Ryan and N. Farrell, *Biochem. Pharmacol.*, 2007, **73**, 1270–1279.
29. I.-S. Song, N. Savaraj, Z. H. Siddik, P. Liu, Y. Wei, C. J. Wu and M. T. Kuo, *Mol. Cancer Ther.*, 2004, **3**, 1543–1549.
30. J. Zhang and X. Zhao, *Eur. J. Med. Chem.*, 2007, **42**, 286–291.
31. J. Goodman, D. Hagrman, K. A. Tacka and A.-K. Souid, *Cancer Chemother. Pharmacol.*, 2006, **57**, 257–267.
32. E. Gabano, D. Colangelo, A. R. Ghezzi and D. Osella, *J. Inorg. Biochem.*, 2008, **102**, 629–635.
33. N. Kitada, K. Takara, T. Minegaki, C. Itoh, M. Tsujimoto, T. Sakaeda and T. Yokoyama, *Cancer Chemother. Pharmacol.*, 2008, **62**, 577–584.
34. L. Martelli, F. Di Mario, E. Ragazzi, P. Apostoli, R. Leone, P. Perego and G. Fumagalli, *Biochem. Pharmacol.*, 2006, **72**, 693–700.
35. M. Gemba, E. Nakatani, M. Teramoto and S. Nakano, *Toxicol. Lett.*, 1987, **38**, 291–297.
36. Z. Yang, L. M. Schumaker, M. J. Egorin, E. G. Zuhowski, Z. Guo and K. J. Cullen, *Clin. Cancer Res.*, 2006, **12**, 5817–5825.
37. A. J. Giurgiovich, B. A. Diwan, O. A. Olivero, L. M. Anderson, J. M. Rice and M. C. Poirier, *Carcinogenesis*, 1997, **18**, 93–96.
38. O. A. Olivero, C. Semino, A. Kassim, D. M. Lopez-Larraza and M. C. Poirier, *Mutat. Res.*, 1995, **346**, 221–230.
39. E. Lindauer and E. Holler, *Biochem. Pharmacol.*, 1996, **52**, 7–14.
40. G. Samimi, K. Katano, A. K. Holzer, R. Safaei and S. B. Howell, *Mol. Pharmacol.*, 2004, **66**, 25–32.
41. G. Samimi, R. Safaei, K. Katano, A. K. Holzer, M. Rochdi, M. Tomioka, M. Goodman and S. B. Howell, *Clin. Cancer Res.*, 2004, **10**, 4661–4669.

42. A. V. Klein and T. W. Hambley, *Chem. Rev.*, 2009, **109**, 4911–4920.
43. G. L. Beretta, S. C. Righetti, L. Lombardi, F. Zunino and P. Perego, *Ultrastruct. Pathol.*, 2002, **26**, 331–334.
44. M. U. A. Khan and P. J. Sadler, *Chem.-Biol. Interact.*, 1978, **21**, 227–232.
45. Y. Kiyozuka, K. Takemoto, A. Yamamoto, P. Guttmann, A. Tsubura and H. Kihara, *X-ray Microscopy: Proceedings of the Sixth International Conference*, 2000, 153–158.
46. C. Meijer, M. J. A. van Luyn, E. F. Nienhuis, N. Blom, N. H. Mulder and E. G. E. de Vries, *Biochem. Pharmacol.*, 2001, **61**, 573–578.
47. R. Ortega, P. Moretto, A. Fajac, J. Bénard, Y. Llabador and M. Simonoff, *Cell. Mol. Biol.*, 1996, **42**, 77–88.
48. M. D. Hall, C. T. Dillon, M. Zhang, P. Beale, Z. Cai, B. Lai, A. P. J. Stampfl and T. W. Hambley, *J. Biol. Inorg. Chem.*, 2003, **8**, 726–732.
49. M. D. Hall, R. A. Alderden, M. Zhang, P. J. Beale, Z. Cai, B. Lai, A. P. J. Stampfl and T. W. Hambley, *J. Struct. Biol.*, 2006, **155**, 38–44.
50. J. P. Berry, P. Galle, A. Viron, H. Kacerovská and A. Macieira-Coelho, *Biomed. Pharmacother.*, 1983, **37**, 125–129.
51. R. G. Kirk, M. E. Gates, C.-S. Chang and P. Lee, *Exp. Mol. Pathol.*, 1995, **63**, 33–40.
52. T. Makita, S. Itagaki and T. Ohokawa, *Jpn. J. Cancer Res.*, 1985, **76**, 895–901.
53. J. P. Berry, P. Brille, A. F. LeRoy, Y. Gouveia, P. Ribaud, P. Galle and G. Mathé, *Cancer Treat. Rep.*, 1982, **66**, 1529–1533.
54. C. Molenaar, J.-M. Teuben, R. J. Heetebrij, H. J. Tanke and J. Reedijk, *J. Biol. Inorg. Chem.*, 2000, **5**, 655–665.
55. G. V. Kalayda, G. Zhang, T. Abraham, H. J. Tanke and J. Reedijk, *J. Med. Chem.*, 2005, **48**, 5191–5202.
56. K. Katano, R. Safaei, G. Samimi, A. Holzer, M. Tomioka, M. Goodman and S. B. Howell, *Clin. Cancer Res.*, 2004, **4578**, 4578–4588.
57. R. Safaei, K. Katano, B. J. Larson, G. Samimi, A. K. Holzer, W. Naerdemann, M. Tomioka, M. Goodman and S. B. Howell, *Clin. Cancer Res.*, 2005, **756**, 756–767.
58. X.-J. Liang, D.-W. Shen, K. G. Chen, S. M. Wincovitch, S. H. Garfield and M. M. Gottesman, *J. Cell. Physiol.*, 2005, **202**, 635–641.
59. D. Fink, S. Nebel, S. Aebi, H. Zheng, B. Cenm, A. Nehm, R. Christen and S. B. Howell, *Cancer Res.*, 1996, **56**, 4881–4886.
60. R. L. Hoffmann, *Tox. Env. Chem.*, 1988, **17**, 139–151.
61. P. Perego, F. Zunino, N. Carenini, F. Giuliani, S. Spinelli and S. B. Howell, *Mol. Pharmacol.*, 1998, **54**, 213–219.
62. A. K. Holzer, G. H. Manorek and S. B. Howell, *Mol. Pharmacol.*, 2006, **70**, 1390–1394.
63. O. Rixe, W. Ortuzar, M. Alvarez, R. Parker, E. Reed, K. Paull and T. Fojo, *Biochem. Pharmacol.*, 1996, **52**, 1855–1865.
64. H. C. Harder and B. Rosenberg, *Int. J. Cancer*, 1970, **6**, 207–215.
65. Y. Jung and S. J. Lippard, *J. Biol. Chem.*, 2006, **281**, 1361–1370.
66. P. Jordan and M. Carmo-Fonseca, *Nucl. Acids Res.*, 1998, **26**, 2831–2836.
67. S. Tornaletti, S. M. Patrick, J. J. Turchi and P. C. Hanawalt, *J. Biol. Chem.*, 2003, **287**, 35791–35797.

68. W. H. Ang, M. Myint and S. J. Lippard, *J. Am. Chem. Soc.*, 2010, **132**, 7429–7435.
69. G. E. Damsma, A. Alt, F. Brueckner, T. Carell and P. Cramer, *Nat. Struct. Mol. Biol.*, 2007, **14**, 1127–1133.
70. K. Rijal and C. S. Chow, *Chem. Comm.*, 2009, 107–109.
71. E. G. Chapman and V. J. DeRose, *J. Am. Chem. Soc.*, 2010, **132**, 1946–1952.
72. D. Draper, *RNA*, 2004, **10**, 335–343.
73. M. Hägerlöf, P. Papsai, C. S. Chow and S. K. C. Elmroth, *J. Biol. Inorg. Chem.*, 2006, **11**, 974–990.
74. T. W. Hambley, *Dalton Trans.*, 2001, 2711–2718.
75. S. K. C. Elmroth and S. J. Lippard, *Inorg. Chem.*, 1995, **34**, 5234–5243.
76. V. Monjardet-Bas, M. A. Elizondo-Riojas, J. C. Chottard and J. Kozelka, *Angew. Chem., Int. Ed.*, 2002, **41**, 2998–3001.
77. A. S. Snygg, M. Brindell, G. Stochel and S. K. C. Elmroth, *Dalton Trans.*, 2005, 1221–1227.
78. T. Cheatham and P. Kollman, *J. Am. Chem. Soc.*, 1997, **119**, 4805–4825.
79. K. Chin, K. A. Sharp, B. Honig and A. M. Pyle, *Nat. Struct. Biol.*, 1999, **6**, 1055–1061.
80. P. Acharya, S. Acharya, P. Cheruku, N. V. Amirkhanov, A. Foldesi and J. Chattopadhyaya, *J. Am. Chem. Soc.*, 2003, **125**, 9948–9961.
81. P. Legault and A. Pardi, *J. Am. Chem. Soc.*, 1997, **119**, 6621–6628.
82. D. Rhodes, P. W. Piper and B. F. C. Clark, *J. Mol. Biol.*, 1974, **89**, 469–475.
83. J. R. Rubin, M. Sabat and M. Sundaralingam, *Nucl. Acids Res.*, 1983, **11**, 6571–6586.
84. J. C. Dewan, *J. Am. Chem. Soc.*, 1984, **106**, 7239–7244.
85. Y. T. Yu, P. A. Maroney, E. Darzynkiewicz and T. W. Nilsen, *RNA*, 1995, **1**, 46–54.
86. P. Fabrizio and J. Abelson, *Nucl. Acids Res.*, 1992, **20**, 3659–3664.
87. S. A. Woodson, *Curr. Op. Chem. Biol.*, 2005, **9**, 104–109.
88. R. Dalbies, D. Payet and M. Leng, *Proc. Natl. Acad. Sci. USA*, 1994, **91**, 8147–8151.
89. R. Dalbies, M. Boudvillain and M. Leng, *Nucl. Acids Res.*, 1995, **23**, 949–953.
90. K. M. Comess, C. E. Costello and S. J. Lippard, *Biochem.*, 1990, **29**, 2102–2110.
91. M. Escaffre, J. C. Chottard and S. Bombard, *Nucl. Acids Res.*, 2002, **30**, 5222–5228.
92. M. Boudvillain, M. Guerin, R. Dalbies, T. Saison-Behmoaras and M. Leng, *Biochem.*, 1997, **36**, 2925–2931.
93. B. Alguero, J. L. de la Osa, C. Gonzalez, E. Pedroso, V. Marchan and A. Grandas, *Angew. Chem. Int. Ed.*, 2006, **45**, 8194–8197.
94. K. Aupeix-Scheidler, S. Chabas, L. Bidou, J. P. Rousset, M. Leng and J. J. Toulme, *Nucl. Acids Res.*, 2000, **28**, 438–445.
95. M. Hagerlof, H. Hedman and S. K. C. Elmroth, *Biochem. Biophys. Res. Comm.*, 2007, **361**, 14–19.
96. M. Hagerlof, P. Papsai, H. K. Hedman, U. Jungwirth, V. Jenei and S. K. C. Elmroth, *J. Biol. Inorg. Chem.*, 2008, **13**, 385–399.
97. S. Bombard, J. Kozelka, A. Favre and J.-C. Chottard, *Eur. J. Biochem.*, 1998, **252**, 25–35.

98. J. R. Thomas and P. J. Hergenrother, *Chem. Rev.*, 2008, **108**, 1171–1224.
99. J. Boer, K. F. Blount, N. W. Luedtke, L. Elson-Schwab and Y. Tor, *Angew. Chem. Int. Ed.*, 2005, **44**, 927–932.
100. C. N. N'soukpoe-Kossi, C. Descoteaux, E. Asselin, J. Bariyanga, H. A. Tajmir-Riahi and G. Berube, *DNA and Cell Biology*, 2008, **27**, 337–343.
101. J. S. Saad, G. Natile and L. G. Marzilli, *J. Am. Chem. Soc.*, 2009, **131**, 12314–12324.
102. H. Schöllhorn, G. Raudaschl-Sieber, G. Müller, U. Thewalt and J. Lippert, *Am. Chem. Soc.*, 1985, **107**, 5932–5937.
103. G. Natile and F. Cannito, in *Metal Complex-DNA Interactions*, Eds. N. Hadjiliadis and E. Sletten, Wiley-Blackwell, New Jersey, 2009, pp. 143–145.
104. P. J. Sanz Miguel, M. Roitzsch, L. Yin, P. M. Lax, L. Holland, O. Krizanovic, M. Lutterbeck, M. Schurmann, E. C. Fisch and B. Lippert, *Dalton Trans.*, 2009, 10774.
105. S. E. Sherman, D. Gibson, A. H.-J. Wang and S. J. Lippard, *Science*, 1985, **230**, 412–417.
106. A. Gelasco and S. J. Lippard, *Biochem.*, 1998, **37**, 9230–9239.
107. G. N. Parkinson, G. M. Arvanitis, L. Lessinger, S. L. Ginell, R. Jones, B. Gaffney and H. M. Berman, *Biochem.*, 1995, **34**, 15487–15495.
108. C. C. Correll, A. Munishkin, Y.-L. Chan, Z. Ren, I. G. Wool and T. A. Steitz, *Proc. Natl. Acad. Sci. USA*, 1998, **95**, 13436–1344.
109. F. M. Jucker, H. A. Heus, P. F. Yip, E. H. M. Moors and A. Pardi, *J. Mol. Biol.*, 1996, **264**, 968–980.
110. S. Gelbel, S. Banckenko, M. Engell, E. Lanka and W. Saenger, *Proc. Natl. Acad. Sci. USA*, 2009, **106**, 7810–7815.
111. P. S. Klosterman, S. A. Shah and T. A. Steitz, *Biochem.*, 1999, **38**, 12784–14792.
112. K. Hindmarsch, D. A. House and M. M. Turnbull, *Inorg. Chim. Acta*, 1997, **257**, 11–18.
113. M. F. Osborn and V. J. DeRose, unpublished.
114. J. A. Cowan, *J. Inorg. Biochem.*, 1993, **49**, 171–175.
115. D. E. Draper, D. Grilley and A. M. Soto, *Annu. Rev. Biophys. Biomol. Struct.*, 2005, **34**, 221–243.
116. M. Roychowdhury-Saha and D. H. Burke, *RNA*, 2007, **13**, 841–848.
117. A. Y. Keel, R. P. Rambo, R. T. Batey and J. S. Kieft, *Structure*, 2007, **15**, 761–772.
118. E. Ennifar, P. Walter and P. Dumas, *Nucl. Acids Res.*, 2003, **31**, 2671–2682.
119. D. M. Calderone, E. J. Mantilla, M. Hicks, D. H. Huchital, W. Rorer Murphy, Jr. and R. D. Sheardy, *Biochem.*, 1995, **34**, 13841–13846.
120. F. R. Keene, J. A. Smith and J. G. Collins, *Coord. Chem. Rev.*, 2009, **253**, 2021–2035.

Met. Ions Life Sci. **2011**, *9*, 379–391

Subject Index

A

B

Metal Ions in Life Sciences, Volume 9 Edited by Astrid Sigel, Helmut Sigel, and Roland K. O. Sigel

Published by the Royal Society of Chemistry, www.rsc.org DOI: 10.1039/9781849732512003379

C

D

E

F

G

H

I

K

L

M

N

O

P

Q

R

S

Y

Z

Bei Fragen zur Produktsicherheit wenden Sie sich bitte an:
If you have any questions regarding product safety,
please contact:

Walter de Gruyter GmbH
Genthiner Straße 13
10785 Berlin
productsafety@degruyterbrill.com